Biomimetics for Architecture & Design

Göran Pohl · Werner Nachtigall

Biomimetics for Architecture & Design

Nature—Analogies—Technology

Springer

Göran Pohl
Stuttgart
Germany

Werner Nachtigall
Scheid
Germany

"The photography on the cover page is courtesy of Alfred Wegener Institut (AWI), Bremerhaven; Claus Kiefer, Becker & Bredel, Saarbrücken; and Göran Pohl, Pohlarchitekten, Stuttgart"

ISBN 978-3-319-19119-5 ISBN 978-3-319-19120-1 (eBook)
DOI 10.1007/978-3-319-19120-1

Library of Congress Control Number: 2015943315

Springer Cham Heidelberg New York Dordrecht London

Printed on acid-free paper

Springer International Publishing AG Switzerland is part of Springer Science+Business Media (www.springer.com)

Preface

From the foreword to the 1st edition:

> It should be stated in advance: This is not a book that directly enables one to build and construct. It is a book that broadens the horizon.

Building biomimetics is a field of biomimetics. The classical definition states:

> Biomimetics as scientific discipline concerns itself systematically with the technical implementation and application of structural systems, processes, and development principles of biological systems.

Building biomimetics would then be correspondingly classified under the subject area of "structural biomimetics," or also possibly under "process biomimetics." There are, however, some points to consider.

First, one must be cautious when translating inspirations from the living world to the world of technology and should not expect the impossible; a direct copy never leads to the goal. However, when the architect or engineer grasps a *fundamental idea* from nature—for example, the environmentally neutral, thermoregulating ventilation systems using solar effects, as practiced by termites, for example—these inspirations can contribute to bolder technological–biological adaptations of these aspects and their biomimetic applications in the engineering sciences. No more, but certainly no less. One must understand that nature presents no blueprints for its structures, and its processes are not always simple to appreciate, let alone to implement. Nonetheless, they are available for our observation.

Second, this book would like making inroads into *analog research.* The previously mentioned ventilation systems of termites and those systems of technology are analogous systems. Such systems can always be principally developed in two manners. Either nature actually provides the driving stimulus for the development of a certain technology, in which case the technical structures develop further under the umbrella of the engineering science disciplines. Or the development of the technology occurs without the knowledge of the biological nature to such structures. In this case, one establishes *a posteriori* a functional similarity, establishes *analogous structures.* On this basis of comparison, nature can be better reconstructed and more subtly observed.

With the application of technical know-how, natural structures can often be much better understood than without such cutting-edge sciences.

The final consideration was an essential reason for the composing of this book. It would not have been written in vain, even if it merely inspires awe in the structures of nature. This inspiration keeps the technological spirit alive for the linking of technology and nature, a link which could be much stronger than is customary today. And without nature always being at the forefront, alone from the understanding that nature and technology must not necessarily be alien to one another.

Foreword to the 2nd edition

The first edition, published only in the German language, was well received and quickly out of stock. It contained the perspective of Werner Nachtigall as subject biologist with a major interest and a certain fundamental knowledge of the concerns of building and design. As a structural biology-oriented text, the first edition contained an illustrated collection of biological precedents.

In the meantime, the extensive book by N. W., "Biological Design—Systematic Catalogue for Biomimetic Design" appeared with Springer Publishers, which integrated this collection of illustrations. The newly freed pages allowed the possibility of a completely new orientation for the 2nd edition: Alongside the biological fundamentals, which a biologist can describe, the book would now also contain illustrations for practical applications of building and design, a task for which an architect is better suited. Both of the composers endeavored to develop a sound and encompassing work, without raising the claim to comprehensiveness. A series of technological analogs, which had been only briefly covered in the biological sections, were grasped once again in the technological chapters and more extensively represented with structural physics and architectural aspects.

The authors coordinated closely on this book and intensively discussed how a new edition could be structured using the basis of the 1st edition. It appeared important to intensify the viewpoint of the architect Göran Pohl and incorporate current examples of biomimetics for buildings in particular. Furthermore, important changes in relation to definitions and standards in biomimetics had occurred during the contributions of G.P. with the VDI. In this regard, this present work is a—hopefully perceived as successful by the reader—coproduction of the biologist W.N. with the architect G.P.

The following chapters are the writings by the individual authors: Sections authored by W.N. are Sect. 1.2 Historical and Functional Analogies to Sect. 2.1.5 Panel Structures; Chap. 4 Natural Functions and Processes as Prototypes for Buildings; Chap. 5 Biological Support and Envelope Structures and their Counterparts in Buildings; Chap. 7 Brief Information to Biological Structures. Sections authored by G.P. are Sect. 1.1 The Term “Biomimetics”; Sect. 2.1.6 Structures of Folds; Chap. 3 Biomimetics for Buildings; Sect. 4.5.4 Example for Ventilation and Air Conditioning: Incorporation of Biomimetic Inspirations in the Structural-Architectural Planning Process; Sect. 5.6.4 Tensegrity—Connecting the Systems of Tensegrity and Pneu; Sect. 5.8 Moving Structures, Chap. 6 Products and Architecture—Examples of Biomimetics for Buildings.

This new edition should offer reliable information to architects, engineers, designers, and urban planners, as well as to teachers and students in all of the stated subject areas, and—possibly—also offer a certain reading enjoyment.

The architectural and engineering aspects of biomimetics have been far more distinctly developed in recent times than the biological aspects. That will certainly be strengthened in the future, and is good so. Biology serves as the initial basis for comparison and understanding of biomimetic principles; biomimetics for the built environment will then work its way into the actual practice and realization of future architectural and urban designs. Therefore, it only appears sensible to place the further development of this book primarily in the hands of professionals and practitioners of the architecture field. For this reason, we have changed the order of authors from the previous German edition of this book.

Acknowledgement

Many thanks to Sam Wesselman, who undertook the translation of this work from German into English.

Contents

About the Authors

Prof. Göran Pohl is professor for design, structural design, and urban planning at the School for Architecture, University of Applied Sciences HTW Saar, Germany. After his studies at the University of Stuttgart, he and his wife Julia Pohl founded the office of Pohl Architects and the Lightweight Structures Institute in Jena, the latter of which has since become Pohl Architects' research center, taking part in a number of projects on biomimetics and lightweight structures. Their works have been published in numerous reference books and magazines, and endowed with national and international awards. Prof. Pohl developed his understanding of lightweight construction and biomimetics as well as his knowledge of the structural aspects of architecture during his studies at the University of Stuttgart, Germany, under Frei Otto and Peter C. von Seidlein, among others, and during his doctoral studies at the TU Delft in the Netherlands under Ulrich Knaack. He is the editor and author of *Textiles, Polymers, and Composites for Buildings (2010) Woodhead Publishing, Cambridge*. He is also author of numerous technical lectures and publications in the areas of building materials and systems, natural and artificial fiber composite materials, and biomimetics as well. In recent years, he has been teaching at several international universities and has participated to national and international research projects. In 2011, he founded the B2E3 Institute for Efficient Buildings at the HTW Saar, which he has been leading since then, and is a founding member of BIOKON International. Besides being a member of the panel committee for biomimetics of VDI (Association of German Engineers), he is also chair of the guidelines committee VDI 6226 for Biomimetic Architecture, Industrial Design, and Structural Engineering.

Prof. em. Dr. rer. nat. Werner Nachtigall studied biology, physics, and the fundamentals of structural engineering and architecture history at the Ludwig Maximilian University (LMU) in Munich and at the Technical University of Munich. With his pioneering insights on technical biology and bionics and the founding of the "Society for Technical Biology and Bionics," he has made great contributions to the convergence of biology and technology, and has become an internationally respected authority on the "study of nature." He is author of numerous books that have set the standards for studies in bionics. His latest book on *Biomimetics for*

Architecture & Design, coauthored with Göran Pohl and published by Springer in 2015, is the first English translation of the 2nd edition of their German book on *Bau-Bionik,* published by Springer in 2013. He has published, among others, *Bionik—Grundlagen und Beispiele für Ingenieure und Naturwissenschaftler (2nd edition, 2002); Biologisches Design—Systematischer Katalog für bionisches Gestalten (2005); Bionik als Wissenschaft—Erkennen, Abstrahieren, Umsetzen (2010); and Bionics by Examples: 250 Scenarios from Classical to Modern Times (2015),* which he coauthored with Alfred Wisser. Prof. Nachtigall is also the author of more than 300 technical scientific papers. He is a member of two academies and his work has been honored with several awards.

Chapter 1
Technical Biology and Biomimetics

Practicing biomimetics means learning from nature for the improvement of technology; in the various technical subject areas it is practiced with varying intensity. Of course it can be interesting or even fascinating for the engineer and the architect to dare a peek over the fence into the wealth of living nature. One must only then be cautious of a too direct interpretation. Inspirations from nature for building engineering or architecture will not function if they do not follow the in between step of abstraction. The approach of biomimetics is then a three-step process: Research → Abstraction → Implementation (Nachtigall 2010). There will repeatedly be occasions to point out this process chain, but first it is necessary to introduce some fundamental questions. How did the term "biomimetics" come into existence? Are there definitions? Why does analogue research lie at the basis?

1.1 The Term "Biomimetics"

The view that "BIONICS" is an artificial word, combined from BIOlogy and techNICS, is unavoidable. Since the 1950s this description has existed; at that time it was formulated during attempts to study the echolocation of bats for yet-to-be developed radar technology. Recently, a different terminology has been found: "BIOMIMICRY", which literally means the "imitation of life" and does not match the goal of this book. "BIOMIMETICS" is the more recent terminology and is professionally accepted. For this reason this term will be used in this book.

The term "biomimetics" implies the understanding of biological structures and processes and their comparable technological applications, methods, or procedures.

Biomimetics is not the mere imitation of nature, neither in material and functional nor in creative regard, rather the grasping of natural principles to aid in the comprehension of analogous, technological questions, which could then be solved by the applications of optimized technologies. The term "technological application" contains all applications of the present time, be they of machine or computer technology. The term covers materials, applications, modes of operation, entities,

G. Pohl, W. Nachtigall, *Biomimetics for Architecture & Design*,
DOI 10.1007/978-3-319-19120-1_1

design, or management. In biomimetics, it is thus about the discovering of the wealth of experience of nature to be utilized for man-made products, a practice of virtual "industrial espionage" of the most experienced researcher and developer on Earth.

In Germany, the pioneers of this field were Heinrich Hertel and Ingo Rechenberg. Werner Nachtigall performed substantial research in the areas of technical biology and biomimetics and promoted the use of "precedents in nature" for technology and economics for decades. Engineers and architects such as Richard Buckminster Fuller and Frei Otto had concerned themselves since the 1950s with "natural structures" and developed structures that have not lost any of their fascinating appeal. Otto linked "natural structures" with the aesthetic and functional expressions of buildings so that they appear logical or "natural," and with the aid of technology they accomplish similar tasks as they do in nature.

1.2 Historical and Functional Analogies

Historically, the biomimetic process developed from the comparison of results from functional morphological research with the requirements of technical constructions. Initially, this process occurred naively, as is customary when a new subject field gropingly develops. Around 1500, Leonardo da Vinci, the closest observer of bird flight of his time, developed flapping wing mechanisms, which were supposed to have functioned according to the principle of flight feathers overlapping during bird flight. One could already speak here of a "functional analogy," if the entire wing structure had not been designed so-to-speak against principles of static structure and aerodynamics. In this case and in a myriad of other "inventions" well into the twentieth century one can today remark that these inventors had paid too close attention to the *similarity of form* and neglected *functioning principles*, which represents the actual missing link for their failed or too simplified abstractions. Philosophical, epistemic approaches speak in any case of the "precedent of nature" and the "imitating technology." W.N. synthesized these issues in his 2010 book *Bionik als Wissenschaft* ("Bionics as Science"). However, earlier, more obvious attempts to integrate the analogy principle with the application of natural precedents also exist. One example is the invention of reinforced concrete.

The Parisian Joseph Monier was a "horticulturalist, paysachiste"; therefore concerned himself heavily with landscape problems. Owing to annoyance with how expensive and fragile large stone or clay planting pots were and to the clever observation that the weathered, branching sclerenchyma structures of *Opuntia* give rigidity to its leaf masses, the idea emerged in 1880 to produce pots with a multicomponent structure. A wire basket, corresponding to the sclerenchyma network in plants, gives tensile strength and simultaneously holds the pressure-resistant cement mass, corresponding to the parenchyma of plants, in shape. At the same time the cement stabilizes the wire basket form.

The fundamental idea of this application appears typically biomimetic: A principle of nature is abstracted; however no forms were slavishly copied. The natural principle would be: *Mechanical synergy of a tension-resistant cylindrical network of sclerenchyma with a pressure-resistant parenchyma matrix.* The technical principle would be accordingly: *Mechanical synergy of a sclerenchyma-analogous steel reinforcement with a parenchyma-analogous cement medium.* A new industrial branch had thus been invented, the reinforced concrete structure. Incidentally, the imaginative gardener lives on in the expression "Monier iron."

1.3 The Form–Function Problem

However, the above-sketched fundamental concept of "functional analogies" was later lost. In 1905, C. Lie gave his mechanically driven "pilot fish" (which was supposed to have hauled one line) the form of an actual fish, with all the corresponding fins at the "biologically correct" locations. An actually efficient hauling device with the fish as precedent would look different in its essential details. The form–function problem is depicted in two well-known examples, the Sony robot dog AIBO and Frei Otto's tree columns (1988).

Behind the popular Sony robot dog, though looks cute, wags its tail, and can pee, lies no biomimetic concept. It is simply the technical copy of a natural form (which is not a negative critique; it sells well, but it is not biomimetic). Otto's "tree columns," as one can observe in form in the Stuttgart Airport and under some highway bridges, do *not* look like trees yet comprise nonetheless an analogous biomimetic concept of the "structural tree." Before their design, studies were performed on branching angles, thickness proportions, and other aspects of tree branches. Also observed was the structure of such a column, which should support a given load over a given area while having least possible mass—the functional goal of the dimensions to be optimized.

1.4 Biomimetics and Optimization

The development of the so-called "tree columns" represented an optimization problem. A further possibility to apply biomimetics for solving such problems, the evolution strategy, also exists. I. Rechenberg and his colleagues had already shown in the 1960s that one can translate the principles of biological evolution for optimizations in technology, by integrating accidents (mutation, recombination) and subsequent testing strategies (selection) in design development. The arithmetic techniques of their "evolution strategy" (Rechenberg 1973) have since been used in an increasingly important manner in the area of technology, in particular when theories for application are impedingly complex or if no basis for the optimization of certain systems exists at all. C. Mattheck (1993) also used the principles of accidents and biological optimization for his processes of "computer-aided design" (CAD) and "computer-aided optimization" (CAO). He had gained inspirations for the development of these very successful and much-used computer processes from his observations of the functions of tree forms.

1.5 From Accidental Discoveries to the Entry into the Market

Sometimes taking the dog for a walk in the forest pays off, or at least that is what happened to Swiss engineer and inventor G. de Mestral. In 1980, the journalist D. Dumanowsky described in the *Boston Globe* the invention of hook and loop fasteners as the outcome of one such walk through the forest in 1941, after de Mestral and his Irish setter had been coated in burs: "It was barely possible to get them out of his wool pants and his dog's fur. Out of curiosity, de Mestral looked at one of the burs under the microscope. Hundreds of fine hooks appeared when enlarged. As such the bedrock for the idea of hook and loop fasteners was laid. With the use of modern production techniques arose eventually the product "Velcro". (The name comes from two French words, "velour" (wool) and "crocher" (hook)." Although barely out on the market, the distributor made a yearly profit in the tens of millions in America alone.

Today it is almost impossible to imagine everyday life without Velcro. But one should not forget that, as a rule, a thorny path lies between a patentable idea and market implementation. With de Mestral it lasted 20 years and initially cost him a lot of money, before the product was established and became financially worthwhile. With their discovery of the Lotus effect, W. Barthlott and Ch. Neinhuis (1997) had to similarly learn the hard way, or at least over a similar timespan. Likewise, it had lasted 20 years from the first microscopic studies of the nub structures on the lotus leaf to the successful façade coating "Lotusan," which has now been provided for hundreds of thousands of houses.

Biomimetic ideas and biomimetic products are simply two different things. Who attempts such an endeavor requires patience, a good patent attorney, and some money. In recent history, interested firms have been unwilling to stick money into the development of a nature-based concept, which is patented and made ready for the market for a high cost, only for the idea to be quickly stolen after a few years. They develop something instead in concealment and throw it onto the market, where it can redeem its cost over maybe 2 years, before cheap(er) copies flood the market.

1.6 Nature and Technology—Antagonistic?

W.N. has, since he began concerning himself with biomimetics in the 1960s, always differentiated between "Technical Biology" and "biomimetics in the actual sense," which he demonstrated in numerous publications; a selection can be found in the literature appendix. Fundamentally, they are only two different perspectives that connect nature and technology. Both belong inseparably together.

Technical Biology investigates the structures, processes, and evolution principles of nature from the viewpoint of the technical physicist and related disciplines. *Biomimetics* attempts to project these base results backwards to technology and to give inspirations for modern solutions better suited for people and the environment.

As already mentioned in the foreword, there is no reason today why nature and technology should be considered so separate, as before. Exactly the opposite: Only when we overcome the boundaries with a meaningful integration, when we realize that the biology-oriented and the technology-oriented disciplines can learn from one another, progress can be achieved.

The engineer should no longer only simply take note of an entire world of structures, processes, and development principles, but use the wealth of knowledge found in nature, wherever it is suitable and meaningful.

The biologist, on the other hand, should no longer be content with simply collecting data and letting himself disappear behind the books in a library. He should be empowered to engage with the structural engineer and offer him insights and perspectives. This encounter should be allowed to reach the limits of reasonableness: Only then can we break out of gridlocked, seemingly unalterable, predefined paths.

G.P., since he began his work on biomimetics as a young architect in the late 1980s, has been deeply influenced by Frei Otto and his ideas when they met each other as teacher and student at the University of Stuttgart. G.P. has worked as an architect since then, using biomimetic inventions when the benefits promise a positive outcome for his building designs. Biomimetics functions as one design tool among other various possibilities of gaining knowledge within a holistic design process.

1.7 Classical Definitions of Biomimetics

The discipline of "bionics" or "biomimetics" is established within the realm of nature sciences, and the term should be therefore scientifically and clearly definable. Particular definitions always reflect the zeitgeist; they gain, however, more precision through the ongoing process of knowledge, as to be found in the following three definitions.

From the beginning of the 1970s W.N. defined bionic/biomimetic work as follows: "Learning from nature for self-sufficient, engineerable design." Nature provides inspirations that the engineer should not simply copy, but incorporate into the structural design—in the art of his or her science. One can also state, "Nature delivers no blueprints for technology," and therefore underline the viewpoint that general stimuli from the most diverse sources can have influence on technical design. However, direct copies never lead to the ultimate goal.

In a convention of the Association of German Engineers (VDI) for the "analysis and evaluation of future technologies," Düsseldorf 1993, which stood under the motto "Technology Analysis Bionics," the attending technical biologists and biomimetics scientists agreed on the clause, quoted earlier in the foreword (Neumann 1993):

> Bionics/Biomimetics as scientific discipline is concerned with the technological implementation and application of structural, procedural, and developmental principles of biological systems.

Bionics is then accordingly a discipline of applied science. The profit of insights and each aspect of biomimetic interpretation always have their bases in the essence of biological systems.

In recent years, the understanding has been established that the VDI definition from 1993, which was intentionally narrow on the grounds of precision and differentiation, should be broadened. In particular, it could not bear one important fundamental aspect of biomimetics, namely influencing technology, so that it can provide a stronger connection between humans and environment. W.N. then suggested the following condensed alternative:

> Learning from structural, procedural, and developmental principles of nature to form a positive network of man, environment, and technology.

This formulation then also encompasses interactions between environmental influences and living beings.

The German VDI set up a work group that further considered such questions and developed specifications for standards of the biomimetic process. However, science can by definition never reach an end point. The current insights from the work on the VDI guidelines, on which G.P. had collaborated, can be found in Chap. 3.

1.8 Biomimetic Disciplines

The subjects of biomimetics can be summarized by the three fundamental disciplines of structure biomimetics, process biomimetics, and development biomimetics.

Structure biomimetics pertains to issues of substances, materials, prosthetics, and robotics.

To *process biomimetics* belong the corresponding viewpoints of climate and energy, construction and possibly architectural design, sensor technology, and ultimately kinetics and dynamics of machine construction.

Development or evolution biomimetics ultimately encompasses areas of neurophysiology, the already implied aspects of biological evolution, and also corresponding viewpoints of procedural and organizational methods.

Therefore, building and architecture biomimetics can be sorted in the broader framework of biomimetic disciplines. However, these subdisciplines must not be strictly held under the banner of "process biomimetics," although there they have their main position, as building and design are processes. Naturally, they encroach into structural biomimetics, especially when it comes to building and insulating materials. Ultimately, they also play an important role in development biomimetics, when a building structure—which in view of drastically more complex structures such as sport halls happens increasingly often—must be processed again and again to produce new variations with a trial-and-error method on a computer.

1.9 Biomimetics for Architecture and Design: Basic Aspects

Biomimetics offers no methods with which one can directly implement into our technical processes. Biomimetics for architecture and design may be translated from the German expression "Bau-Bionik" to "building biomimetics," meaning biomimetics that aims on aspects of architecture and/or design. "Building biomimetics" will then still not be a method to directly build houses or design Items. However, the large range of natural precedents certainly offers the potential of finding new ideas. The difference lies in the fact that the idea generating process in this field can both lie away from the technical paths, more with the natural precedents, and still lead to concepts based on synthetic and technical aspects. In the end, both methods are often mixed. It will be therefore difficult to find a pure "biomimetic" structure, and often only parts of structures are biologically inspired (thus "biomimetic"). If the defining components of a building or building part are biologically inspired, then the building as a whole can then be designated as "biomimetic."

Architects, building engineers, and designers use the research results of biomimetics as a design approach; they actively employ biological insights as design methods or design tools. Biomimetic work itself is defined by its methods; biomimetics is then actually not to be seen as a discipline of the sciences.

Certainly, biomimetics broadens the horizon and offers an incomparably detailed basis for the abstraction of natural precedents, which could or does already enter into the creative design processes of building engineers and architects in various modes. Chimney structures of termites for example have provided inspiration—and also more broadly and to a larger extent—for solar-driven thermoregulating ventilation systems in Europe and Africa. One recent, well-known example is the ventilation system designed by the firm Arup for the East-Gate Hall in Harare, Zimbabwe.

With the translation of inspirations from the living world into technology world must—and we will always be referencing and are addressing here once again the foreword from the first edition—be cautious and cannot expect the impossible. A direct copy never leads to the goal. If, however, a fundamental idea from nature is grasped, for example, the environmentally neutral thermoregulating ventilation from solar effects, then the inspirations can provide for stronger technological–biological handling of these aspects and their biomimetic application in the engineering sciences. One must only understand that nature delivers no blueprints and that their structures and processes are not easy to appreciate or behold much less implement. However, they are present in multitude.

Of course, it cannot hurt to remember once again the principle of biological–technological and technological–biological analogies. Ventilation systems of termites and those of technology are *analogous systems*. Such systems can always be developed in principle in two manners. Either nature provided the driving stimulus, in which case technical structures are further developed under the umbrella of the engineering science disciplines or the development occurred without the knowledge of the nature to such structures. In this instance, one establishes a posteriori a functional consistency, inventing analogous structures. On this basis of comparison nature can be more subtly observed.

With the insertion of technical know-how, natural structures can often be much better understood than under biological viewpoints alone. A better understanding of this kind in turn offers a more advantageous basis for implementation and so forth.

Thus, a discipline is then able to learn from the other.

1.10 Nature and Technology as Continuum

In the end, all research activities mean nothing other than chipping away at a large continuum, even if it is at different corners and with different tools. Natural evolution has lead to fantastic structures, processes, and developmental principles long before there were humans on this planet. Ultimately, evolution is also the source for the human physiological–mental capacity and only from that could the idea of human technology even be conceived.

Thus, technology is nothing other than the continuing of natural evolution with another means. Therefore for us technology is, epistemically speaking, not something "principally different." We see, aside from pragmatic needs for differentiation, no compelling reason why nature and technology should then be considered as opposites, as it has occurred in the past.

Rather, technology and nature form parts of a continuum. This fact can either be statically understood, or it can be further developed and used. The tool for that is biomimetics. Not the only and surely not the most important.

But in many aspects the best.

Chapter 2
Buildings, Architecture, and Biomimetics

The juxtaposition of structural sciences and biology leads to a multitude of—sometimes surprising—analogies. It shows primarily that the fundamental principles in both disciplines are comparable throughout. It is therefore worthwhile to peer over the fence, not simply in one direction but both.

Ecological, structurally functional, and esthetic viewpoints additionally demand a return to the old principles of construction. An "organic" shape of building is not intended, instead one that incorporates and uses natural properties. Architects of antiquity have already noted that their building volumes were embedded in a preexisting environment, compelling them to construct structures oriented to the prevailing winds (structurally functional aspect) and ultimately yielding a convincing and harmonic impression (building esthetic aspect). So-called primitive cultures followed these rules as well up until recently (ancient Iranian architecture) and still today (native architecture in some parts of Africa). These ancient cultures are therefore interesting, as their building design is "biomimetic" so to speak, namely it is completely analogous to the process of natural evolution according to its trial-and-error methods. One could not pre-calculate a complete, comprehensive structure even in the Middle Ages; Gothic domes essentially arose from trial-and-error methods.

Concrete possibilities for comparison can be found between the technological dwellings of humans and other living organisms and their structures; aspects of temperature regulation, as they are embodied in polar bear fur or solar-driven climate systems, or as they are constructed by termites, belong to these observational categories.

G. Pohl, W. Nachtigall, *Biomimetics for Architecture & Design*,
DOI 10.1007/978-3-319-19120-1_2

2.1 Technical Biology and Biomimetics of Building and Load-Bearing Structures

In the following sections, forms of building structures in nature and analogous technical concepts will be juxtaposed to one another, as they have occurred in historical, physical, functional, or ecological observation.

The juxtaposition of these analogs will consist of the following seven sections, from dome-forming node-and-rod structures to the question of whether one has actually completely understood the honeycombs of honeybees and if they are in fact "technologically optimal." In the frame of these analogies, the architect B. Kresling and the biologist W. Nachtigall wrote short summaries on several subjects that could shed light on biological structuring and self-organization processes in different aspects. They are reproduced here in italicized quotations.

2.1.1 Dome-Forming Node-and-Rod Structures

Structures of this type are composed of rod members (pressure and tension rods) and nodes (joints). An optimized structure works with a least possible amount of members, which ideally form a triangular mesh network and regulate the flow of forces so that the individual members are relieved of bending stress and bear only pressure and tension stresses.

The basic forms of equilateral structures of this type are three of the Platonic forms, the tetrahedron, the octahedron, and the icosahedron (Fig. 2.1a). The nodes of these structures all lie on an imaginary spherical shell. Each node is surrounded by the same number of equilateral triangles. Three members of a tetrahedron, or four in the case of the octahedron or five in the icosahedron ("basic frequency," "frequency one"), connect to one node. If one were to subdivide the resulting triangle further (Fig. 2.1a), the resulting connecting members would no longer lie on the same sphere but on an "inner" sphere.

In such domes several members surround a node, namely five or six ("higher frequencies"). One can also say that the base triangles are subdivided into several meshes and these are "exploded" onto a spherical form. Analog biological structures possess up to seven members meeting at one node (Fig. 2.1b).

The sphere form as such is of course completely symmetrical. In contrast, if one were to lay a fine mesh network over it, two types of nodes would emerge and therefore a reduced number of symmetry planes. Particularly irregular meshes with a relatively large number of members per node are found in biology. These are often interpreted as "mistakes;" they can however also imply that dynamic self-organization processes have taken place, which would then suggest a functional or mechanical meaning.

In contrast to technical, spherical meshworks, which are from the beginning "rigidly" arranged (Fig. 2.1c) and cannot be expanded in volume or easily modified,

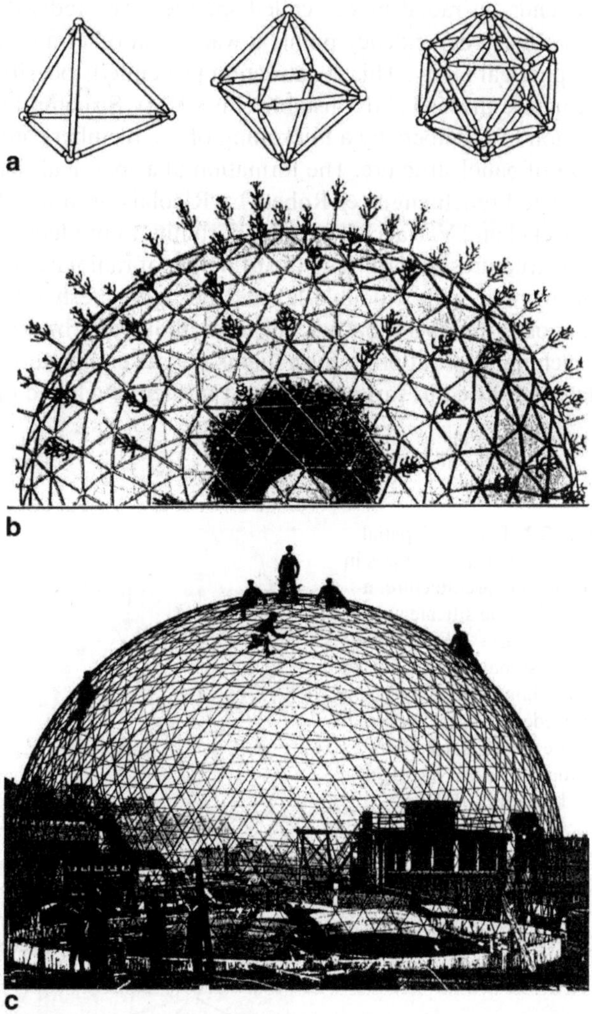

Fig. 2.1 Dome-forming node-and-rod structures in nature and in architecture. **a** Platonic forms, members of the same length complete a triangle. **b** Biological sphere network with dissimilar member lengths: silicate skeleton of the radiolarian *Aulosphaere spec.* (Haeckel 1899). **c** Architectural sphere network with members of equal length: first planetarium of Zeiss, Jena

a "natural" spherical form—for example, that of the radiolaria—must be able to morph and adjust. It rotates conceivably around a center of gravity that is often not quite centered. When that is the case, it slightly deviates from the spherical form and becomes somewhat irregular and instable.

2.1.2 *Special Forms of Spatial Node-and-Rod Structures*

The spherical-appearing radiolaria often carry one to several hollow spheres within one another, which had been formed earlier. In the formation process each new shell depends on radial braces called spicules. The individual members grow outward from these dependency points toward each other and ultimately fuse together into a spherical entity. This construction principle is possible only with a node-and-rod structure that is subtly instable (Fig. 2.2a). Structural stability is reached after the fusion of members by a thickening of the members and nodes, transforming into a sort of panel structure. The formation of a spherical shell is then complete.

The French engineer Robert Le Ricolais used the drawings of radiolaria by E. Haeckel and V. Haecker as an opportunity to produce experimental models for spatial structures according to the principle of radiolaria skeletons. In his first designs, he worked with a double-layered hexagon mesh grid, which is strengthened by diagonal members that jut out from above and below a middle layer (Fig. 2.2b); reaching a sort of proto form that is not yet completely stable. This structure can be later modified in various ways and further developed into a fully stable structure.

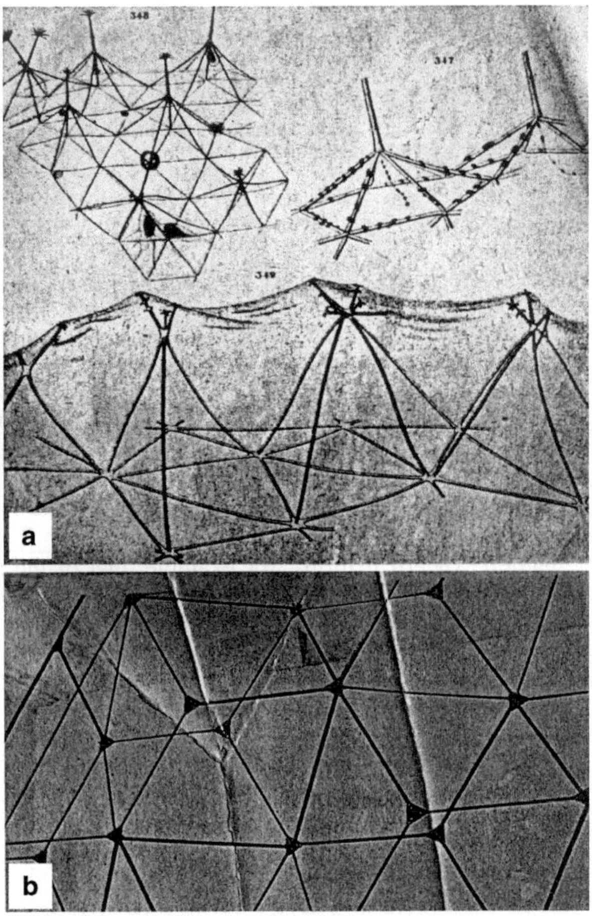

Fig. 2.2 Forms of spatial node-and-strut structures in nature and architecture. **a** Detail of the silicate skeleton of radiolaria. **b** Early three-dimensional dome modeled according to the *Sargoscena* precedent, original photo: R. Le Ricolais, ca. 1935 (Adapted from Nachtigall and Kresling 1992a)

2.1.3 *Self-supporting Structures ("Tensegrity Structures")*

Le Ricolais had already suggested that the structure of radiolaria does not represent a pure truss framework but a structural hybrid of a frame and supportive cladding. One designates structures that support themselves as "tensegrity" structures (R. B. Fuller), in French as "structures auto-tendantes" (D.G. Emmerich). They consist of building elements that are supported on either tension (pull wires) or pressure (freely suspended and untouching pressure rods) (Fig. 2.3b) but not both. A. Chassagnoux, a student of Emmerich, suggested that the smallest irregularities in the tensions of the cables result in a warping of the structure. Theoretically, several shifted variations are possible for a spatial entity, which means differing from the ideal geometrically defined form based on the center of mass. Instead they oscillate so to speak around the center.

Analogous biological structures are represented, for example, by sea radiolaria from the group of the Acantharea (Fig. 2.3a). Tension elements are here again

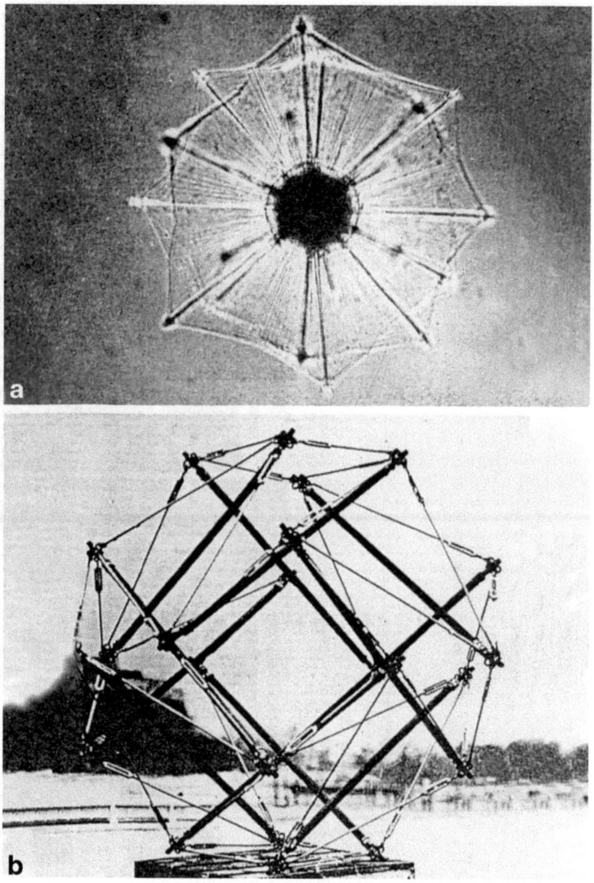

Fig. 2.3 Self-supporting structures ("tensegrity structures") in nature and architecture **a** Sea radiolarian of the group Acantharia with skeleton of strontium sulfate (Courtesy of C. Carre). **b** Tension wire-pressure rod "tensegrity" structure by G. Emmerich

braced with radial, compression-resistant spines, which can also be augmented. The outer membrane in its totality forms the biological equivalent to the tension elements. For this purpose, the tension work performed by the "cables" automatically "adjusts" to the straight growing, pressure-bearing spines.

2.1.4 Orthogonal Lattice Structures

One finds stunningly consistent and—which concerns each idealized axis—nearly rectangular lattice structures in the walls of the tubelike glass sponge (Fig. 2.4a). They consist of membranes in which star-shaped spikes are suspended. These spikes bear six arms in the directions of the three spatial axes toward which they can grow to meet other arms and fuse together into the orthogonal lattice structures of the matured sponge. Before fusing, the spikes often shift and orient themselves repeatedly anew; they "wander" in the rhythm of the active tensing and slackening movements of the membrane. As soon as the spikes have organized into an orthogonal grid network however, bending stress occurs in the nodal points, which causes the nodes to strengthen themselves. Additional spines are also formed afterward

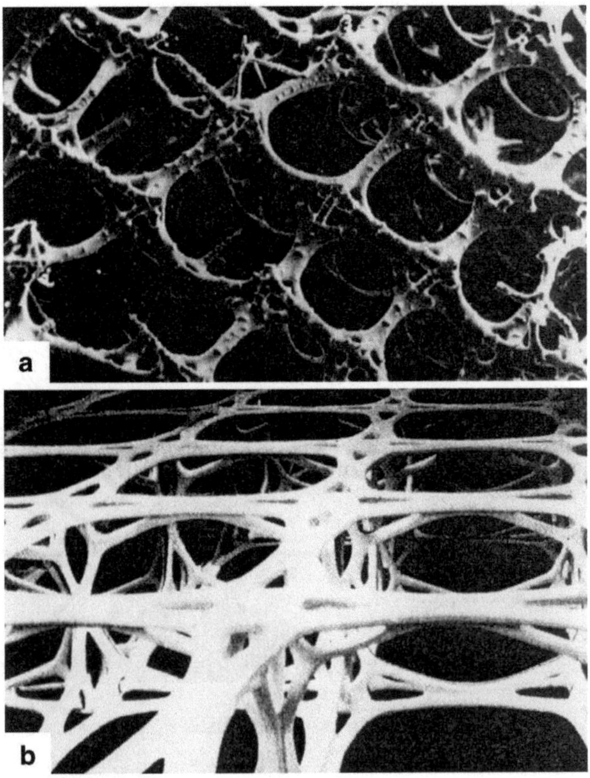

Fig. 2.4 Orthogonal lattice structure in nature and architecture. **a** Glass sponge Aulocystis spec. **b** Experimental node-and-rod structure with rigid nodes by Frei Otto, 1962 (Adapted from Nachtigall and Kresling 1992a)

for further stabilization of the network. As soon as this process is complete, the formation of the next layer begins. In the ontogenesis of the sponge, one tubelike, closed, orthogonal lattice is layered on top of another; the outermost layers being the youngest.

Orthogonal lattices consisting primarily of flexibly connected members are of course not stable in themselves. Why would nature then work with such systems?

The architect Frei Otto designed similar orthogonal lattices (Fig. 2.4b). In his design, the nodal points could no longer be articulated and had to be formed as rigid nodes so that the system remained stable. The structure is used mostly as a load-bearing floor system, therefore for bearing loads in the horizontal direction, and as it is planar, it needs to be supported from below in small enough frequencies to avoid bulging.

In contrast to technical structures, material in biological structures is accumulated—and later hardened—in locations where bending stresses arise. These stresses are thus functionally used and simultaneously dissipated by the growth processes induced by them: The tensing movements by the membrane are co-responsible for the forming of pressure-resistant spines, from which the tension system is suspended. The linear growth of the spines increases in turn the tension in the membranes and is thereby co-responsible for their development.

In organisms, which form a structural framework from precipitated, or in other words, initially viscous and then hardened materials, two-formation systems cooperate in feedback to one another.

The comparison of biology and technology yielded the following insight for this structural form: Nature clearly does not work according to the technological principle of pre-calculated, measured, and stably prefabricated structural elements. Because natural structures must be able to grow, they must work with "preformed deviation," meaning the admission of slight instabilities and resultant accidental variations. This insight signifies: Optimizations in a biological structure do not require reaching a form with an ambitious margin of safety. Rather, a structural form that is sensitive with respect to variations yet still precisely efficient is reached. Simultaneously, the partially self-evoked tensioning from the growth process is simultaneously used for the stimulation of this process, resulting in a network of building processes, function, and adaptations to specific structural loads.

Such self-organizing processes are understandably unable to be reenacted with large-scale building technology. They could however lead to, for example, experimental constructions for the fabrication of innovative materials. Engineers search for means to be able to consistently test the structural behavior of a building for potential failures or even to let the structure correct itself. Studies of micro-vibrations could in this instance, as they occur in the construction of the mentioned biological structures, provide worthwhile inspirations. It could also be that the inverted process is pursued; namely someone, who acquired knowledge about similar processes in technology, would a posteriori correctly describe or even correctly understand the natural processes. That would be "technical biology" par excellence.

Technical biology can also lead at the same time to insights that are and are not immediately usable in biomimetics (that does not devalue the technical–biological process by any means).

2.1.5 Panel Structures

Figure 2.5a shows the base forms of regular volumes, which one could construct from panels bound at their edges. They are three of the Platonic forms: tetrahedron, cube, and dodecahedron. The characteristic of a stable panel structure is accordingly the meeting of edges in a "Y" formation.

In 1984, the Danish engineer T. Wester found that there are strict, formal, and mechanical correlations between the network forming node-and-rod structures and panel structures. It is related to dual symmetries. Consequently, the computer programs developed for geodesic dome structures could be reformulated and utilized for panel structures as well.

One can construct a panel structure in such a manner that the panels are flexibly joined to one another at the edges. Shear forces (which try to shift the panels against one another) occur as a result. One can form the edges as linear joints (i.e., in the form of a piano hinge) or with dovetails: Such structures are also stable due to the Y configuration of the vertices of the panels—as long as no more than three panels meet at one vertex. If the joint lines of four panels intersect (then in the form of an "X"), one obtains a foldable structure as a rule.

In the first mentioned case, the complete structure finds itself in equilibrium when the sum of all occurring torques is equal to zero. A spatial structure can be composed from such panels; an example from T. Wester is shown in Fig. 2.5b, namely a building structure from load-bearing glass panels. It is almost certain that many biological structures, for example sea urchin shells, follow this structural principle (Fig. 2.5c). In these shells and in the shells of other organisms, the individual panels—with slightly dovetailed edges or seams—also meet in a Y form. One can actu-

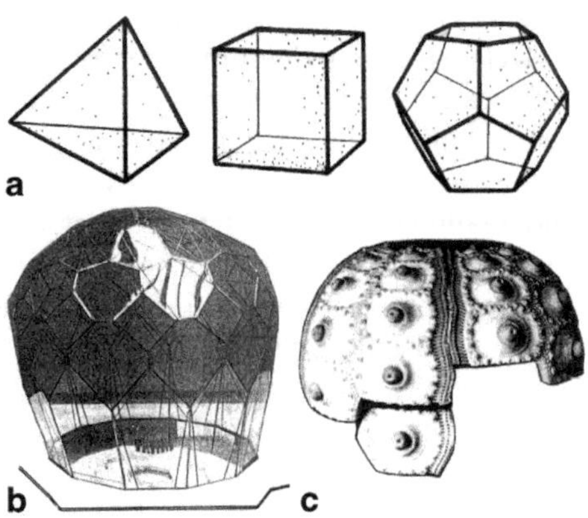

Fig. 2.5 Panel structures in nature and architecture. **a** Platonic forms: maximum of three panels around one vertex. In stable panel structures the edges meet in the form of a "Y". **b** Project for a museum building by T. Wester and K. Hansen (1988). **c** Australian sea urchin *Phyllacanthus imperialis* from the collection of MNHN, Paris (Adapted from Nachtigall and Kresling 1992b)

ally remove the individual panels in older, completely dried-out specimens and insert them back in as well ("clipping together"), obtaining once again a stable shell. The sea urchin appears to integrate this ability as a growth principle. New growth marks always form along the edges of the panels and remain parallel to each other. "Neighboring panels grow so-to-speak at a right angle to their edges towards each other, so that theoretically only tangential shear forces can occur within" (Wester 1984).

Ute Philippi, a doctoral candidate under W.N.— in collaboration with the Institute for Structural Mechanics at the University of Stuttgart—concerned herself for some time with the finite element (FE) modeling of sea urchin shells within the frame of the SFB 230 ("Natural Structures").

The studies yielded, among other findings, that the peculiar "apple-shaped" shell form is particularly well adapted to the tension forces caused by the tube feet and the undirected forces acting on the exterior. The shell presents no weak points. How is it formed then and how can it be statically functional even during its formation process?

To this question, the structural panel approach mentioned earlier can provide food for thought. However, it does not completely explain the essence of sea urchin shells; they possibly belong to technical hybrids, which one can understand only if one has understood "purely technical" entities and can combine two ideas: Perhaps the sea urchin shell behaves simultaneously like a panel structure (shear forces) and like a shell structure (bending-induced forces). Such structures are also not completely stable during growth but subjected to shear forces, which cause the panels shift slightly against each other, and "bending forces," which are directed over the seams. These forces are however—as indicated by the glass sponges—functionally used: As panel structures, the sea urchins could use the anticipated shear forces on the interlocking edges of the panels expected in such a structural form for the accumulation of calcite crystals, as a stable shell structure it could use the deformations elicited by the shifting panels for its construction. In such a construction process, the biological shells could grow both longitudinally and latitudinally. Therefore, it could offer an interesting solution for a difficult technical problem, namely volume enlargement or diminishment, which is always linked with tension points in one direction or another along the surface. Combined linear and volume growth could be used for technological purposes, possibly for an assembly process.

The use of two seemingly contradictory structural principles by one biological entity suggests that this form does not occur in a static but in a dynamic equilibrium condition.

One could thus formulate the underlying model concept as follows:

> In certain biological building processes oscillations are used in order to reach an equilibrium state for any given case. The form that results from this dynamic process contains the characteristics of two antagonistic structural principles.

Both authors of the quoted article have noted in various discussions in the struggle to find the most appropriate approach that a completely typical characteristic has been addressed in the comparison of biology and technology.

They found it in its quintessence: "technologists and biologists should toss arguments and counter-arguments back and forth like ball. In a fair game it's the playing itself that is most important—with the passing of the ball between biologists and technologists the alternating learning from one another should be an end in itself. In general, a result—like the final outcome of a game—from scientific discoveries and likewise from technological achievements is always only tentative: the game is never won, it is only postponed: the process is the goal."

2.1.6 Fold Structures

As illustrated by the Japanese paper-folding technique ("origami"), one can fold paper into complex forms. The innovative Japanese physicist K. Miura technically implemented such fold structures, which are analogous to biological systems such as deciduous tree leaves, flower petals, folding insect wings, and possibly bee honeycombs or plant cells as well. This technique focuses on spaceship design and ultra lightweight design.

The previously described panel structures and Miura's fold structures allow comparisons. In both cases only tangential shear forces occur at the fold edges. A surface cannot be warped along a fold (or generally speaking, along the axis of a cylindrical or conical curve). This property functions for isotropic, thin-gauge materials, such as paper, due to their inelastic deformation behavior. The formal characteristic for a stable panel structure is, as stated, the arrangement of no more than three panels around one vertex. The panel edges functioning as linear joints therefore form a Y-fold structure. Because they should be light and rigid, they must be produced from the thinnest possible planar surfaces and be able to be folded together in space-saving manner (Fig. 2.7).

Structural applications require, for example, a large spanning width of the panels of a folding structure or additional folding elements, such as heat insulation. The construction of thicker fold elements is necessary for this purpose. These elements are limited in their ability to be folded together, but nonetheless retain the essential characteristic of fold structures, that is, the distribution of loads in third dimension. Owing to their complex geometries, folded structures have been hitherto difficult to produce. At the ETH Zürich and the EPFL Lausanne, Switzerland, research teams have occupied themselves with the construction of fold structures using wood (Fig. 2.6). As an outcome of the research results from the Lightweight Structures Institute Jena, Germany, a fold structure was developed by the architect team Steinmetzdemeyer/Pohl for the new convention center in Luxembourg that is supported by a lightweight and well-insulated wood construction and—as with the naturally multifunctioning capabilities of leaves for example—supplies solar energy and is partially transparent to allow light in. This structure is planned as a zero-emission convention center and as such can predominantly provide enough energy for itself (Fig. 2.8). At the University of Applied Sciences, HTW Saar, Germany, architects with G.P. are researching on comparable fold structures with the goal of achieving simple constructability using only woodshop and carpentry machinery.

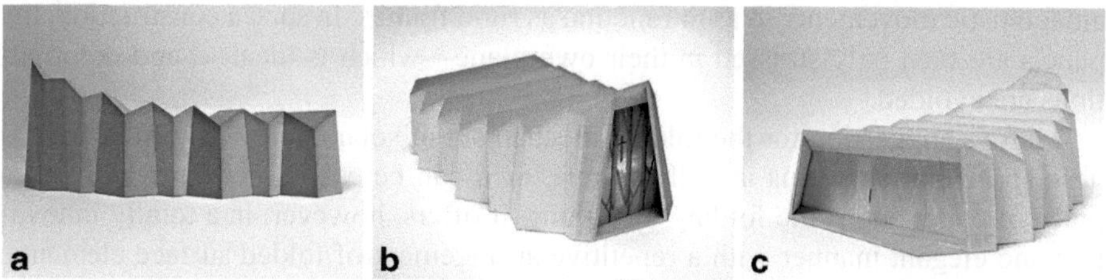

Fig. 2.6 **a, b** Rigid fold structure for a chapel in St. Loup, Switzerland, EPFL Lausanne, Switzerland (MFB Architects, fig. M. Keller)

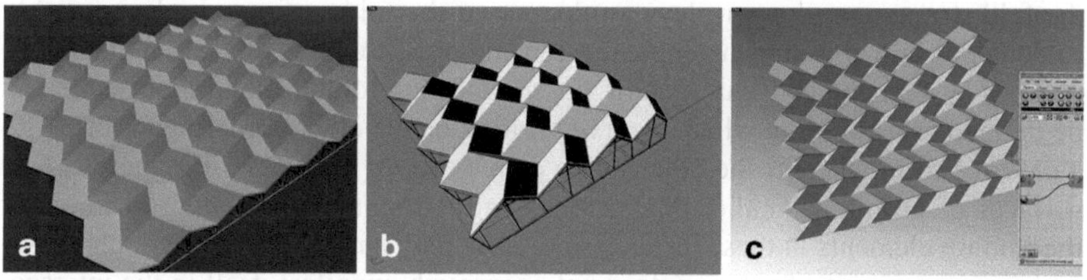

Fig. 2.7 **a, b, c** Geometry and fold studies of the Leichtbau Institut Jena (Institute for Lightweight Structures Jena)

Fig. 2.8 Convention center Luxembourg, Model, Pohl Architects

The formal characteristic for such requirements is the X-shaped vertex, which assumes the role of linear joints; a mirrored arrangement of the surfaces around a vertex and a stress solely in the planes of the surfaces.

If the mechanical behavior of the fold structure is to remain controllable, the fold panels are best coordinated when the neighboring panels automatically perform antagonistic movements so as to function as rigid frames. In such a construction, the panels are then only stressed in their own plane—which is ideal—, and deformations are avoided.

Miura's suggestion for the folding mechanism of solar energy panels on a Japanese space station seems initially simple, as it can be easily reconstructed with a piece of paper and some folding directions. It offers, however, in a totally innovative and elegant manner with a repetitive arrangement of folded surface elements, the technological possibility to unfold the entire structure in the three spatial directions in one single movement and then re-fold into a single closed entity (with "classical" letter folds, as we practice it, each fold occurs in a chronological order). The surfaces are interconnected as a kinetic chain. In a vacuum, where the mass of the fold structure no longer plays a role, one of the fold surfaces can be smoothly folded open and closed by a simple tension and pressure force along two diagonals.

Each system and configuration of zig-zag forming foldlines of this structure can be defined as a curve, in which a curved fold pulls the other bordering surface into a concave form whereas the other into a convex form. Although, in principle, fold surfaces of infinitely smaller thicknesses can be designed, the actual thickness of the folding elements in these instances does not play a role: The fold is replaced by a hinge, whose axis of rotation does not necessarily manifest itself in the intersection of two planes.

In the fan-folded field of the hind wings of many insects, curved veins transmit a muscle-induced movement from the base of the wing over the entire folding surface. Owing to this pretensioning, the fields are interconnected. The wing can automatically open itself, although it has no muscles in the actual folding axis (!).

2.1.7 *Honeycombs of the Honeybee—Still Somewhat Puzzling*

It may surprise that one can also consider bee honeycombs from the perspective of fold structures. The hexagonal honeycombs (Fig. 2.9a) are in no way inherently stable. As a linked chain with more than three members, the walls take up neither lateral pressure nor tension forces. During construction, two layers of honeycomb cells slightly inclined toward horizontal are arranged on both sides of a shared, perpendicular, middle lamella. From flattened beads of wax this middle lamella emerges as a surface with rhombus-shaped folds, from which the walls of the hexagonal cells on both sides of the lamella are constructed. This waffled middle lamella does not however strengthen the honeycomb structure in the same way as the flat floor layers of a sandwich structure do. From a geometry point of view, these fold surfaces consist of parts of rhombic dodecahedrons (Fig. 2.9b, c), on the vertices of which the edges of three surfaces meet alternatingly Y-shaped (stable panel structure) and X-shaped edges (instable fold structure).

The Hungarian mathematician Fejes Tóth described in a humorously titled article "What the bees know and what they don't know" an alternate arrangement of honeycomb cells from the middle layer that would theoretically save 0.35 % surface area. This "ideal" honeycomb of Fejes Tóth illustrates from a mechanical

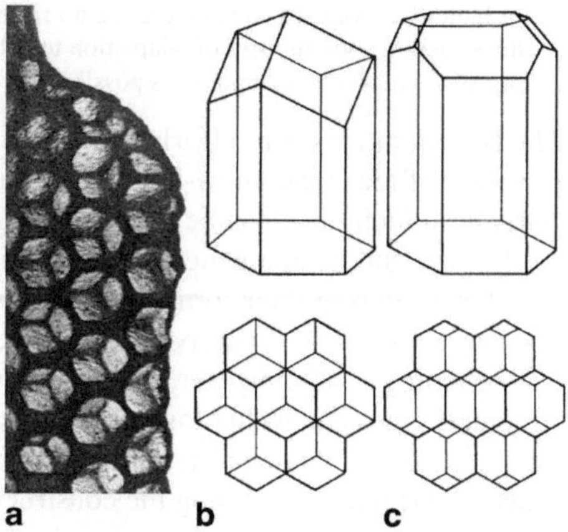

Fig. 2.9 Honeycomb—a naturally optimized structure to be "technologically" improved (?); compare to text (Adapted from Nachitgall and Kresling 1992b). **a** (*dark*), **b** (*light*)

standpoint, with its symmetry of surface edges meeting in a Y, the characteristics of a pure, or stable, panel structure. That signifies, however, that a honeycomb structure, as honeybees construct it, which is partially formed as a fold structure, cannot be stable. On the other hand it does not readily collapse; meaning the stability must lie on mechanisms that emerge from the building process and static-structural conditions of the vertically suspended position and are compatible with its comparably heavy live load. Following the principle of a perfect equilibrium, we know that the cell walls, smoothed and polished by the bees to a fine lamella, do not bulge. That would immediately be the case with similar plastic building material if they were placed under only the slightest bending forces. In this case it could result from an equilibrium state that arises from complex symmetry relationships and bistable formation.

Bee researchers H. Martin and M. Lindauer have reported that bees obtain information about the thickness of the wax walls with smoothing motions of the mandibles and touching with the feelers due to the "aperiodic vibration behavior," and they can then apply the finishing touches accordingly. However, this appears possible only if the evoked modifications are distributed not from each cell, where the individual bees work completely independent from one another, but as sinusoidal vibrations disruptive over the entire structure. This perception must be modified by newer findings, which show that there are places in the honeycomb that vibrate in phase and others that vibrate in opposite phase (Tautz 2007).

A difficult to solve contradiction presents itself here. In a stable structure each local deformation would theoretically have to be "picked up" by an instantaneous compensation. It also needs to be explained how bee honeycombs, with their coarse framing, thick walls, and round bulges, have reached their perfect geometric form in equilibrium. The question must further be answered as to why the bees—if the purpose is not material thriftiness, but the result of some mechanical necessity—reduce their cell walls to a minimum. Perhaps the paradox of a hybrid of two antagonistic structure types in equilibrium provides an answer. From these considerations one can abstract a further model concept:

> In biological fold structures there are no rigid guidelines to follow. From the elasticity of the structures possibilities of adaptation to actually occurring forces emerge for stationary organisms, for mobile organisms possibilities of adaptation to several functions.

The considerations very clearly show how cautiously one must proceed in the interpretation of technical know-how to the understanding of biological structures. The "structural intention" can be thoroughly different.

Biology and its ontogenesis, which builds systems that must be somehow functional even during their formation processes, possess a certain autonomy that one does not understand if purely technical viewpoints are "imposed" on it.

Technology can, however, essentially help to clarify these questions. Biological structural types and technological structural types are to be compared with caution. In this case, the analogue research does not lead to functional similarities but to functional differences during the construction phase. These can certainly disappear after the fabrication process so that the finished system appears to be completely describable by technical-static aspects.

Both authors formulated model concepts at the end of their deliberations (keywords are stated in parentheses for the hitherto established concepts that perhaps should be newly reconsidered):

1. Biological structures cannot be described by "pure" technical–structural types. (*complexity?*)
2. Biological building processes simultaneously proceed according to laws of mechanically antagonistic structural principles. (*compromise solution?*)
3. In biological building processes stable forms can only be approximated. A biological structure is, therefore, always only efficient under completely certain circumstances. (*Instability?*)
4. In ontogenesis, chronological and already "anticipated" partial problems are solved, which in each case determine the conditions of the following development step. (*form–function adaptation?*)

If one further pursues the reasoning for individual structural types as formulated here, it will yield the "generalized model concept of a biological structure as one of an at least bivariate system with mechanical feedback."

That would mean that a biological organism does not completely owe its organization or its mechanical efficiency to genetically predetermined and driven processes, but essentially to the capability to reach a dynamic state of equilibrium by the use of physical processes in all growth stages.

2.1.8 Do Tensegrity Structures have a Fundamental Cytomechanical Meaning?

Questions of cell biomechanics have been—one could almost say criminally—neglected to benefit of cell chemistry and genetics for quite some time. D. Ingber regarded the cytoskeleton (Fig. 2.10a) as a tensegrity structure. Such structures have

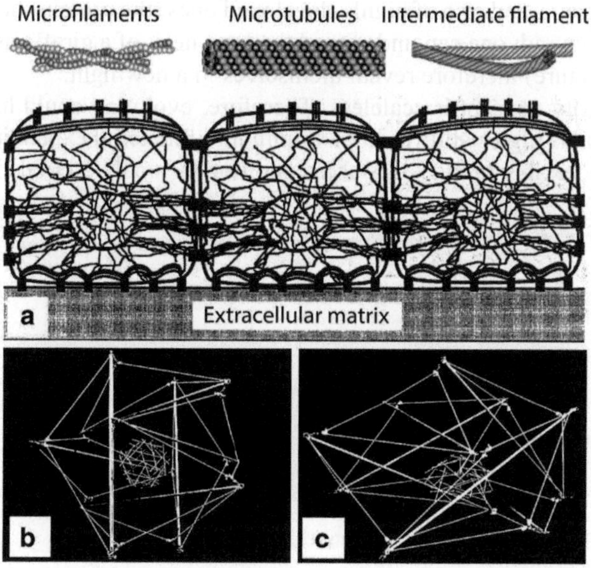

Fig. 2.10 Tensegrity and the cytoskeleton. **a** Schema of the cytoskeleton, in principle consisting of the three elements illustrated above, and its extension through three cells with intercellular junction complexes and focal adhesion complexes to the extracellular matrix. **b** A two-level, "free-floating" tensegrity model, which also models the nuclear envelope, built from pressure-stressed aluminum struts and tension-stressed bands. Connective bands between the "*outer*" and the "*inner*" levels are *black* and therefore not visible in a *black* background. **c** The model from **b** is flattened and the *inner level* is correspondingly shifted when it is laid on a *solid* base

been described in the previous section. They consist of a continuous system of tension-resistant elements (model: rubber bands) that work mechanically together with a discontinuous system of pressure-resistant elements (model: aluminum struts) and stabilize themselves, dependent on the external and internal force proportions, in each specific location. Ingber regarded the microfilaments of the cytoskeleton as tension pairs, and the microtubules (and the compartments generally set under pressure) as pressure pairs, whereas the intermediate filaments were seen as "tension braces," (mechanical integrators) particularly in the region of the nuclear envelope (Fig. 2.10b, c). The cytoskeletons of individual cells are connected with one another and via the extracellular matrix with the "outside world" through the intercellular junction complex and basal adhesion complexes to the extracellular matrix. So they form, also in mechanical respect, a continuum. Forces that work on the extracellular matrix should therefore be directed deep into the cell.

Therefore, the principle of tensegrity as well as the architectural basis of a cellular mechanotransduction can be observed. If mechanical energy is thereby led to molecular transductors and induces this biochemical process, particularly on the cell wall, then the fundamental questions of mechanochemical transduction and "cell response" to stimuli could also be approached with this model. Questions of the emergence of active structures (in porous clay materials), of the evolution of the simplest life forms, and also of highly developed ones (the protein shell of viruses is tensegrity structured; one can understand the long neck of a giraffe as pretensioned tensegrity structure) therefore reveal themselves in a new light.

The tensegrity concept is scaleless. Therefore, evolution could have favored it because it obtains high stability with a minimum of material expenditure but can at the same time flexibly modify its stable condition by small internal changes.

Chapter 3
Biomimetics for Buildings

Our technology-oriented economic world has only relatively recently discovered the functionality and esthetics of natural forms and structures. However, people of all cultures have already intuitively used the functioning principles of nature: storing water in terraced fields, using wind for the separation of chaff from the wheat, or natural climate control in living spaces in earthen houses with updraft cooling in hotter regions. In addition, most functions in nature are closely linked with a sign-emitting, physical manifestation. Success by attraction (i.e., flowers and their coloration, which lure insects by attraction for pollination) depends on these functions.

> …In spite of this, we still build buildings counter to nature from the past epochs. Our times demand lighter, more efficient, mobile, adaptable, or, in brief, natural houses. … This consequently leads to the further development of the lightweight structure, of the building of cells, shells, sails, and airborne membranes. (Frei Otto)

What does nature teach about form and function? Natural life forms distinguish themselves by their multifunctional conceptualization of building parts and functioning groups for the most various demands: envelopes, warmth, thermoregulation, energy production, structure, enclosure, unfolding processes, transportation, movement, and growth. These are only some of the tasks that are fulfilled by nature among innumerable examples and variations. These "technological" developments of life are responses to laws of physics and chemistry, to the necessities of growth and reproduction, and to reaction and utilization. Nature has accordingly created and optimized products whose marketability has been tested and whose "product profile" has been honed and suitably configured for its niche. The results of their developments would be listed in a book of examples or guidelines for successful management, if nature indeed needed one.

Architects must also navigate a multitude of demands comparable to nature, but additionally they must deal with the difficulties of creative implementation as well. Constraints to design result from materials, which can merely fulfill *one* purpose (either structure or shelter), or even from the clients and purchasers of buildings themselves, whose demands ("only beautiful") sometimes further restrict the process. Integrative design means handling a wide variety of requirements within the

G. Pohl, W. Nachtigall, *Biomimetics for Architecture & Design*,
DOI 10.1007/978-3-319-19120-1_3

project with intelligently behaving materials. For architecture, this process also means fulfilling of esthetic sensibilities as part of their general public duty for finding an appropriate design for the collective urban environment.

3.1 Architecture and Biomimetics from the View of Architects, Engineers, and Designers

Architecture must serve the people. All structures—perhaps with the exception of monuments—have to fulfill a functional purpose; monuments fulfill at the very least the purpose of remembering. As a result, a social duty is conferred onto every structure. They perform this function more or less dutifully, be it in consideration of the function to be provided or in its form. For the consideration of the issue of "design" the following essential questions could be posed: What is the effect in relationship to the beholder? Which duties belong to the artificial interpretation of our environment? What consequences arise from understudied designs and what kind of self-image is implanted by such neglect?

From the global changes to cities due to the lack of social awareness emerge "Simcity"-like urban spaces without vision or quality, with pre-programmed potential for violence and questionable sustainability. Global change does not only stop at the built environment. The global change in cities and, as a result, their formation is not only an economic, social, urban challenge but also above all a cultural one. In his conclusion from "Aus Krise zu Innovation" (From Crisis to Innovation), the architecture theorist Philipp Oswalt pointed out:

> The debate over the crisis of shrinking cities is currently an impulse for the development of new concepts and models. The starting point of the classic modern movement was quite similar.

For classical modernists it was not only about the development of a new architectural style or urban typology, but also about the future-oriented understanding of design and "ultimately a new model for society." The approach of biomimetics similarly aspires to a new societal model: Technology is not to be used as an end in itself, but it must be integrated into a cycle, in which the efficiency of energy use and material application is considered as a given, as nature would teach us.

It would then be too easy to simply relate the precedents of nature to the forms of our structures. Biomimetics is not merely a stylistic form in which one visually perceives a quasi-natural origin in building shape—often represented by rounded forms ("biomorphic"). Biomimetics instead implies the previously formulated structural and functional chain of "abstraction, interpretation, and application of insights" from biology to technology; only then can a form emerge.

In what manner then have the structures of humans found themselves in relation to natural structures? Human beings are accustomed to suitably adapting themselves to the conditions of their environment. In this manner they cannot wholly differentiate themselves from other living beings, such as beavers, which can form entire lakes with their dams, or termites, which construct complex structures with thermoregulating functions. What differentiates human beings from these organisms is then, among many other aspects, the deliberate construction and continual development of their technological products. A prominent product of their creative ability is of course the building of houses, the development of architecture. The human being consciously erects structures that must serve a whole variety of selected purposes. They fulfill the requirements of protection from weather, protection from enemies, and security. The dwelling also offers, as locale for communication, spaces for meeting, and, as place of protection, the possibility for retreat from society. Human beings have adapted their dwellings for further purposes: for the purpose of communal activity, for the purpose of recovery, of enjoyment, for the storage of supplies, for the practice of religion, and so on. The complete, differentiated design of these spaces is an artistic act of creativity that only humans can accomplish. For this purpose they use their ability to develop technologies. Technologies and their devices were initially developed by humans to overcome physical shortcomings: Spears and stone axes were needed for hunting, as they are not fast enough, or to defend themselves, as they are not strong enough. Owing to the lack of effective fur against the cold, humans at first considered the use of primarily hunted furs from the animal kingdom. Later they began to produce their own material for the desired effects of warmth and moisture protection. They used manufactured materials as a second skin—as replacement fur—and as a third skin as well, their dwellings. However, it is due to lack of technological advancement in the design of this third skin that places humans right at the beginning of development compared what nature has already performed, namely the energy and substance-sparing implementation of multifunctional materials.

Today houses that tend to distance themselves from natural conditions are still being planned and built: instead of reduced sun infiltration and better air circulation, they are regulated with central air conditioning; and instead of conserving the same solar warmth in the winter, they are heated. Many can only perform one or the other. Often the building must be completely sealed to function efficiently with central heating or cooling, or it is too drafty. Natural structural precedents are in every sense more “intelligent.” They combine several functions in one structural method. For multifunctioning capabilities biomimetics can provide many inspirations and precedents, so that in the future modern building skins will be able to combine many functions, each according to needs such as protection, warmth, and light (as with polar bear fur). In this manner, building envelopes will be able to produce energy and convey materials similar to leaves, and additionally be structural and able to grow like the skin of an elephant.

The “new concepts and models,” of which Oswalt wrote, are to be interpreted within the general belief that technology must not only be developed for the sake of “forward” advancement alone, but should also have the ability to rediscover the multitude of preexisting ideas from nature with the application of biomimetic ideas. The design world can in turn be provided as an enormous technological and artistic spectrum and, if nothing else, the meaningful relationship to their precedents from nature.

3.2 Historical Background and the Origins of Building

Nature offers an inexhaustible arsenal of examples for "living structures" that recall architecture in form and function. These structures have not only inspired architects today but also in the past centuries. According to his story, Brunelleschi (1377–1446), the famous architect, builder, and artist of the Renaissance, was inspired by the form of a chicken egg for the design of the dome of Sanra Maria del Fiore in Florence. Leonardo da Vinci (1452–1519) is also naturally assumed to have more or less investigated nature and have attained his inspirations in this analytic manner.

At the beginning of the twentieth century the structures of the Art Nouveau era had begun to imitate nature with floral patterns and curved building volumes. Artists and architects were inspired by the publications of the biologist Ernst Haeckel in Jena. Haeckel had intended with his studies and publications to argumentatively support Darwin's theory of evolution. However, he developed no functional morphology and instead depicted the various forms of nature more in the manner of an art historian and in the order of ornamentation. Haeckel's sensational books, such as *Art Forms in Nature* and *Art Forms of the Ocean*, managed to obtain global influence even in America and are today in their original form sought-after rarities. An example for the application of "natural" art forms and the orientation to ornament is the entrance gate to the 1900 World Exhibition in Paris by René Binet, which is based on a radiolarian skeleton. Architects today have overcome this glorified natural romanticism, which had then been singled out and characterized in the literature

Fig. 3.1 Planetarium Jena, 1942–1925. (Design: W. Bauersfeld, Dyckerhoff and Widmann, Photo University of Jena, Planetarium)

as so-called "bionic building art" and superficially correlates to ornamentation and formalism. Architects increasingly understand the complexity of natural structures and how they can function as sources of inspiration.

The American engineer Buckminster Fuller is one of the most well-known engineers, who had already occupied himself in the 1950s with mechanisms of biological systems and their effects. Fuller's most well-known trademark was the geodesic dome, one of which he erected for the 1967 World Exhibition in Montreal. This construction typus also recalls, with its delicate, with their delicate, materially minimal construction, skeletons of single-celled radiolaria, which Haeckel had previously investigated. An earlier dome, forerunner so to speak and inspiration for Fuller's lightweight structures, was constructed for a planetarium building for the firm of Carl Zeiss in 1924–1925 at the former workplace of Haeckel in Jena, Germany. The three-dimensional (3D) spatial framework was subsequently poured out with concrete (Fig. 3.1).

3.3 Definitions and Methods of Biomimetics for Buildings

Different definitions of biomimetics and building biomimetics circulate in publications that often pertain but also sometimes lead to major misunderstandings. To combat this confusion, guideline committees have been set up by the VDI—the Association of German Engineers—that have described the differently used terms. These and current definitions, such as the term "building biomimetics," are subsequently recorded here. Various methods as to the application of biomimetics are also used, the most of important of which are listed here.

3.3.1 Definitions from the VDI

The VDI, the Association of Gerrman Engineers, does its work in a certain manner, where one is to investigate into regulations that are, in the European countries, usually to be seen as state-of-the-art. The VDI has occupied itself for some time with the issue of biomimetics in its important committee work for the definition of standards. VDI guidelines for biomimetics have recently been developed, the first of which appeared in 2010/2011. The framework guidelines for biomimetics VDI RL 6220 and the VDI RL 6226 Architecture, Engineering, Industrial Design, both of which were developed with the participation of G.P., chairman for the VDI 6226, define biomimetics as

> The interdisciplinary combination of Biology and Technology.

The goal of biomimetics according to the VDI definitions is the

> Abstraction, transfer, and application of knowledge gained from biological models.

This occurs, according to the VDI definition, in interdisciplinary collaboration.

3.3.2 Methods of Biomimetics

The scientific community has solidified two approaches as methods of biomimetic process for buildings, which differentiate themselves by their starting points, as well as a third approach that represents a combination of the two.

One of these methods considers a course of development from biology: Technological developments are stimulated from insights of biological research, push started so to speak ("*Biology Push*").

The other method is driven by a technical scope, in which approaches to solve technical problems are sought within biology, thus extracting then the biological approach to improve an already mostly existing technological product ("*Technology Pull*").

A third method, which makes use of already pursued insights, must also be emphasized within the context of this publication. Building and architecture fields, which are oriented on a quick generation of knowledge, cannot afford research before the construction of each individual building. These fields differ from more linear industries that follow the development process of research → development → serial production and better suited for the continual construction of new prototypes. They use a constantly changing combination of thousands of different solutions to materials and functions for the purpose of developing new design ideas. Biomimetically interpreted, this means that what is considered as advantageous for building and architecture emerges from a pool of "pre-researched" biological and analogous technological mechanisms. The evaluation of which is the methodological approach of "*Pool Research.*"

W.N. used the following labels in his epistemology-based book *Bionik als Wissenschaft* (2010) ("Biomimetics as Science"):

"Biology Push": "a discovery in biology is the starting point." ("What could one improve in the area of technology with the help of a certain biological finding?") (p. 196, 197)

"Technology Pull": "The posing of a problem from technology is the starting point." ("Which findings from the living world could help solve a technological problem?") (p. 198, 199).

"Pool Research": "pooling of information." Filling of the biological data reservoir, from which one can draw information for a technological problem. (p. 156).

These labels are also borrowed for W.N. and A. Wisser (2012). The three brief descriptions selected here implement these exact labels as key phrases.

3.3.3 Biology Push and Technology Pull as Methods of Biomimetics

The processes of development according to biomimetics can either be pushed by biology or pulled by a technological product. These processes can then be referred to as "Biology Push" ("Bottom Up") and "Technology Pull" ("Top Down"), respectively.

The contrast between these two approaches drives research in both fields. With "Biology Push," the biological discoveries are the basis for the development of new, technological products. The direction of development runs then from the knowledge and data of biology to the formulation of an idea and development of a technological product. With "Technology Pull," an existing technological product obtains new and improved qualities by the interpretation and application of biological principles. In this case, the direction of development results from a request from the technical world for biology—in the form of the question "Are there comparable approaches in biology to solve this problem?" On the basis of this question, "ideas" are sought in living nature, which could then give an impetus to the technological side and help lead to new or improved products.

3.3.4 Pool Research as Method of the Biomimetic Process for Architects, Civil Engineers, and Industrial Designers

This third method comprises the collection of fundamental knowledge from biology to better understand biological functions with the goal of technical application, without however needing to immediately determine what that application might specifically be (compare Fig. 3.2). In the discussions for the VDI guideline VDI 6226, the following was ascertained:

> For this reason, the design and the development of solutions in building construction and industrial design seldom follow predictable combinations of rational and subjective aspects. Biological models can be integrated into this process at various stages, as a result of which the fields of building construction and industrial design differ significantly from most other technical disciplines with a more unidirectional orientation.

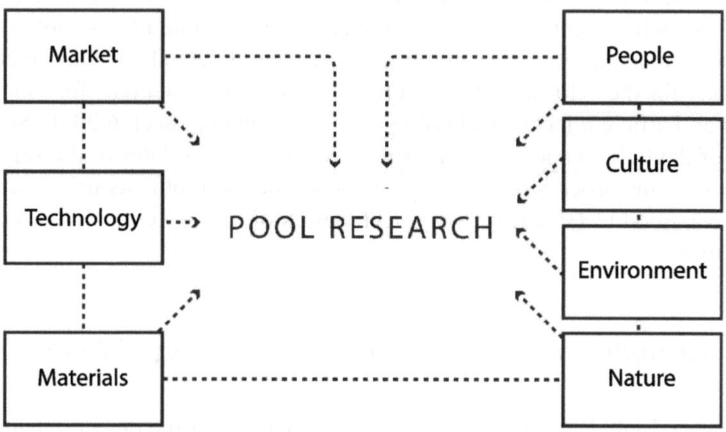

Fig. 3.2 Pool Research analog VDI guideline 6226. (Courtesy of Göran Pohl, Pohl Architects) Figure Translation: Market, People, Culture, Environment, Nature, Materials, and Technology

In this respect, this method does not befit the definition of either of the two previously addressed work processes. Biomimetics is not looked upon in the building and design worlds as a distinct discipline but as a *creative tool*. The insights and knowledge can both be abstracted and described in great detail as well. Frei Otto attempted to understand the functional bases of natural forms and building processes in collaboration with biologists, for example, the botanist Johann-Gerhard Helmcke or the zoologist Werner Nachtigall and other experts, and demonstrated as well how essential the basic fundamentals can be for the further, and often much later, development of biologically inspired structures. In contrast to the "*Biology Push*" and "*Technology Pull*" methods, the "Pool Research" method does not necessarily or immediately underlie an interest for abstraction and application. Instead, the knowledge generation itself is the core of the process and conduct. This result can then be directly or indirectly followed with a technical application or as a whole lead to a discovery of an area for a potential application.

The process of "Pool Research" can follow different paths. One possibility is extraction of knowledge from an in-depth study of a biological precedent from the "pool," which could at some point drive a biomimetic development. A driving inspiration often only emerges in the linking of knowledge to functions in biology; sometimes even years later, namely when the technology has "matured" and the question or issue can be properly formulated (comp. Hill, B. 1998).

Often with the "Pool Research" method analogous technological functions have already been compared with the insights into biological processes, resulting in a tabular register of (both technical and comparable biological) examples and function details, a so-called morphological box. The insights are then evaluated according to these morphological categories and can lead to new "biomimetic" solutions.

In this manner, "Pool Research" is a strategy for abbreviating the duration of the development process. The insights derived on the basis and means of "Pool Research" are of interest for architects, engineers, and industrial designers who can generate ideas on a broader basis and determine potential courses for realization.

The research project "BioSkin" of the Austrian Institute of Technology (AIT), under the leadership of S. Gosztonyi, explored the potentials of biomimetics in a basic study for the "House of the Future." The research potentials for biomimetic-inspired and efficient facade technologies are described in Sect. 6.29. In Sect. 6.30, the use of daylight is addressed, as well as shading as it relates to the ridge forms of cactuses. This process of investigating fundamental problems using natural examples is a good testament for the effectiveness of "Pool Research," essential for architecture.

3.3.5 Evolutionary Light Structure Engineering (ELiSE)

Evolutionary Light Structure Engineering is a method for the development of optimized lightweight structures.

At the Alfred Wegener Institute in Bremerhaven, Germany, a method that enables a systematic and effective development of form-optimized geometries based

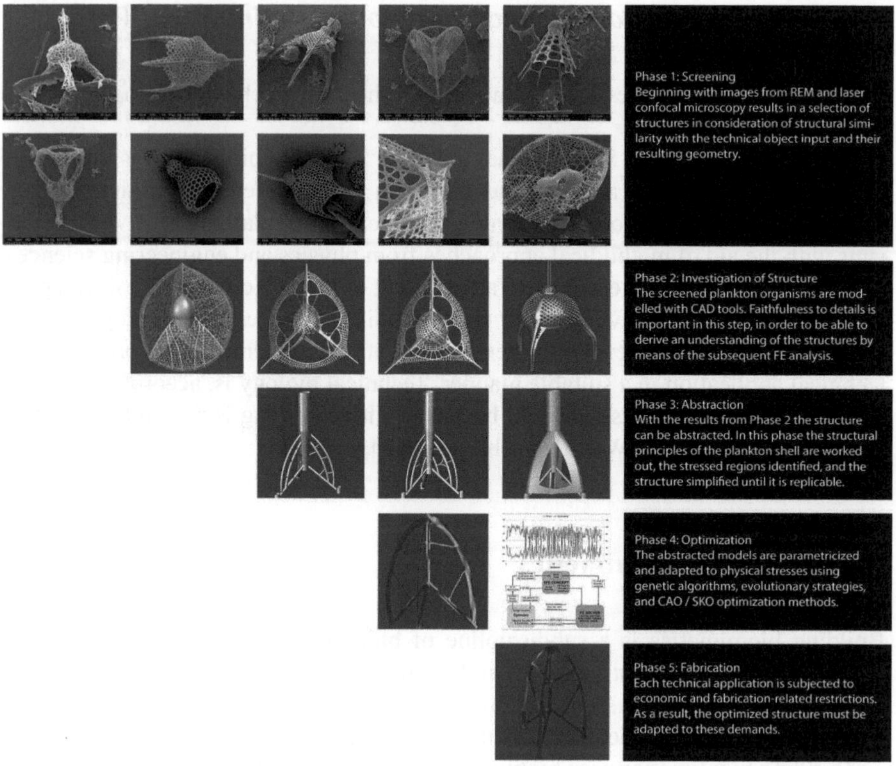

Fig. 3.3 Phases of ELiSE. From above to below: screening, structural study, abstraction, optimization, and fabrication using the example of an off-shore mast. (Courtesy of Christian Hamm, IMARE)

on studies of plankton shells was developed (Fig. 3.3). In this method, the plankton shells are identified with the aid of various search techniques, and after biomechanical examinations and finite element (FE) simulations and calculations, they are evaluated, abstracted, and adapted to structural stress situations and restrictions due to fabrication with the help of methods such as parametric optimization of genetic algorithms.

Using ELiSE, information and data are drawn from unique collections of samples and preserved specimens (Hustedt Laboratory for Diatom Research), as well as a 3D databank of concrete, pre-optimized lightweight structures (parametric computer-aided design (CAD) models and microscopic data). The technical application is supported by the foundational research in the areas of evolution, biomechnanics, diatom taxonomy, and genetic algorithms.

A function-optimized derivation of a completely new and diverse structures and geometries for the development of technical lightweight structures results in the process. The ELiSE process has been implemented in the areas of automobile design, air and space industry, medicine, offshore structures, civil engineering, industrial housings, and consumer goods.

3.3.6 Technical Biology, According to the Definition of VDI

Next to the terms "bionics" or "biomimetics," the term "technical biology" is often used in the literature.

VDI 6220 explains in Sect. 5.4 that the term "technical biology" was introduced by Werner Nachtigall as a complementary term to biomimetics. Technical biology comprises the analysis of form–structure–function correlations of living organisms with the aid of methodical approaches from physics and engineering sciences. "Technical biology" is thereby the starting point of many biomimetic research projects, as it allows us to have a deeper understanding of processes of biological precedents on a qualitative level and then can initiate an implementation process for a technical application in a suitable manner. Technical biology is, according to N.W., an essential facet of basis research, because "where nothing is researched, there is nothing to implement" (Werner Nachtigall 2010, p. 198).

3.4 Building Biomimetics

Building biomimetics is a subdiscipline of biomimetics and covers the areas of building design and construction of architecture and civil engineering.

On the basis of core essence of building biomimetics and its methods, building biomimetics is also applicable to industrial design.

Building biomimetics uses biomimetics as tool for creativity.

Building biomimetics connects the classical definition of biomimetics with the analogy researches and with technical biology.

Building biomimetics uses next to the "traditional" methods of "Biology Push" and "Technology Pull," in particular the method of "Pool Research."

3.5 Classification of Building Biomimetics

According to the VDI guideline 6220, a product is considered biomimetic when it fulfills these three criteria:

1. Biological precedent
2. Abstraction from biological precedent
3. Transfer and application

The VDI definition implies that all three criteria must be fulfilled. If it is only consistent with one or two of the criteria, then it cannot be described as biomimetic.

In his book *Bionik als Wissenschaft*, which applies the theory of cognition to biomimetics, Werner Nachtigall (2010) signified this process with the subtitle: "Knowledge → Abstraction → Application."

A definition always becomes difficult in the context of complex products, whose development was shared by several lines of knowledge, among them from the area of biomimetics. On the basis of the dimensions of products from architects and civil engineers and the resulting complex relationships therein, it would be difficult of ascertain a perfect classification according to the precedent of the VDI. There will never be houses completely inspired by biology. Nonetheless, they can be differentiated according to the degree of inspiration, abstraction, and technical application and, furthermore, the significance of biomimetics for the development of a building can be underlined. One can term technical products as biomimetic when the prominent characteristics are biomimetic.

The following *classification for buildings* is to be understood on the basis of an analysis of the development lines and the degree of biological inspiration in the architecture. It facilitates the understanding of building biomimetics. This classification of building biomimetics represents the influence of architectural understanding. For improved understanding of building biomimetics, the following categories are conceived:

Similar to nature: buildings as sculptures similar in appearance to nature
Nature analog: building methods analogous to nature
Integrative: biomimetic principles as components of architecture

3.5.1 Similar to Nature: Buildings as Sculptures Similar in Appearance to Nature

So-called landmarks still play a major role in architecture today, particularly when a building is supposed to be established as a special attraction. Buildings can be used then as built exclamation points, when they set themselves apart from the common perception of the built environment, when they are different from the traditional experience that is taught to us: A house has to be built with angles, with windows, and a pitched roof or, if necessary, a flat roof.

Sculptures that are formed in the "appearance" of nature follow the precedent of Binet (Fig. 3.4) and other realized precedents and are developed as a direct image or as loose interpretation of natural forms. Examples begin with the artistic embellishments for the Casa Milà by Antonio Gaudi and today with the TGV station at Lyon-Satolas by Santiago Calatrava (Fig. 3.5) and Parasol for Sevilla by Jürgen Mayer H. (Fig. 3.6).

Particularly, Santiago Calatrava, the brilliant contemporary Spanish engineer and architect, pushes supporting elements to the limit and stages buildings as sculptures derived from nature, which serve the purposes of function and beauty in equal regard.

Fig. 3.4 Entrance building for the 1900 World Exhibition in Paris, Georges Binet. (Courtesy of Haeckel House, University of Jena)

Fig. 3.5 TGV station at Lyon-Satolas, architect Santiago Calatrava. (Courtesy of A. de Luz Mendes)

Fig. 3.6 Metropol Parasol, architect Jürgen Mayer H. (Courtesy of K. Köhler)

3.5.2 Nature Analog: Building Methods Analogous to Nature

J.G. Helmcke, biologist, and Frei Otto, architect, discussed in the 1950s and 1960s whether similarities between living and built structures are coincidental or whether conformity to laws also underlies living structures, similar to built structures. From 1970 to 1985, with the collaborative research center (Sonderforschungsbereich, SFB) SFB 64 "Weitgespannte Flächentragwerke" ("Long-Spanning Surface Structures"), studies were conducted under the leadership of Frei Otto on natural structures, which received high international recognition. This research concerned itself, among other things, with networks in nature and technology, expandable structures in living nature and technology, and biology and construction. From 1984 to 1995, it was followed by the collaborative research center SFB 230 "Natürliche Konstructionen" ("Natural Structures"), a research program intended for architecture, urban planning, building structure, and design. With this program, self-forming and self-organization processes in all areas of inorganic and organic nature and technology including house and settlement construction were considered. Self-optimization, form finding, and origination of form in technology and art were likewise researched alongside behavior mechanisms of animals and humans in relation to house and city and the esthetics of natural and technological structures. To the understanding of biomimetics at the time, Frei Otto wrote in SFB 230: "For the streamlining of the topic, biomimetics, or the using of living structures as technological precedent, was not incorporated into the original program. Understanding nature was more important for us than using nature." Obviously this exclusionary

Fig. 3.7 Olympia Stadium in Munich, architects Günther Behnisch and Partner. (Courtesy of G. Pohl)

approach could not be sustained. The study of nature inevitably caused reactions in building form and design. Many architects and engineers have been influenced by the results of the research work from the SFB 64 and SFB 230 to investigate natural structures and gain insights for their creative processes. They discovered the issues of lightweight construction and unlocked the secrets of natural, minimal structures step by step. In the modern pursuit for energy-saving and materially efficient structures, the consideration of findings from research into natural structures becomes more and more essential (Fig. 3.7).

3.5.3 Nature-Integrative: Biomimetic Principles as Components of Architecture

This case involves the integration of biomimetic principles in architectural structures. The integratively termed "structural architecture" helps visualize the real effects of the forces and loads in relation to the used materials.

The relationship with nature will become clear with an example. If one studies the effects of loads on a thigh bone in a cross-section of the bone, he or she will recognize the material reaction to these external conditions in the course of the spongy bone and its density. The internal structural stress leads to a redistribution of the material and to a structurally optimized arrangement of mass. Comparable approaches to structure in architecture have led to developments of minimal surface structures and shell and expanding constructions, for example. In nature the structural system never assumes only one function but fulfills a multitude of various demands simultaneously. Exterior skins perform the tasks of load distribution, substance exchange, and attraction, and occasionally defense against enemies as well (e.g., sea urchin shells).

In architecture, the considerations of complex structural systems of nature can lead to construction of effective and sustainable buildings that would then have to assume a multitude of tasks: structural support, moisture control, acoustics and sound insulation, heat insulation, advantageous forms for internal air flow, advantageous forms for external air flow, volume reduction, material optimization or minimizing, and additionally, the task of design esthetic. A building facade could reach this level of complexity with the use of sophisticated material and functional components, with the use of delicate, low-input structures as well as its physical realization as an updraft facade, combined with storage and cooling masses within the building.

3.6 Potentials of Building Biomimetics

Building biomimetics with its access to the reservoir of ideas and inspirations found in nature can offer the serious potential to better develop technical products.

Observing nature as prototype will not always be crowned with success. However, the failure to consider the potential gain of knowledge from building biomimetics will lead from the start to a reduced palette of possible solutions.

3.6.1 Demands of Modern Buildings: Modern Architecture with the Use of Biomimetic Insights

The demands of modern buildings are, as they have been for the hundreds of years of technological development, characterized by the pursuit for efficient discourse with our resources. It is then to be noted in the current era of computers and the Internet that the diffusion of "new" materials and constructions for ever newer products has been accelerated. The multitude of possibilities has become unimaginable, and this always to the advantage to the task of construction. However, the individual elements of increasingly commonplace technical composite materials are usually inflexibly bound to one another as a complete package. Nature follows a different course in this instance. Its composites decompose after the death of the creature again into its individual parts. But with the aid of computers, technological work methods can resemble the processes of biological genetics sometimes to an astounding degree: Our computer technology enables, for example, "genetic design processes." On a broader front, analogies to biological processes, biomorphic architecture, and biomimetically developed detail solutions, which are being incorporated into building structures, are emerging.

Therefore, it can never be repeated enough that it does not matter in which way the building arrives at its final form, biomimetic or not. The secret of a well-designed building lies in the skillful combination of all creative tools and in the knowledge of technology and praxis.

3.6.1.1 Energy Efficiency, Material Efficiency, and Functionality

The transition from lightweight structures, designated by Frei Otto as "natural structures," to integrative buildings is smooth. The more complex the solution is for the fulfillment of several demands in one structural element, part or building, the more one can speak of "integrative biomimetic principles." The necessity of understanding complex systems often leads to consideration of individual aspects and thereby better explaining the function as a whole.

In nature, material efficiency means the effective discourse with the "expensive" materials produced from metabolic processes. Nature has developed particularly efficient and light shell and fold structures that can grow and be nonetheless stable. Their potential can certainly be fathomed for technological applications. Natural structures have developed building processes in plants and in animals that, on the one hand, negotiate the use of locally accessible raw materials in the form of efficient shell and fold constructions and are structurally optimized and in many regards multifunctional but, on the other hand, can be constructed and expanded with growth processes. Examples are not only the shell structures of mussels and sea urchins, but also the folded structures of leaves: hornbeam, palm varieties, and so on.

After studies of natural growth and optimized forms, the knowledge of bone mineralization and structurally optimized fiber arrangements was abstracted and implemented for the speed skating hall in Erfurt (architects: Julia and Göran Pohl), where they were re-interpreted as pressure-bearing steel struts, so-called bone struts. As components of a combined enclosure and structural system they consist of several spatially linked, arched frames with a superstructure and a substructure. The superstructure is supported with the "bone struts" only at the most structurally necessary positions.

Naturally occurring shell forms served as additional inspirations, such as those of the sea urchin (Fig. 3.8). The strengthening ridges found in the sea urchin shell are represented in the architectural interpretation by individual spanning elements.

Fig. 3.8 Shell structure of the sea urchin *Echinus esculentus* as inspiration for an effective building method. (Courtesy of G. Pohl)

Strictly speaking, the structure of the speed skating hall follows a shell-like beam construction that consists of an arched framework, which exhibits a span-length of 83 m and is radially arranged at the ends. The enclosure as such spans approximately 20,000 m^2, which can be compared with the area of an entire soccer stadium.

The structure of the "bone struts" was preceded by direct parallel studies on bone and tree growth systems, whose findings were abstracted from the structures of natural systems. The technical interpretation followed on the basis of the performance requirements, namely using simple, industrial fabricated, and commercially available prefabricated products (T profiles) to produce an affordable lightweight structure with the least amount of material waste. This lightweight profile is optimized like its natural "precedent," optimally adapted to its purpose of application (Fig. 3.9).

Comparisons of the sea urchin shell with the shell-like, arched frame structures of the speed skating hall in Erfurt reveal the differences and similarities between the two (Fig. 3.10) .

The ribs can be clearly seen in the image of the sea urchin shell. The similarities between both construction methods arise with spatial merging of the rib structures, which gives the shell construction in both instances its 3D form (although the

Fig. 3.9 Biomimetically developed, so-called bone struts of the speed skating hall in Erfurt, Pohl Architects. (Courtesy of C. Bansini)

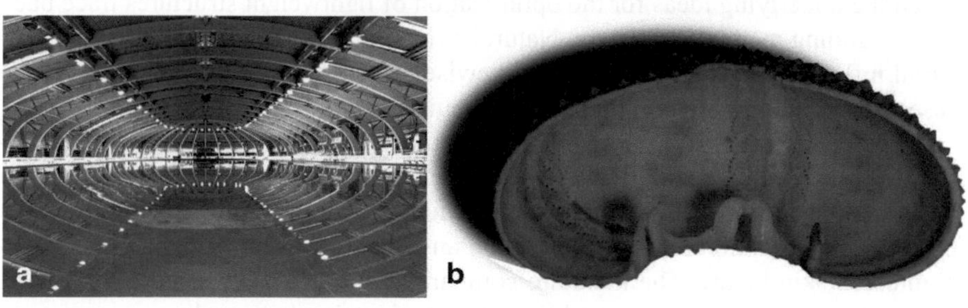

Fig. 3.10 **a** *Left* Interior view of the speed skating hall in Erfurt, Pohl Architects. (Courtesy of G. Pohl) **b** *Right Echinus esculentus*, interior of a shell. (Courtesy of G. Pohl)

individual ribs for the speed skating hall in Erfurt are built as frameworks). The difference is based on the fact that the rib supports the sea urchin shell fused with the shell envelope. In contrast, the structure of the speed skating hall is more skeletal and the shell envelope is not fused with its ribs. The bond is only produced as such so as to distribute external stresses from the envelope to the load-bearing elements, avoiding compression loads (which in theory can also occur within the sea urchin shell, but due to the small dimension of its structure, they would not be able to exert a serious influence on the entirety).

The cited example of the pressure-strut structure for the speed skating hall in Erfurt shows how an optimization and technical application can be reached with the "traditional" material of steel by analysis of natural efficiency strategies. The limitation of the structure consists mostly of the questions of cost-efficient use of prefabricated products to realize an open structure of this type and of materiality and material costs. Neither a massive single strut in the form of a standardized I-beam, nor a solid laminated timber strut instead of the open steel struts would have been capable of yielding the separation of the spatial envelope from the structural system. The expression of the building elements "arch" and "envelope" and with it the comprehension of the structural system is reached by the overcoming of alleged technical constraints—with help of the knowledge of natural optimization mechanisms.

3.6.1.2 Life Cycle

The life cycle plays a commanding role in nature, whether the matured structures are occupied by new life forms or decomposed into its basic elements, from which new life forms can emerge. Researchers are currently developing materials and building elements in the scientific areas of biomimetics that can integrate themselves with a life cycle, as observed in nature. However, research on this subject is only yet at the beginning.

3.6.1.3 Material-Efficient Construction with "Old" and "New" Materials

Often the underlying ideas for the optimization of lightweight structures trace back to the building methods of nature. Natural structures react to internal as well as external influences, and their forms are likewise influenced by such factors, as is the case with technology-based, human-made buildings.

Lightweight materials for envelope structures, which in turn possess good insulation qualities with a high level of light penetration and diffusion, are similar in construction to the system within polar bear fur. The fur of the polar bear fulfills the purposes of insulation and the redirection and even diffusion of light to the dark-pigmented skin below. The hairs are comparable to parallel-oriented glass fibers, which also perform insulation and light distribution simultaneously.

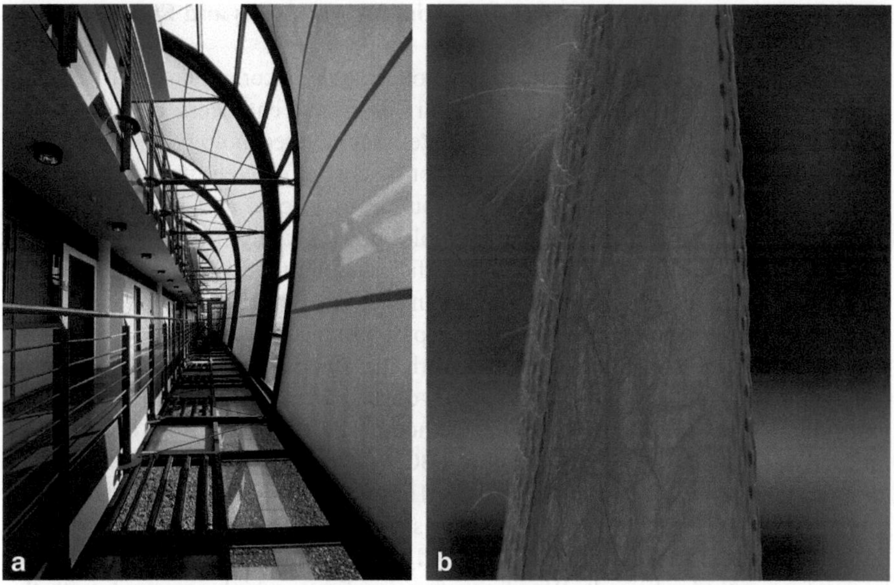

Fig. 3.11 **a** Terminal EF in Erfurt, updraft facade with translucent envelope, Pohl Architects. (Courtesy of G. Pohl) **b** Construction detail of the glass fiber weave for the light diffusion system of Terminal EF, Pohl Architects. (Courtesy of G. Pohl)

The example shown in Fig. 3.11a and b of a facade construction with multi-layered polytetrafluoroethylene glass and interspersed glass fiber weaving fulfills similar functions as the polar bear fur, but adapted to technical demands and without the pigmented underlayer. In contrast to polar bear fur, the weave consists of wool-like, randomly arranged thin glass fibers that scatter the light instead of directing it yet still possess—as with polar bear fur—a heat insulating effect.

This building technique was implemented for a facade structure on the "Terminal EF" building in Erfurt, Germany. It supports the building's cooling system in the summer, conserves warmth in the winter, and provides glare-free work spaces for the entire year. This building skin houses the access and communication areas. The heat insulation value is with 1.1 $W/(m^2K)$ comparable to the then (end of the 1990s) conventional glass facades (compare: Sect. 6.37).

3.6.2 Potentials of Nature-Integrating Building Techniques

Nature-integrating systems combine technological knowledge and insights gained from various sources with those that originated on the basis of natural precedents.

3.6.2.1 Biomorphic? A Research Potential for Architects and Engineers

Shell-like and biomorphic structures have once again become popular in contemporary architecture. The form language of architects and engineers, developed with the help of generative methods and computer software technology as non-uniform rational B-splines (NURBS) models, cannot be cost effectively implemented with traditional building technology. Current building systems and technologies can barely keep pace with modern planning tools and can barely fulfill the consequent demands. Examples for a computer-driven fabrication process for timber construction are provided by the Centre Pompidou in Metz by Shigeru Ban (Fig. 3.12).

Higher costs associated with building part production, assembly of auxiliary structures, and the construction itself absorb the otherwise feasible material cost-saving potential and actually exceed than—despite these savings—the normal cost expenditure with traditional building methods. The research effort Bionic Optimized Wood Shells with Sustainability (BOWOOSS) has taken this problem as an opportunity, initiated together with the HTW Saar and the Bauhaus University Weimar, both in Germany. A numerical translation of the results from 3D structures designed for wood fabrication is processed by a computer (CIM) directly on the basis of an optimized result. Wood construction firms, which can work with this kind of 3D data, have already collaborated with specialists for the production technology. The development of new joint and connection details and flexible modules should, as a corollary, be undertaken for suitable materials in order to yield the

Fig. 3.12 Centre Pompidou in Metz, France, under construction, architect Shigeru Ban. (Courtesy of G. Pohl)

desired sustainable results in symbiosis with composite solutions. An optimization approach comparable in its result can be found in nature with shell structures and should be modeled and investigated as a potential for technical derivation.

3.6.2.2 Hierarchical Structures as Optimizing Strategy

The optimizing strategies of the shell constructions in microorganisms had already fascinated Frei Otto and his team at the SFB 230. Analyses of the formation of diatom shells lay at the focus. Presumptions about their functions and a direct implementation in technology were expressly ruled out of the scope of the proceedings. However, new scientific work has indicated clear mechanical constraints for diatoms and additionally the optimization of mechanical features.

Planktontech, a virtual institute of the German Helmholtz Society, concerns itself with the fundamentals and principles of optimization capabilities of lightweight structures in marine microorganisms, in plankton. The shells of diatoms (Fig. 3.13) and radiolarians stand at the focus, which are distinguished by their strength coupled with minimal material application. With the help of modern microscopic observation tools the shells are analyzed, translated into 3D data, and processed with various calculation and optimization tools, making it possible to study the biomechanical characteristics of ocean organisms as well as principles of evolution.

The 3D data are to be used for industrial applications as well, in the area of lightweight building methods. The emphasis lies in the combination of lightweight structures with composite materials for the development of new products in numerous technology sectors such as architecture, automobile design (→ tire rims), and medicinal technology. For this purpose, C. Hamm developed ELiSE (compare Sect. 3.7.2), a tool that searches for structure elements and new forms for technical lightweight structures and makes them available in the form of finite element method (FEM-calculated base forms in a 3D database (Fig. 3.14).

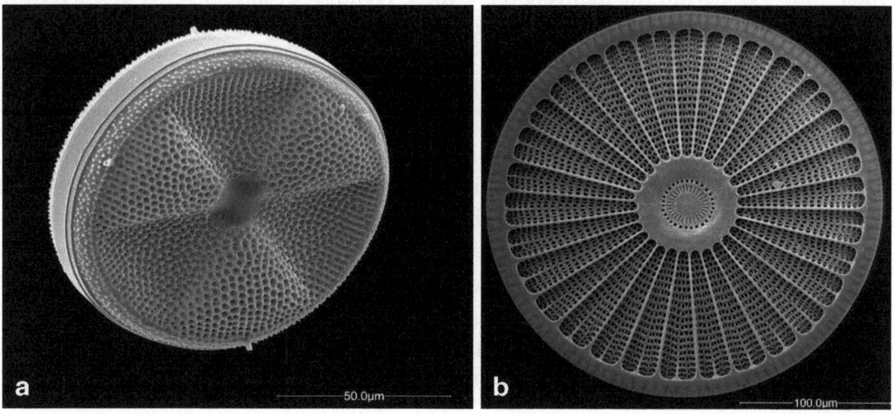

Fig. 3.13 **a, b** Diatoms *Actinoptychus* and *Arachnoidiscus*. (Courtesy of Alfred Wegener Institute Bremerhaven)

Fig. 3.14 Three-dimensional analysis, representation of the form of *Actinoptychus* for the investigation of various load-bearing stress situations. (Courtesy of Lightweight Structure Institute of Jena)

Fig. 3.15 Typical structure of diatom shell, idealized. (Courtesy of Alfred Wegener Institute, Helmholtz Centre for Polar and Marine Research)

Members of PlanktonTech include the German Alfred Wegener Institute; Harvard University; Rutgers University; the Universities of Kiel and Freiburg, Germany; the TU Berlin; the Institute for Textile Technology and Process Engineering, Denkendorf; and the Lightweight Structure Institute of Jena, the research department of Pohl Architects.

Measurements of the strength and stability of diatoms, which had been performed in the frame of the research activities of PlanktonTech, have shown that considerable forces must be applied in order to break their biomineralized shells. With the application of FEM calculations on diatom structures, conclusions could be drawn from the material characteristics of the shells. Silicate withstands applications of high tension and pressure and has elasticity similar to the solid bone. In addition, the diatom exhibits an extraordinarily efficient arrangement of structural members inside the silicate shell (Fig. 3.15). The complexity of the shell structure appeared to be important: a simplification of the shell geometry with the same level of material application led to an essentially lower rigidity.

Another task of PlanktonTech is to discover potential areas for application to structures for support and envelope systems of buildings. Normally findings from the areas of biology cannot be directly translated into a technological dimension. The lightweight shells are in this instance certainly exception, as surface pressure and material section can both be scaled in proportion to each other. According to Hamm et al. (2009) a diatom shell can be scaled, for example, by a factor of 10^6 without having to essentially change its internal relative dimensions.

The functions of the shells can also be essentially focused on the factors of mechanical strength and lightweight construction, often in combination with permeability.

It has longer been known that the formation of the diatom shell occurs in special vesicles, the "silica deposition vesicles" (SDV). These vesicles form a hollow chamber inside of which the precipitation of silicate occurs. On the other hand, how the formation of the SDV is driven, is still largely unexplained.

Diatoms attain maximum stability with a minimum of material and therefore follow the same rules as modern lightweight constructions do. This approach is pursued for many different applications in facade and roof structures.

According to the fat droplet hypothesis from Helmcke (compare Sect. 5.1.1), during the process of their shell formation, the diatoms produce fat droplet molecules on the exterior surface whose negative space forms the shell shape with typical openings filled with liquid silicic acid. But this hypothesis, which had at the time been debated in scientific circles, appears completely illogical in light of the well-known multiplicity of forms and the identical formation of a species.

The roof structure of an enclosure for a new train station at Luxembourg-Cessange (Fig. 3.16; prize winner of an international competition for the new construction of the Europa station Luxembourg-Cessange, Pohl Architects with SteinmetzdeMeyer and Knippers Helbig Engineers, compare Sect. 6.26) was designed according to the precedent of diatom shells. The refinement of the structure is hierarchical, starting from a primary structure system to a supporting secondary system and a tertiary structure, which with its triangular grid pattern forms the glass roof. As the basic module, a hexagonal form was chosen that allows from 15.0 to 21.0 m span lengths. The load-bearing structure consists of welded steel tube profiles. The secondary support system is likewise divided into hexagonal modules and spatially forms a dome, so that the roof system supports itself as a shell structure primarily subjected to pressure loads. Upon this structure lies the triangular frame network in 1.50–2.25 m grid that delicately follows the dome forms of the secondary structure and carries the glazing.

As approaches to the application of hierarchical structures, as they occur with radiolaria, a series of tests were performed in the frame of the collaboration with the research institute Planktontech at the ITV Denkendorf for carbon-fiber-reinforced polymer hexagon structures to help better understand the hierarchical silicate constructions of diatom shells and to generate test structures for facades and roof elements. The sheathing of a water tower in Chemnitz, Germany, is planned as an

Fig. 3.16 Train station roof at Luxembourg-Cessange, Pohl Architects with SteinmetzdeMeyer and Knippers Helbig Engineers. (Courtesy of rendertaxi/Pohl Architects)

application for a hierarchical facade construction (Fig. 3.17a and b, design: Pohl Architects).

The multiaxial, transformable roof for the open-air theater in Feuchtwangen, Germany (design: Julia and Göran Pohl, Fig. 3.18) stands as a glass-fiber-reinforced plastic (GFRP) structure in its test phase. For the lamella-like GFRP wings that are 18.0 m long and 2.6 m wide, subsupport frames of glass fibers are integrated in triangular recesses. The hierarchical structuring is in this case also a further development of the shell principles of diatoms and uses the design approach of using standardized industrial building elements for the advantage of simple assembly.

The shape of the columns for the roof structure (Fig. 6.52, Sect. 6.52) visualizes the features of optimizations, which had already been implemented in an analogous approach with the speed skating hall in Erfurt. The branches are used similar to tree limbs. Welded steel tube profiles are once again used, which corresponds to the location of structural stress.

3.6.3 Evolving Design and Evolutionary Urban Planning

The possibilities of modern computer technology lead to selection processes that can be to a certain degree similar to those of natural evolution. At the Institute for Computer-Based Design, ICD, at the University of Stuttgart, Germany, Achim Menges and his colleagues have researched the possibilities of computer-supported

Fig. 3.17 **a**, **b** Studies for the sheathing of a water tower in Chemnitz, Germany, with hierarchical facade elements with a fiber composite, lightweight method of construction. Pohl Architects. (Courtesy of N. Feth)

algorithms for evolutionary design strategies (Sect. 6.23) and evolutionary urban planning (Sect. 6.27). In cooperation with the Institute of Building Structures and Structural Design, ITKE, at the University of Stuttgart, led by Jan Knippers, building structures are studied and optimized for their engineering–technical function, their structural behavior, and—in special cases—their mobility. From this collaboration, fascinating biomimetic-inspired buildings and structures are developed.

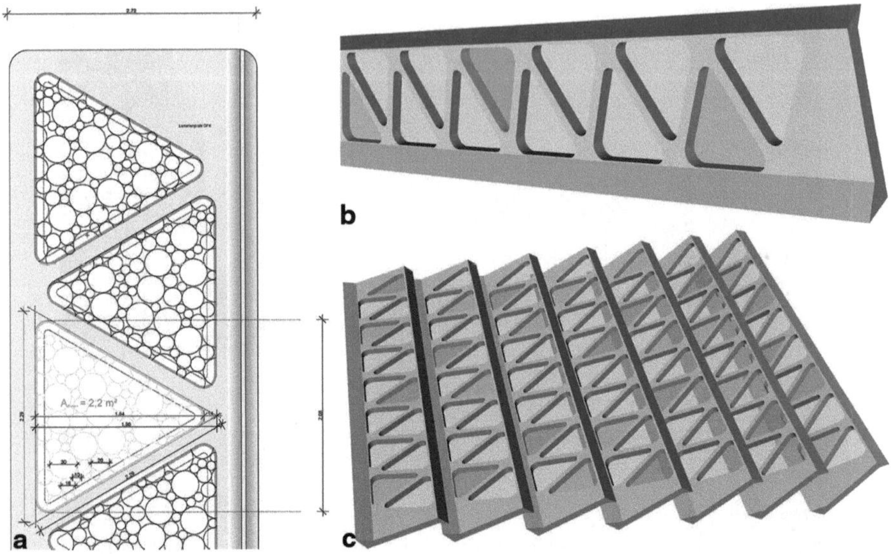

Fig. 3.18 Roof structure of the open-air theater in Feuchtwangen, Germany. Pohl Architects. **a** Hierarchical subdivision of the frame structure with porelike, celled elements. **b** The elements show a design variation with differently colored glass panels that were discarded in the final planning phases, as the colors would have interfered with the stage scene. (Courtesy of Pohl Architects)

3.7 Methods and Approaches Related to Building Biomimetics

Building biomimetics applies to architects and civil engineers. The predominant methods used by these fields have been discussed in the preceding sections. Alongside these methods further tools and approaches exist that follow similar paths and should not remain unmentioned in this context.

3.7.1 Scionic®: Industrial Design and Biomimetics

At the University for Arts and Industrial Design in Linz, Austria, a connection between inductive inspiration and industrial design and serial production is taught under the leadership of Axel Thallemer. According to Thallemer, Scionic® focuses "on heuristic inspirations from nature, virtual model building, and iterative optimization, as well as empirical verification of the found forms. In (natural) scientific base research," and formed products "in interaction with esthetic, technological, scientific, and psychological factors."

The theoretical superstructure was specified in Thallemer and Reese (2010) "Visual Permutations" and Thallemer (2010) "Scionic."

The education program for industrial design at the University for Arts and Industrial Design in Linz describes Scionic® as synergy between the factors described by Thallemer. "As a distinction from the perception of the now loaded term 'design,' the neologism of Scionic® is propagated for the same purpose. This term references the fundamental knowledge database of nature as a sign in the sense of syntax, semantics, and semiotics." Products can emerge as such, whether they are of virtual or real nature, mobile, or immobile.

3.7.2 Methods of Structure Optimization and Self-Organization

At the Karlsruhe Institute of Technology, KIT, Claus Mattheck developed particular methods with the goal of structure optimization according to precedents from biology. At KIT, the Institute for Material Research investigates strategies of nature for the structural optimization and the potential of these insights for the application in technology.

Nature does not principally optimize all of its building processes. On the basis of access to nutrients, energy, and building materials, structures have been developed that either demonstrate little optimization or are exceptionally thrifty in their use of locally accessible resources. These processes use the so-called Soft Kill Option (SKO) method. This method simulates the mineralization in a bone, where high-stressed areas are structurally strengthened and less-stressed ones reduced in mass. Applied to technical structures, the SKO strategy leads to efficient use of materials and therefore a lightweight structure (Fig. 3.14). The SKO method is effective for support structures and defines the directions of stress within the structures so that areas under more stress are supplied with more strengthening material and non-structural areas receive less material mass. According to this principle, very efficient and lightweight construction methods can be developed, predominantly with application in machine and vehicular design.

Another and, in principle, simple and therefore diversely applicable method is Mattheck's "Method of Tension Triangles." This method is a graphic tool for the production of contours along directions of forces. The method of the "tension triangles" is a graphic method for the rounding of transitions with a forking or bending of the material. It describes, with help of simple geometric definitions, branches in the structure that are superior to the perpendicular geometries in common structures. The model contour retained as such is used for areas dominated by shear forces, for example, notches and sectional transitions, to induce mechanically efficient flows of forces. If the form of the component deviates from this flow of forces, the notch stress can be reduced by a local augmentation of the "form contour." Otherwise weight and material can be saved by a diminishment of the form ("shrinking") if there is unnecessarily high material mass (Fig. 3.19).

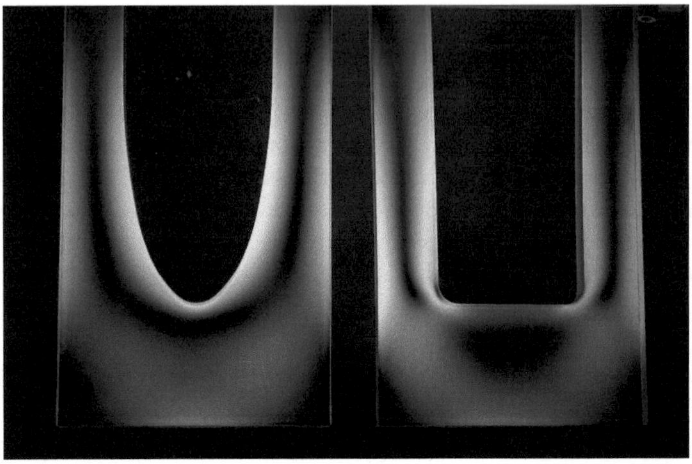

Fig. 3.19 Visualization of the behavior of forces in transition zones from a thick to thin cross-section of two steel components: **a** A tension-optimized construction developed according to the "Method of the Tension Triangles." **b** A traditional design with straight edges and rounded corners, model: KIT. (Courtesy of G. Pohl)

A closer description of the SKO method and its application is found in Sect. 6.21. Section 6.22 subsequently introduces the method developed by Frank Mirtsch for self-organization, a process of producing strength-enhanced sheet metal using indentations.

Chapter 4
Natural Functions and Processes as Prototypes for Buildings

4.1 Polar Bears and Alpine Plants: Transparent Insulation Materials

These examples were chosen as they define a prototype that span a wide variety of similar biological adaptations. Polar bear fur is extensively covered in this chapter due its well-investigated nature.

4.1.1 Polar Bear Fur as Solar-Driven Heat Pump and Transparent Insulation Material

Connections have already been drawn between thermal solar panels and the furs of arctic animals by Grojean and Koautoren in 1980. Electrical engineers and material scientists have since then brought forth a comparison utilizing this well-known data.

On the one side of the comparison stood solar panels, glass houses, double panel converters, and selective absorbers; on the other side the furs of endothermic animals. The authors have formulated five demands for solar energy converters:

(1) Largest possible absorption of energy.
(2–4) Minimal waste by using conduction, convection, and radiation.
(5) Relative independence from angle and direction of sunlight.

With polar bear hairs these demands are fulfilled by a unique, highly reflective cylinder inside the hair follicles and their foundation in a dark layer of skin, as well as the insulation characteristics of the fur. With these characteristics the hairs form the model of an ideal absorber, consisting of randomly arranged, yet nearly parallel cylinders with rough inner cylinders that as a whole function like optical fibers. As such they are calculated to have an average heat flow of 210 W m^{-2} during sunlight absorption.

Detailed studies for this system, as they are referenced in the following pages, were first performed by the physical chemist H. Tributsch and his research group.

G. Pohl, W. Nachtigall, *Biomimetics for Architecture & Design*,
DOI 10.1007/978-3-319-19120-1_4

Based on his work, astounding analogies emerged between the modern building material TIM (transparent insulation material) and the fur of the polar bear. Though the research of the polar bear fur and the development of TIM ran on essentially separate tracks, the eventual application of the TIM principles on the polar bear fur allow on the one hand a better functional understanding of this biological construction (from the standpoint of "technical biology"), and on the other hand the peculiarities of the fur were able to give inspiration for further technological forms (from the standpoint of "biomimetics").

Principle of the Heat Pump Heat pumps correspond to the inverse principle of the Carnot cycle. With the input of work a warm mass is generated for heating purposes; an output that can be much larger relative to the input, because the warmth is drawn from a reservoir of lower temperatures. The quality coefficient of the heat pump corresponds to the reciprocal value of the efficiency of thermal effects from the Carnot cycle. In this sense the polar bear fur is also a heat pump, as it concentrates warmth from the reservoir of sunlight on the skin of the animal.

Polar Bear Fur: Morphology and Radiation Effects The hairs are white and possess a central core cylinder. Figure 4.1a shows sections through these hairs; the central cylinder is visible as a dark sliver. In contrast, the white hairs of other animals, for example of a gray horse, are open, thin-walled cylinders and lack such central structures.

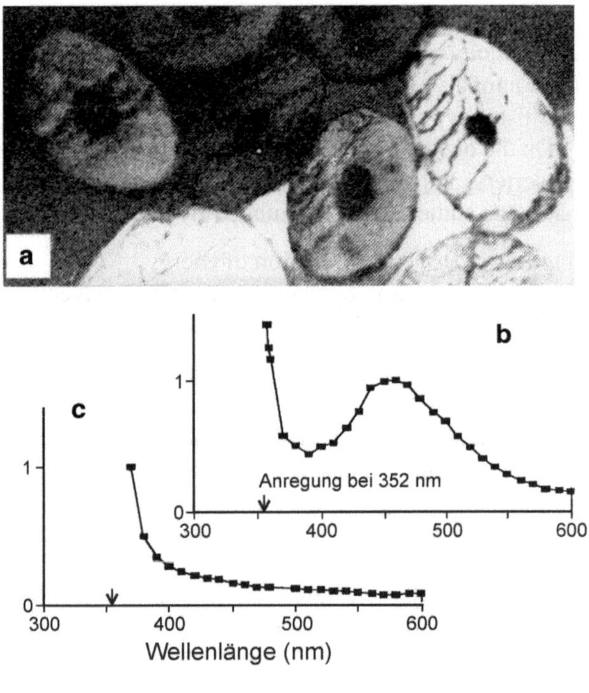

Fig. 4.1 Polar bear hair and luminescence. **a** Section through the white hairs of a polar bear. **b, c** Laser-induced ($\lambda = 352$ nm) luminescence in the hair of a polar bear **b**; no significant luminescence in the hair of a white pony **c**. (Adapted from Tributsch et al. 1990)

The central cylinder contains structures that scatter the light ("scattering centers"; Fig. 4.3a). Together with the total internal reflection in the outer layer, the hair therefore has the capability to function as an optic fiber. Furthermore, it can transform shortwave light into light of longer wavelengths. If one stimulates the hairs with a shortwave ($\lambda = 352$ nm) UV laser, one will find a broad luminescence maximum in the hairs of the polar bear fur around 450 nm; the hair of a white horse displays no such characteristics by comparison (Fig. 4.1b, c). Light scattering, reflectivity, and luminescence are clearly then basic functions of this hair.

The Polar Bear Hair as a Light Absorber and Solar-Driven Heat Pump The applied formulas used in this section are listed in Fig. 4.2.

If the process of light capture were to only rest on light scattering, then the thermodynamic boundary factor K_s would (with a refractive index of air equal to 1) be $K_s = bn^2$ (Eq. (1)) (where β is the geometric factor; $\beta = 4$ for three-dimensional light capture; n the concentration factor).

As the refraction factor of the hair is higher than that of air, diffuse irradiation can be concentrated in the hair without changing its frequency. The maximum concentration factor K_s is in this case 9.72.

The situation is represented differently in a frequency shift as a result from the phenomenon of luminescence (compare Fig. 4.1b). The hair can absorb high-frequency UV light and convey it to a lower frequency by luminescence. In this man-

$$K_s = \beta n^2 \quad (1)$$

$$Q_s = \frac{Q_a(\nu_e - \nu_a)}{\nu_a} \quad (2)$$

$$K_f = \left(\frac{\nu_e}{\nu_a}\right)^3 \left\{ \exp\left[\frac{-h(\nu_a - \nu_e)}{kT_s}\right] - \exp\left(\frac{-h\nu_a}{kT_r}\right) \right\}^{-1} \quad (3)$$

$$K_f^a = (1 - \eta_a) K_f \quad (4)$$

$$T(\nu, L_\nu) = \frac{h\nu}{k}\left(\log \frac{2h\nu}{3} c^2 L_\nu + 1\right)^{-1} \quad (5)$$

$$T_s = \frac{k_p T_a + k_s T_b + S}{k_s - k_p} \quad (6)$$

$$q_s = \left(\frac{k_s}{k_p + k_s}\right)\left(k_p T_b - k_p T_a - s\right) \quad (7)$$

$$\eta = \tau\alpha \frac{q_p^* - q_s}{S} = \frac{\tau\alpha}{1 + \frac{k_p}{k_s}} \quad (8)$$

Fig. 4.2 Equations that can be used for the formulation of the effects of the polar bear hair as a light absorber and heat pump and the polar bear pelt as transparent insulation material. See definition of the variables in the text. (Based on Tributsch et al. 1990)

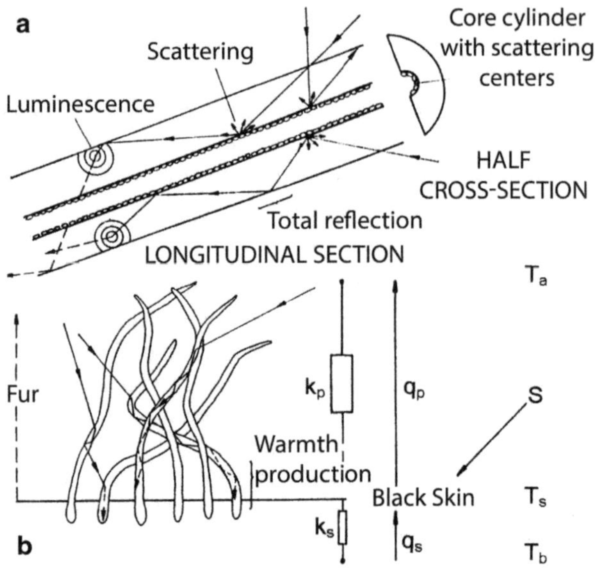

Fig. 4.3 Functions of the polar bear fur. **a** Polar bear hair as a light absorber; mechanisms: scattering, luminescence, and total internal reflection. **b** The fur of the polar bear as transparent insulation material (*k* is the inverse of heat resistance, *q* the heat flow (*arrows* in positive direction), *T* the temperatures, *S* the averaged radiation energy, in relation to the absorption into the *black skin* and transformation into warmth. Suffixes: *p*, fur; *s*, skin and fat layer; *b*, body; *a*, environment). (Adapted from Tributsch et al. 1990)

ner, a portion of low-frequency warmth radiation can be generated that is ultimately led from the hair roots into the black skin; here the warmth is absorbed into the body. The flow of warmth Q_s is in this case linked with the impinging radiation Q_a by the frequency shift $\nu_e \rightarrow \nu_a$ (2) (indices: ν_e frequency of the absorbed light, ν_e frequency of the emitted light). For diffuse solar irradiation on the Earth's surface the highest possible concentration factor K_f is given by relation (3) (T_r is the temperature of the absorbed radiation and T_s the temperature of the emitted radiation).

With a large enough shift in frequency $\Delta\nu$ in the magnitude of $10^{14}\ s^{-1}$, as it is exhibited in the polar bear hair, thermodynamic concentration quotients of several magnitudes can be expected. However, Eq. (3) considers only the pure thermodynamics of solar energy transformation. Under realistic conditions a degree of efficiency η_a should be calculated, which relates the net-power flow with the total flow from solar irradiation. It results in Eq. (4).

A small shift in frequency already results in a relatively high concentration factor; in this regard the polar bear hairs have reached a maximum compared with white hairs of other animals (Fig. 4.1b in comparison with c).

It is advantageous to calculate according to the associated temperature variables of the radiation rather than with the radiation variables; one can calculate then the flows of radiation as one would calculate flows of heat in a temperature gradient.

The temperature of the solar radiation depends heavily on the frequency ν, but only marginally on the irradiance L_v: Eq. (5). Systems that transform the radiation energy from the sun into heat can be favorable or unfavorable according to their effectiveness; unfavorable when they use a high irradiance and a small aperture, favorable when they work with a larger aperture and provided that a frequency shift compensates a rise in radiation. Therefore a minimal frequency shift of 5×10^{13}–10^{14} s^{-1} is required, and the band of emission should not be too small in comparison to the band of absorption. Both conditions are fulfilled by the hair of the polar bear.

One can therefore summarize the effects as follows (Fig. 4.3a). The absorption of the largest part of the light in the hair of the polar bear results from scattering processes on the core cylinder. This absorption process cannot occur with a concentration factor higher than $K_s = 9.72$, though this has the advantage that the diffuse light causes the polar hair to appear white: biologically advantageous in a white environment. If nature had used a more efficient luminescent absorption process, the fur would unfavorably appear in a different color. After the light is once absorbed, nature uses luminescence as an efficient optical principle. As the hair is cylindrically constructed, the light remains on the outer envelope in the hair by its total internal reflection, without it being scattered away (what would more be the case with a planar surface). That is a prerequisite for luminescence effects, as they do not tolerate the additional scattering. In comparison with the white hair of other animals, polar bear hairs are clearly more luminescent.

With two important modifications in the direction to a "luminescence gate" and a broader band of luminescence, the hairs have received the character of heat pumps: the radiation is led to the root region of the hair, transformed, and absorbed by the (dark) skin. On the total reflecting exterior envelope of the hair, blue light and UV light are efficiently collected and transformed into luminescent light.

The viewpoint referenced here is controversially discussed in literature; Koon (1998) has taken up a critical review. He compared the measurements from different authors, some of whom are referenced here, and related them with his own measurements (Fig. 4.4). In this figure the ordinate axis is defined on a $-10 \log (I/I_0)$ scale with I_0 being the intensity of the incidental light and I the intensity of the transmitted light. Thus it would be calculated for a typical 2 cm hair of a polar bear with a

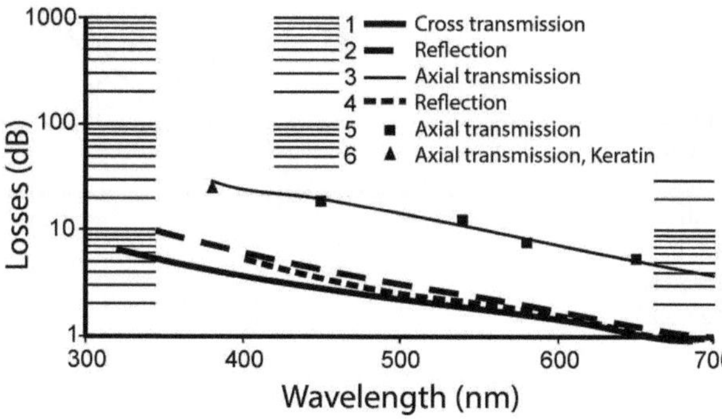

Fig. 4.4 Optical losses with the guiding of light in the hairs of the polar bear pelt, calculated according to different authors, compare to the text. *1–2* hair (from Tributsch et al. 1990), *3* axial light guidance (from Tributsch et al. 1990), *4* reflection on the fur's surface (from Grojean et al. 1980), *5* axial light guidance in a single hair (from Koon 1998), standardized to a 2.3-mm long hair, *6* axial waste in keratin (from Bendit and Ross 1961), standardized as with 5. *Curves*, each standardized to 90, 80, 40 and 82%, with 700 nm. (Adapted from Koon 1998)

reduction of the light intensity of up to an order of magnitude of 20 (!). The author specifies some boundary conditions under which the fiber optic hypothesis could work despite the immense waste (2 dB mm^{-1} in visible and around 10 dB mm^{-1} in UV light) that occurs obviously in the absorption on the "interior side of the exterior envelope of the hair." For details and a critical consideration of the different results one should refer to the original works.

The Polar Bear Fur as Transparent Insulation Material In their totality the hairs form the pelt; this encloses a multitude of air-filled interstitial spaces. It is translucent, but retains warmth due to the insulation effect of the air pockets; therefore it functions as a transparent insulation material.

Scientists have developed a series of such materials that can be compared with polar bear fur based on their underlying theoretical concepts. In simplified form the following can be accepted (compare Fig. 4.5):

If one solves the heat flow equation $q_p = q_s + S$ (with $q_s = k_s(T_b - T_s)$ and $q_p = k_p(T_s - T_a)$) according to q_s, one obtains the Eqs. (6) and (7). As soon as the radiation variable S is larger than the product k_p $(T_b - T_a)$, the heat flow is directed into the body. With the light radiation-induced heat flow q_p* and the heat permeability τ of the fur and the skin absorption coefficient a, the equation emerges for the efficiency of the usage of solar radiation power (8).

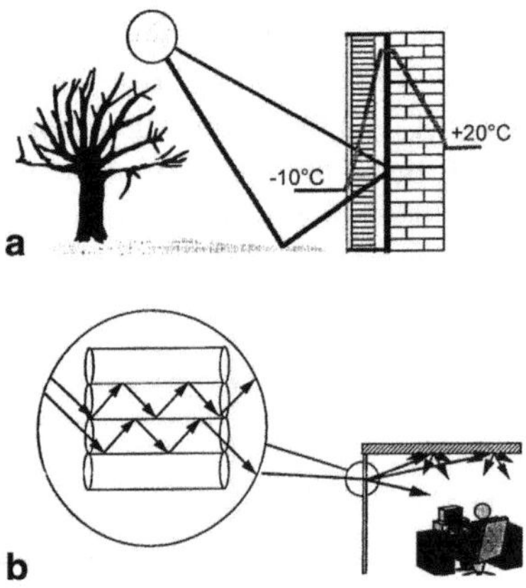

Coefficient of Warmth Transmission	$k = 0{,}7\ W\ m^2\ K^{-1}$
Degree of transmission of visible light	$\tau = 65\ \%$
Total energy passage	$g = 65\ \%$
Net energy gain per area per heating season	$90 > \Delta W < 120\ kWh\ m^{-2}$
Net energy waste of an opaque wall per area per heating season	$-25 > \Delta W^* < -30\ kWh\ m^{-2}$
Energy savings per area per heating season	$115 > \Delta W^{**} < 150\ kWh\ m^{-2}$

c

Fig. 4.5 Translucent heat insulation and typical characteristics. **a**, **b** Kapilux-H, product Okkalux Capillary Glass GmbH. **c** Characteristic values of a typical translucent heat insulation system. Boundary conditions: warmth conductance ITWD=0.1 W m^{-1} K^{-1}, built depth 10 cm, both sides glazed, south facing, Swiss Plateau, wall behind the insulating element 30-cm lime sandstone. K value opaque wall 0.33 W m^{-2} K^{-1}. (**a**, **b** According to Okkalux from Herzog 1996, **c** from Rüegg and Völlmin)

Therefore one can state that an increase in the transparency of the pelt (τ) by the combined scattering and luminescence effects in the hairs and likewise with the optimization of warmth absorption in the black skin (a) increases the radiation absorption. The low warmth conductivity of (dry) fur (k_p) and relatively high conductivity through the skin and peripheral tissues (k_s) can function properly by prevention of large losses of heat and high preference for warmth absorption. One can understand the curious light capturing system of the polar bear pelt as a compromise between biological conditions, the development of white fur, and the physical advantage of the "harvesting" of available light.

Technological Potential of Natural Systems Due to its naturally white environment and the fact that despite the low ambient temperatures it hardly ever falls into an energy deficit due to its long and thick fur and its tremendous body size, the polar bear cannot maximize the warmth capture capabilities of its fur. Rather it uses these capabilities for different, physiological purposes, among them possibly orientation (not considered here).

Technology could borrow two principles from this system: For one, the evolution of innovative TIM's that absorb not only direct but also scattered, diffuse light and ultimately transforms them into heat radiation of longer wavelength for warmth; for the other, the general insight that—in contrast to technology—systems in nature are always optimized as complete systems (the polar bear as part of an entire heat energy system), never as individual, closed elements. Therefore it could be that the inclusion of other qualities (e.g., flow passages for automatic heat convection) in TIMs leads to a lower thermodynamic effectiveness but an overall better, building-technical efficiency.

4.1.2 Transparent Insulation Materials in Technology

To our knowledge polar bear fur has not stood as direct inspiration for the development of TIM materials, though the analogy has been known. Earlier TIM materials were fabricated from plastic tubes, which were however thermally unstable; current TIMs consist of very thin, parallel-oriented glass tubules. Sun radiation penetrates the parallel tubules under total internal reflection, strikes a black absorber panel that warms and radiates its warmth into the room behind. The warmth is not able to excessively escape back outside due to the insulating air content of the system (Fig. 4.5a). With Kapilux-H panels the tubes are in average 3.5 mm thick and covered on both sides by a glass panel so that they are protected from dirt and other contaminants. If the absorber panel is removed, the light is scattered deep into the space improving overall illumination (Fig. 4.5b). The k value of the mentioned panel system amounts to 0.8 W m^{-2} K^{-1} with a total energy transmission efficiency of 80%.

Accordingly, wall systems with transparent heat insulation gain more warmth over a heating period than what escapes through a normal wall of same dimension. They appear opaque; their transmission factor is only roughly 65%. In contrast, equally well-insulated windows (triple-layered heat-insulated glazing with krypton gas) with equal k value are up to 65–70% transparent but their total energy penetration amounts to only 40%; in contrast with the transparent heat insulation it amounts to 65%. "Insulation effects and solar energy yield are opposed to one another."

With a solar power of, for example, 500 W m^{-2} on a sunny day, 30% is lost due to reflection and escaping warmth. With the effective heat power in the absorber (330 W m^{-2}) in a symmetrical construction about half is due to the opaque wall and the other half to the insulation material. The warmed insulating wall emits much warmth externally particularly during the night and unclouded sky conditions, which diminishes the total heat gain.

Regardless of this fact, the technical application of the polar bear principle wins more and more fans. One of the pioneers was the architect Thomas Herzog of Munich, who applied this concept to a youth educational center in Windberg. Two pilot projects by A. Kerchberger at the University of Stuttgart resulted in a positive net energy gain of 113.7 kWh m^{-2} per heating season for a single family house in Ormalingen/Basel Land for 31 m^2 of active translucent insulation surfaces. For the restaurant Hunderwiler Höhe at an altitude of 1306 m, application of the insulation resulted in a positive energy balance per heating period of 138 kWh m^{-2} with an area of 42 m^2. Okkalux translucent insulation materials were used in both cases. This type of energy balance is, aside from the building method, greatly dependent on its location. In high-altitude Davos, the energy conservation per area is about doubled in comparison to Stuttgart, and about one and a half times the value on the Shetland Islands.

On its path through the translucent insulation material visible light is heavily broken and scattered. One can obtain a targeted light distribution by combining fiber optic elements—for example for glare-free illumination of entire office spaces using daylight—and simultaneously the desired positive warming effect. For this concept numerous designs have already been developed and realized.

In comparison to polar bear fur, it is lesser known that plants have also formed transparent insulation, for example the arctic willow, *Salix arctica*. This is a creeping willow that grows little above the earth. The catkins carry a hair-felted coat and are therefore protected against the cold. In addition, the entire plant covers itself with cellulose fuzz "as a carpet" (Fig. 4.6a). The heat radiation of the sun heats the plant's surfaces; due to the enclosed air in the cellulose fuzz the warmth cannot

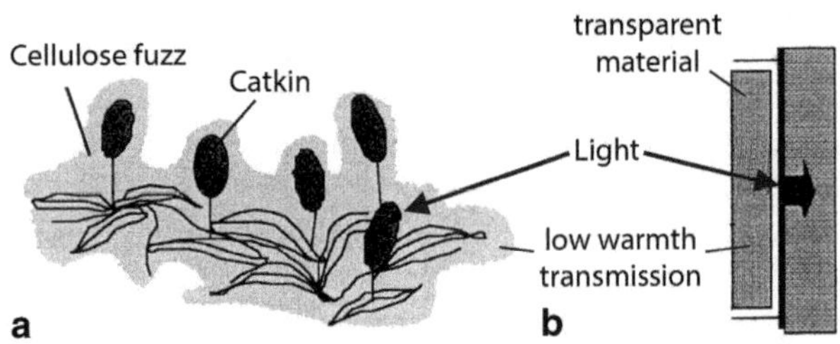

Fig. 4.6 Transparent heat insulation. **a** With the arctic willow, *Salix arctica*. **b** Technical transparent insulation material. (Adapted from Tributsch 2001)

escape easily. The system functions then completely analogous to the technical transparent insulation materials (Fig. 4.6b).

4.2 Termite and Ant Structures: Solar Air Conditioning

We have traditionally insulated houses homogeneously around the entire building perimeter. Animals insulate themselves using different methods simultaneously: each body part is insulated in a particular manner. A tundra goose "bears around 25,000 feathers on its body. In all, 20,000 of them alone are concentrated on the head and neck, where the brain and nervous system are located. The feathers protect the goose from piercing cold temperatures, above all during flight or arctic storms." Here it becomes clear that insulation and the degree of its efficacy in nature are specially modified for each need. "Intelligent solutions for heat insulation should be considered for architecture as well" (H. Tributsch). Well-studied examples for "adaptive insulation" are found in termite structures.

4.2.1 Climate Control in Enclosed Termite and Ant Structures

A classic print from 1781 in Fig. 4.7a shows one complete mound and one longitudinal section through a mound, possibly of a species of the genus *Macrotermes*. To note are the relative thicknesses of layers (related to the average) of the mortar-like structure, and thus their warming capacity as well. Nothing certain is known about the heat conductibility of the materials, though it can be abstracted from the structure that it is not very high, surely lesser than that of concrete, for example. It can be assumed that termites use a principle also utilized by the ovenbird (Fig. 4.13e), at least when the structures stand in partial shade: before the warmth during the day can diffuse from outside in, parts of the exterior wall will already be in shadow causing the flow of warmth to slowly reverse.

The "directional principle" of the compass termite *Amitermes meridionalis* functions similar to this "location principle" as well.

Compass termites construct their large, narrow structures with the length oriented north–south (Fig. 4.7b, c). The broad side then absorbs the heat radiation of the morning and evening sun, which can be advantageous due to the considerable temperature drop at night. When the sun sits at its highest position, the sunlight only strikes the narrow ridge of the mound. The heat absorption of the structure is proportional to the sunlit surface and the sine of the angle between the sun and the longitudinal median.

Solar or metabolic heat-driven air circulation and "pore ventilation" are typical for closed termite structures as a result of the partially permeable material. The Swiss biologist M. Lüscher has already studied the climate balance in termite structures for some time. The African termite *Macrotermes bellicosus* builds a variety of

Fig. 4.7 Termite structures. **a** 1781 print, *left* structure (possibly of *Macrotermes*) sectioned. **b**, **c** North–south orientation for structures of the compass termite, *Amitermes meriodionalis*. (**a** from Henry Smeathman 1781, **b**, **c** adapted from V. Frisch 1974, edited and complemented)

different installations: In the Ivory Coast, the closed mounds have long ducts that run underneath the outer surface and are covered by a porous material; in Uganda, they are open from below, but closed above in broad blind tubes; these are covered likewise with a porous material (Fig. 4.8a). Air circulation is induced in the interior by sun irradiation and metabolic heat; the direction of which depends on the time of day and amount of sunlight.

Cool and moist air is drawn up over the lower chambers (1) into the nest (2) with the Queen's chamber (3), collects in a dome lying above (4) and flows through the outer tunnels (5) and (6) back into the lower chambers. During the passage between

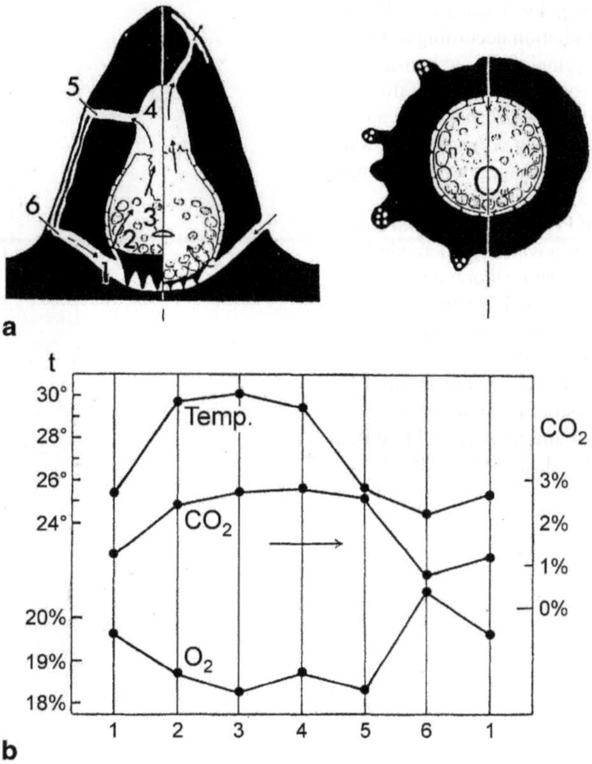

Fig. 4.8 Structural and regulatory processes in the structure of the termite *Macrotermes bellicosus*. **a** Longitudinal and cross-section of mounds of the Ivory Coast *(left half)* and Ugandan *(right half)* varieties. **b** Course of temperature and gas concentration with circulation in Ivory Coast mounds. The numbers in **a** and **b** correspond to one another. (Adapted from Lüscher 1955)

(5) and (6) CO_2 can diffuse out and O_2 can diffuse in. The behaviors of the temperature and gas concentration curves (Fig. 4.8b) reflect again the altogether beneficial effects.

Ants, which can also construct large mounds but in cooler climate regions (with the large red wood ant around a meter tall), use solar orientation somewhat differently. For ants, capturing the sun's warmth is more important. The nest is located in the earth under a hill and extends slightly into the hill. The sun-warmed area under the hill would only use sun-rays (a) (inset image in Fig. 4.9a) with the hill additionally using sun-rays (b). The usage is most effective when the hill side slopes about perpendicular to the sun's location, whose springtime location benefits the (otherwise partially shaded) nests the most. This is approximately the case.

A similar principle can be applied to plant conservatories (Fig. 4.9b): In springtime the sun-facing glass panes should stand about perpendicular to the average angle of sunlight.

Ants use another peculiar method in order to drive up the nest temperature in springtime: The individuals spread out onto the sides of the hill, warm themselves in the sunlight, disappear back into the structure and emit their warmth ("elementary heat storage and transport"). One can construct improvised heat storage structures for small greenhouses according to this principle. Here single rows of arranged, water-filled glass or plastic bottles can be implemented as elementary heat storage devices (Fig. 4.9c).

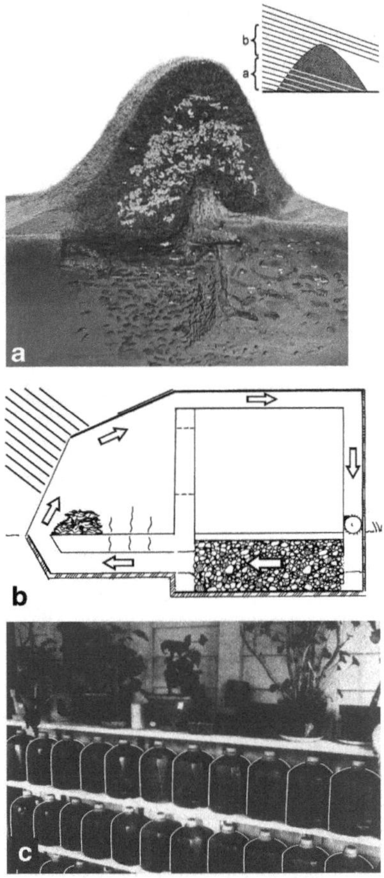

Fig. 4.9 Utilization of heat radiation according to the "principle of ant construction." **a** Structure of the hill of the *small red* wood ant, *Formica polyctena*, section. Inset image: radiation usage; compare to the text. **b** Principle of optimal slope for a winter garden window and stone storage mass. **c** Water-filled bottles as heat storage devices. (**a** from V. Frisch 1974, based on a diorama from the natural history museum of the TU Braunschweig, and **b**, **c** from Yanda and Fisher 1983)

4.2.2 Solar Chimneys in Termite Structures and Buildings

Energy Balance in Buildings The bar graph in Fig. 4.10 shows the percentages of heat loss for different building parts for a cooler climate region. Ventilation systems clearly represent the lion's share. It amounts to about 27% in average for conventional houses. With low-energy houses it lies by 57% (due to the lower significance of the other areas of heat loss). It is therefore worthwhile to focus serious attention on these ventilation effects.

Ventilation Channels in Termite Structures and Their Technological Interpretation Many termite species, such as members of genus *Macrotermes*, equip their mounds with chimney-like structures that rapidly warm up under direct sunlight (Fig. 4.11a). The planar, elongated structures of compass termites (Fig. 4.7b, c), whose upper portion is penetrated by chimney-like ducts. When the heated air rises, negative pressure develops below that draws cooler air up from the base of the structure. The base may have access to groundwater, sometimes with tunnels more

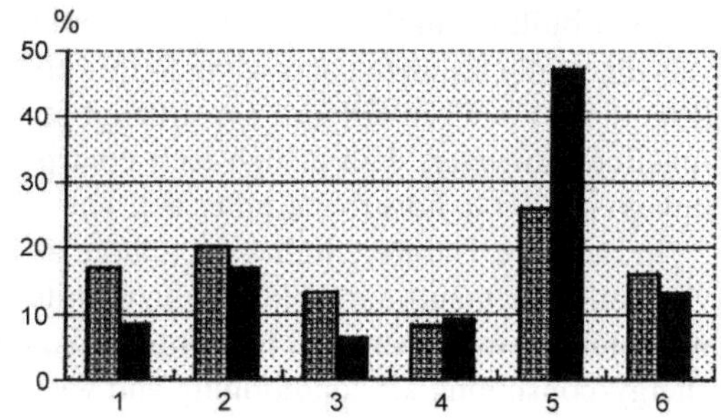

Fig. 4.10 Energy loss in houses. *1* Walls, *2* windows, *3* roof, *4* basement, *5* ventilation, *6* heating. *Gray bars*: standard house. *Black bars*: low-energy house. (Adapted from Bundesminister für Wirtschaft 1996)

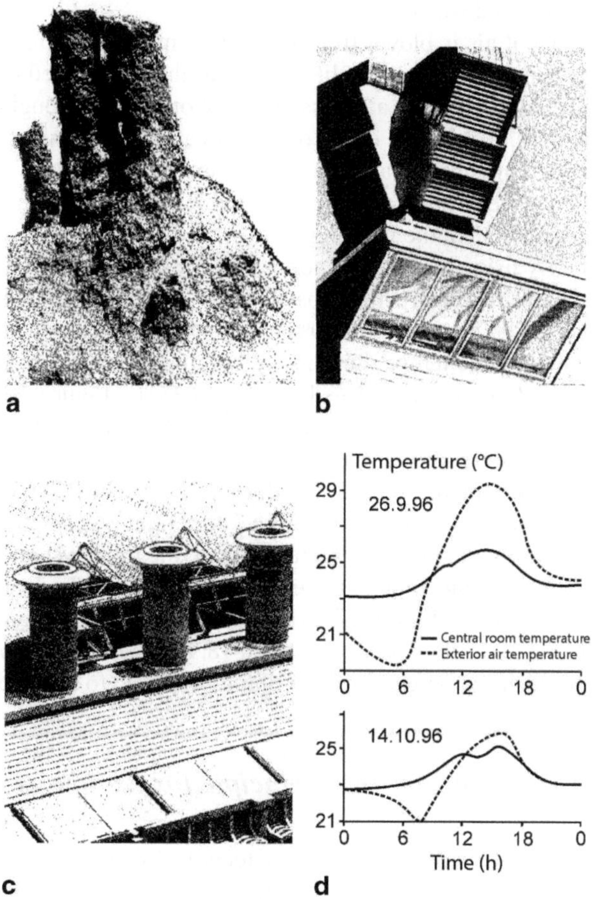

Fig. 4.11 "Termite-like ventilation systems." **a** "Chimneys" on the structures of the termite *Macrotermes sp.*, Avash National Park, Ethiopia. **b** Analogous ventilation chimneys on the building for the machine engineering department of Leicester University, England. **c** Analogous ventilation chimneys for an office building in Harare, Zimbabwe. **d** Temperature variation over the course of a day in the office building from **c**; compare to the text. (**a** from a photo by A. Sielmann, edited, **b** from Spiegel 1994, edited, and **d** from F. Smith from ARUP Journey 1997, edited)

than a dozen meters long. Compass termites can maintain the interior temperature around the queen's chamber at a constant 31 °C, even when the outer ambient temperature fluctuates between 3 and 42 °C. Under extreme exterior conditions they must then change the diameter of the chimney; they can accomplish this by accumulating or removing building material.

As a built example, a departmental building at Leicester University, England, was thermally regulated according to this prototype. The attached ventilation tower (Fig. 4.11b) is 13 m tall and constructed of bricks. The low-tech system can only function, however, with a high-tech control system that regulates the air supply, though similar to how the termites act. The system is more closely described below.

In Eastgate, Harare, Zimbabwe, a large office building was also constructed with ventilation elements according to the termite principles. The architect Mike Pearce was ordered to construct a building in which one could be comfortable without energy-consuming air conditioning and without practically any heating. He solved the problem in collaboration with the climate engineer Ove Arup using air shafts that form a cohesive system in the building, integrated in double-layered roofs, floors, and walls.

Cool air is blown from the atrium into this system and transported to the individual rooms through slits in the baseboards. Based on the systems of the termite, heated air masses are passively siphoned out through the altogether 48 chimneys (Fig. 4.11c) by the effect of solar heated and rising chimney air alone. The heat is stored in concrete and remains for the night and early morning: The city of Harare, which lies some 1500 m above sea level, reaches temperatures at night that barely remain above the freezing point.

By using biomimetic concepts 10% of construction costs were able to be saved; the entire construction costed only 36 million dollars. The monthly energy consumption lies at almost 50% lower other comparable buildings in the city. The average daytime temperature in the building lies by a mild 23–25 °C. Without the suction of cooler air it rises however to 35 °C. Figure 4.10 shows comparison curves for temperature variation in a day. On September 26, 1996 it was a hot day with a temperature differential of about 10 °C; the previous night was cool. The cooling effect amounted to 4.5 °C. The day of October 14, 1996 was preceded by a warm night (around 20 °C); the temperature differential was small, only 5 °C. The cooling effect still amounted to 2 °C. As mentioned, the total system functions only with the utilization of fans, though the energy balance is still advantageous. The relative power amounts to 9.1 kW/am^2. Six other similar, but mechanically ventilated buildings in Harare consume between 11 and 18.9 of these units, so that in comparison 17–52% of electrical power for the building was able to be saved.

4.2.3 The Termite Principle for Buildings

Figure 4.12 illustrates further technical examples for "termite ventilation"; three examples for the inclusion of solar-driven, termite-analogous "thermoregulating chimneys" as well. The engineering firm Arup & Partner was involved in all projects. The buildings are located in England and built in the first half of the 1990s.

The previously mentioned Queens Building at the University of Leicester (Fig. 4.12a) already represents a classic in design, whose concept had been significantly covered by the press at the time. With a footprint of 10,000 m^2 it houses the

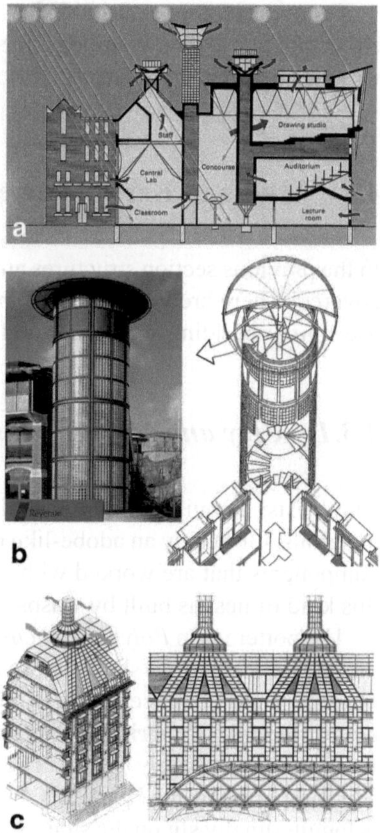

Fig. 4.12 Ventilation according to "termite chimney principle" for three buildings constructed in England during the 1990s, each with the participation of Ove Arup & Partner. **a** Combined "natural lighting" and "natural ventilation," Queens Building, University of Leicester, England. Short and Ford, London 1981–1993. **b** Inland Revenue Center Nottingham, England. Michael Hopkins & Partners 1992–1995. **c** New parliamentary building, Westminster, England. Michael Hopkins & Partners, London, from 1992. (Adapted from Herzog (Ed.) 1996)

department of mechanical engineering. It combines an almost entirely natural system for lighting with "natural" ventilation using cross-ventilation with the chimney effect. The ventilation chimneys are with their hood attachments in total 17 m tall and provide the building with a unique architectural feature.

An addition for the Inland Revenue Center in Nottingham (Fig. 4.12b) possesses a large, solar ventilation chimney as well, which additionally houses a staircase. Daylight can penetrate far into the structure; the interior flooring is not cladded and serves as a heat store and light reflector. Additional heating can be provided by long-distance, district heating; the total energy consumption remains under ideal conditions less than 110 kWh m^{-2}.

The design for the new parliamentary building in Westminster, London (Fig. 4.12c) includes no less than 14 towering thermoregulating chimneys; each converged upon by ventilation shafts in the exterior walls. These shafts supply fresh air and remove already circulated, stale air, for which they use a thermal wheel (left portion of Fig. 4.12c). The flooring system follows the same principle as shown in Fig. 4.12b. Groundwater is implemented for cooling. The energy requirements are resultingly low, around 90 kWh m^{-2}.

Thomas Herzog, as a publisher of the instructional book "Solar Energy in Architecture and Urban Planning," has given a series of further examples for different varieties of solar-driven air conditioning.

4.3 Mud and Earth: Ancient Materials

In the previous section structures and building methods of termites were extensively covered. There are yet numerous examples of shelters of other animal species that use similar building substances and methods though on a smaller scale.

4.3.1 Clay and Mortar Nests

Clay nests, as built by many swallows, are always a mixture of mud and fibrous materials, ultimately an adobe-like material. One can describe animal mortar as clay components that are worked with a saliva secretion. Figure 4.13a–4.13d illustrates this kind of nest as built by wasps.

The potter wasp *Polybia emacinata* sheathes its hanging honeycombs with an almost spherical mortar shell, whose diameter-thickness ratio is about 30:1. The form bears a round entrance on the side. The potter wasp *Polybia singularis* builds its nest differently. It fabricates thick-walled "ceramic nests," which are around 30 cm long and weigh barely 1.5 kg. The honeycomb structures are formed entirely with soil, supported by the side walls. In the middle it bears an exit hole, and the entrance is longitudinally slit on the side. Nests of this type become hard as stone. Indigenous potter wasps of the genus *Eumenes* build fingernail-sized, urn-shaped nests in the form of "construction ceramic" (as a potter would use for the production of large vessels). They bring their captured prey into the nest, which are then loaded with an egg, whose larva develops in this climatized clay shelter.

The nest of the South American ovenbird, *Furnarius rufus*, is particularly impressive. The genus bears the name "*Furnarius*" due to the oven shape of their nests (*furnus*: oven), which have already been mentioned in the introduction to this chapter. They are built from adobe as mud mixed with plant parts. For the fabrication of the 5–10 kg nests, the birds support the walls with around 2000 mud clumps each weighing 2–5 g. With an interior partition they separate an antechamber from the actual brood chamber. The diameter is measured at around 25 cm, and the walls are quite thick: the diameter-wall thickness ratio is calculated up to 7.5:1. The warming capacity of the nest is correspondingly large. By the time the exterior envelope is fully heated through from the early afternoon sun and the heat slowly begins to transfer into the interior, the outer wall will already be in partial shade and therefore able to release the heat. This principle is used for thick-walled adobe structures, utilized by the Pueblo in North America, for example.

Less noticeable than the structures of ovenbirds are the strategies of beavers to preserve warmth during hibernation periods. They spread moist mud, which freezes

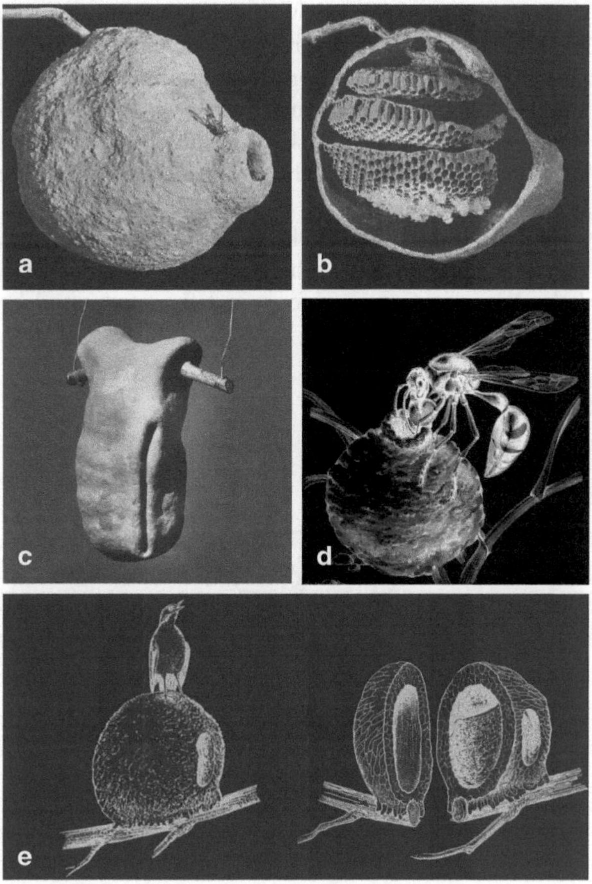

Fig. 4.13 Mud and mortar nests. **a** Potter wasp. **b** Section of a. **c** Potter wasp *Polybia singularis*. **d** Potter wasp *Eumenes* spec. **e** Ovenbird *Furnarius rufus*. (**a**, **c**, **e** from v. Frisch 1974 and **d** from Freude 1982, edited)

and efficiently insulates similar to the igloo of the Eskimos, onto the interior surface of the den. The unfreezing, dammed pool of water has the capacity to store warmth and is used in turn as a source of warmth for the den. Snow traps air and can accordingly function as insulation as well.

The only heat source for the beavers is their fat reserves in their bodies, which is slowly burned through winter. They rest close to one another to concentrate their body warmth and to reduce the amount of body surface exposed to cold. All other strategies are attempts to keep heat loss from the structure at a minimum.

4.3.2 Construction with Adobe

Adobe—essentially mud strengthened with additives—has advantages and disadvantages. This material has been consistently used by cultures since primordial times wherever it was accessible. The first Neolithic houses in the Euphrate–Tigris delta region, ca. 8500 BC, were thick walled, mud rotundas with integrated peaked roofs (Fig. 4.14a). In dry regions one can build multistory buildings with it. Because of the modest tension strength of the material, the walls are reinforced with tree branches which protrude and serve as the scaffold for the always necessary repairs after rainfall (Fig. 4.14b). Spherical or paraboloid rotundas can thereby be fabricated for purposes such as grain storage as they are in the Chad region of Africa (Fig. 4.14c).

It is little known that the mud structure also has an old tradition in northern climates, and not only with structurally uncritical, low-lying structures. Along the Lahn River in Germany four- and five-story mud houses have existed since the

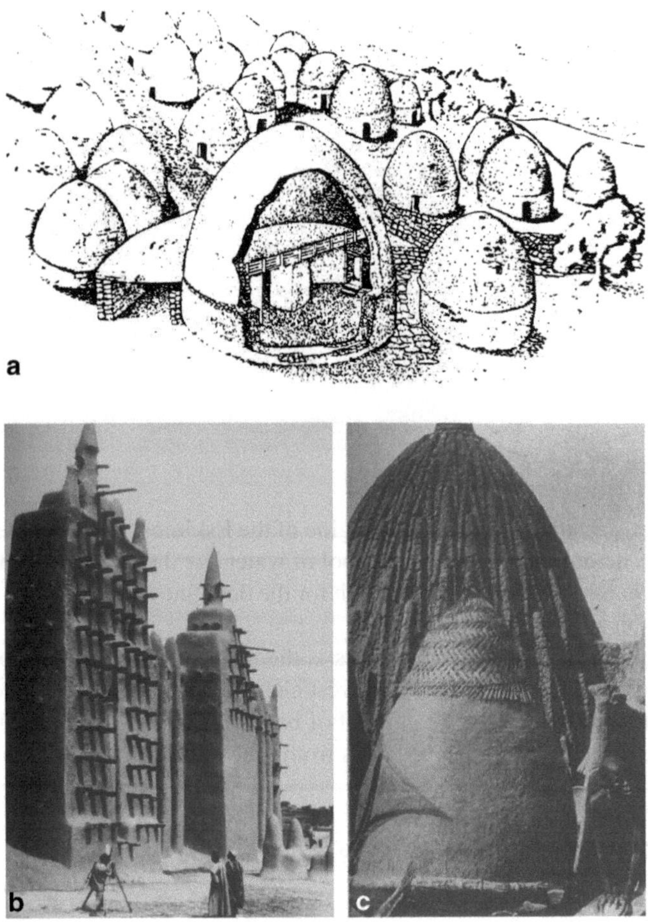

Fig. 4.14 Adobe structures. **a** Neolithic mud rotunda structures, Euphrate–Tigris delta, ca. 8500 BC. **b** Mosque Mopti, Mali. **c** Grain store, Musgu, Chad. (**a** from Müller-Karpe from Brandt 1980 and **b**, **c** from Brandt 1980)

Middle Ages; however, their meter-thick walls are cladded and therefore inconspicuous.

Where mud and manpower are available but the population is poorer, for example in the highland regions of Peru, self-built adobe structures built with competent guidance can be the better alternative for affordable housing. In Peru one can consecutively produce bricks or entire walls of adobe within wooden forms; for reinforcement one uses locally available Ichu grass, which grows at an altitude of 3500 m. For security against earthquakes one includes parts of wood or metal—also bamboo—or one places stones on the separating surfaces that generate additional friction against the tendency to shift, increasing the shear resistance between individual layers.

Teaching projects, which have awoken interest in the workmanship of indigenous materials and pride in the achievements attained by self-built structures, were frequently led by, among others, Volker Hartkopf in Peru and Balkrishna Doshi in India. The projects of the latter are depicted in Fig. 4.15. "The building should demonstrate the total integration of form, enclosed space, structure, and simple building technology. The people who had built it were so excited by this construction technique, by the building form, and by the capability to easily and 'naturally' apply alterations to their own house, that they had the feeling they could carry on the old rites of Pithora Bava."

Figure 4.16 shows how the inhabitants of an adobe house can assert their own influence onto the form of the building over time with self-design, mending, adding, and remodeling. In this case the house was inhabited by a large family from Mali. The photos were taken in 1993 and 2001, respectively.

Building-Climatic Peculiarities of Adobe The necessarily thick walls of adobe structures for their structural stability cause the adobe material to function as a heat reservoir. By the time the thick walls have been completely heated through under the tropical sun, it would have already become evening, allowing the rooms remain cool the entire day.

During the relatively cool nights the warmth is then released into the interior as desired. The humidity due to the respiration of the inhabitants is absorbed by the drier interior walls and diffuses to the exterior wall where it evaporates.

The material is denoted by a high resistance to compression forces, but relatively low resistance to tension and shear forces. As compensation for the latter, length-oriented supplementary materials such as tree branches, twigs, stalks of grasses, and even bamboo pieces are added. More recently, technical components such as plastic elements and metal wires and stakes have also been used for this purpose. Due to their relatively high water content, these types of structures should be securely shielded from electrical currents.

Figure 4.17 shows two traditional mud mosques from the Niger region in Africa, one that is relatively thin-walled with distinct exterior decoration and demonstrates architectural design possibilities (a); and one that is very unique, thick walled, and illustrates particularly well the physical and structural properties of adobe (b).

Adobe architecture is often combined with specific air circulation systems for cooling and air conditioning. One can then attain a thermally effective adobe

Fig. 4.15 Balkrishna Doshi in Ahmedabad, India. **a** Longitudinal section. **b** Finished weavings. **c** Almost complete coating of mud. **d** Mud brick backing for the weaves. **e** Interior. (Adapted from Balkrishna Doshi 1995)

structure, as illustrated for example in Fig. 4.18a, b. Important to note is the stable and relatively low interior air temperature in relation to exterior temperature and above all in relation to the roof surface temperature (c).

Adobe structures can become as hard as concrete, yet their building and climatic effects are completely different due to their composition and microscopic structure. Figure 4.19 illustrates this difference using recorded temperatures in the region of Cairo. The interior air temperature in an adobe structure remains within the comfort zone of this region; in a similar concrete structure, not at all.

Typical Questions and Answers for Adobe Construction Under the website "AdobeFragen," mentioned in the literature appendix, a question–answer catalog for adobe construction was published as information for questions that are repeatedly asked.

Fig. 4.16 Alteration of an adobe house over 8 years. People and textiles (hangings) removed by photo editing. (Original photos: Peter Manzel/Agentur Focus, Material World; www.menzelphoto.com)

What does adobe mean?

The word comes from Arabic Atob: Sludgy, gloopy earth or Atubah: Mud brick). The word means muddy soil as well as mud brick, mud pavement, and buildings of mud brick or spread mud with reinforcement materials and ultimately an architectural style.

Can adobe only be used in climate regions with little rain?

No; it is obviously well suited to those regions, but is widespread in countries all over the world, where one can construct earth architecture. The material itself is essentially the same from place to place; only the construction method differs by region.

What are mud bricks?

Mixed mud is spread with additive materials into a simple wood form and dried under the sun, which in sunny climates lasts about a week.

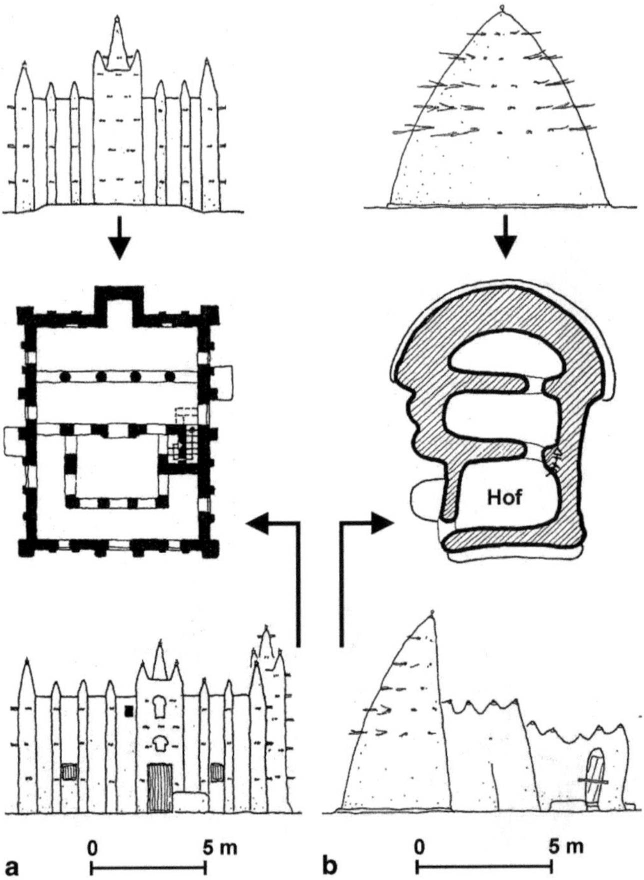

Fig. 4.17 Two traditional mud mosques from the Niger region, Africa. **a** Sirétaga Basariconta 1937. Area 91 m^2, tower height 8 m. **b** Conea. Unconventional, *thick*-walled architecture. End of the nineteenth century. Large Khaya tree in the courtyard. Interior partitions function as support walls. Area 52 m^2, tower height 7 m. (Adapted from Gruner 1990)

Must one mix straw into the adobe bricks and is it suitable for all mud or soil types?

It is suitable for all types, as the actual mud functions as binding agent, whose portion has classically measured in 150-year-old adobe buildings) been up to 32%. An admix of straw improves the rigidity.

Do adobe structures tolerate rain?

In principle, little. Vertical surfaces in regions with up to 60 cm rain per year per square meter erode at only about 1 cm per 10 years, horizontal surfaces faster (5–8 cm per year). They must be annually reworked or protected by linings.

Is adobe a good insulator?

It does not function particularly well for the passage of heat, but rather for the storage of heat ("fly wheel effect"). In an adobe structure the current average

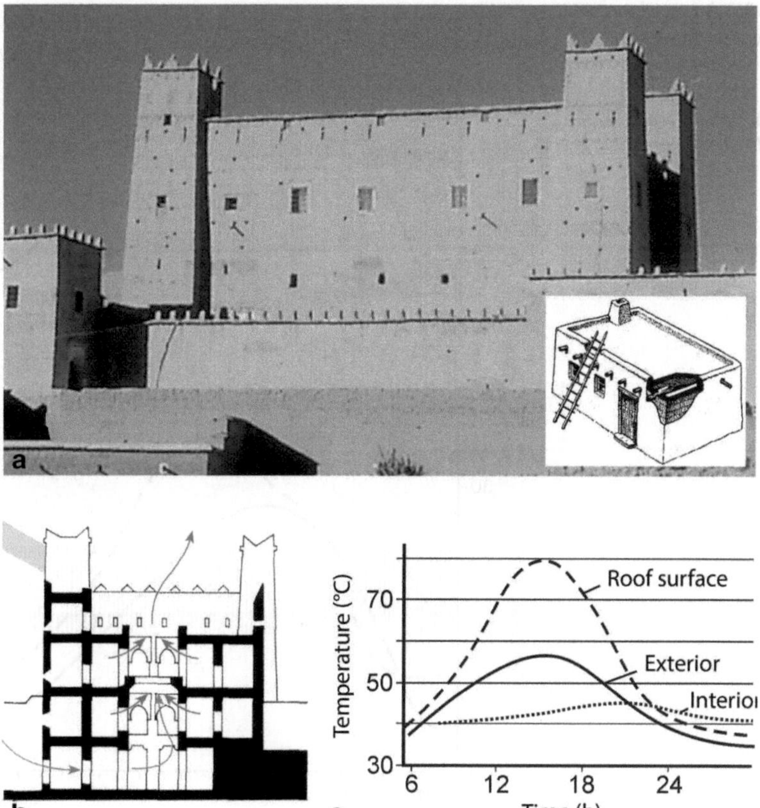

Fig. 4.18 Thermally effective adobe structures. **a** Casbah, Draa Valley, Morocco. Inset picture: Pueblo structure, southwestern USA. **b** Section of a. Heat absorption in *thick* walls and forced ventilation. **c** Typical temperature behavior in an adobe structure. (Adapted from Behling and Behling 1996)

interior temperature corresponds to the middle between the highs and lows of day and night temperatures of a few days before ("heat delay effect"). When the ambient temperature in 24 h period fluctuates between 15 and 30°, the fluctuation of the interior temperature amounts to only a few degrees. When over a few days the daytime temperature measures at 45 °C and the nighttime temperature 30 °C, the interior temperature would adjust to 37 °C: a higher comfort factor that dampens the major changes in exterior temperature.

Can one waterproof exterior walls of adobe?

Yes, with a cement lining, for example. The disadvantage: The lining hinders the passage of water vapor. Or by the application of moist, surface drying earth. More natural coatings are currently being researched.

Why is adobe so little used?

Fig. 4.19 Changes in temperature over the course of a day for structures with "*thick* walls." **a** Mud brick vault (adobe). **b** Prefabricated, similarly shaped concrete test model

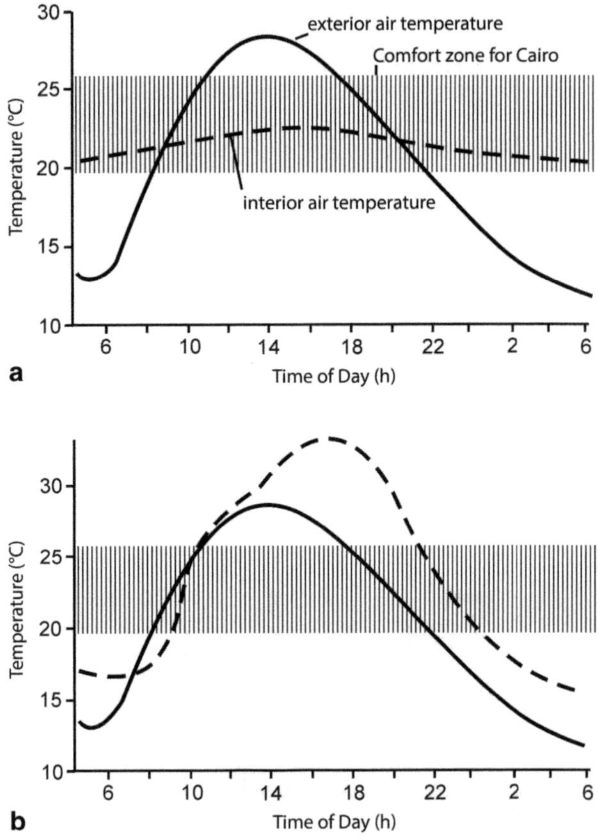

Perhaps because the building material has "poor people" image. In New Mexico there are on the other hand expensive adobe structures for wealthy customers, who have turned adobe into a status symbol. That can be hoped: In the USA adobe structures are commercially produced in Tucson and Albuquerque; there are total 0.2 million of these types of buildings of which 97% are in the Southwest.

How long have adobe structures been fabricated?

Adobe has been used for ages in every world region and climate. Jericho dates back to 8300 BC, buildings in Iraq up to 8000 BC, the first mud brick structures in Iran originated from 5600 BC, and in Peru and Ecuador from 3400 BC. In North America the oldest continually inhabited adobe structures are centuries old. In the sixteenth Century the forest stand in Germany was drastically reduced, as wood had been heavily burned for heat or used as building material. Decrees to build structures from earth substances originated from this time period in order to save wood. Based on similar reasons this occurred again in the eighteenth and nineteenth centuries, even in the periods after the World Wars. Since 1970 this type of construction for public buildings is no longer admitted in Germany.

Small Hospitals Built from Adobe Adobe structures must be adaptable to the particular conditions of the tropics. The prototype of "central hospitals" in urban centers of the industrialized nations cannot be indiscriminately applied to developing nations particularly in rural regions. For basic hospital care for these regions, as well as for urban slum populations, decentralized "basic health-care facilities" are being planned, which must cater to each need (visitor behaviors, cooking customs, time of use of the single functional units). "The decision process for construction and building materials is determined by a catalog of selection criteria that considers altogether the local capacities in view of durability, maintenance, costs, savings devices, etc." The Institute for Tropical Building, Starnberg, Germany performed studies and ultimately provided vaulted structures for certain tropical regions. These structures can be formed from adobe or other materials. Figure 4.20 shows a 120-bed hospital in Kaedi, Mauritania financed by the EU, which has been in operation since 1988. The construction of the hospital entails vault structures made from fired laterite stones and a foundation of natural stones.

Water Resistant Skin for Mud Bricks Mud brick or spread adobe surfaces tilt after the absorption of water from downpours, and the subsequent re-drying causes the formation of cracks; a problem that occurs in many semi-arid regions.

The Technical University in Melbourne developed an emulsion containing silicon that penetrates about a centimeter deep into the adobe surface and thereby binding with the material. A solid layer emerges, which hinders water uptake by around 99%. The emulsion is water based, affordable, and environmentally friendly. Mud houses in Port Moresby, Papua New Guinea have been experimentally treated with this coating.

Self-Built Projects "with Mud, Wood, and Straw." Because low-lying mud structures are structurally simple, they are well suited for do-it-yourself groups

Fig. 4.20 Part of a hospital facility in the tropics, vault construction of fired laterite stone (photo: Lippsmeier + Partner)

and communities, which have become increasingly popular in recent years. One of which has appeared on the Köllertalstraße in Saarbrücken, where a small, new settlement with 14 living units and community center was built. A structural wood skeleton was loaded with a mixture of straw and mud using a labor force of long-term unemployed and welfare receivers as part of an employment program. After an unpaid work effort of 400–600 h for each house, they were able to move into them. The connection of work, qualification, social integration, and housing construction creates a special feeling of motivation and achievement, which by far exceeds the usual employment measures.

4.3.3 Earthen Materials and Dwelling in Earthen Structures

Rammed earth is another ancient building material that has been used since ca. 7000 BC in the Indus River region of Pakistan and for the Great Wall of China. In 1937 a five-story hotel was built in Germany with this material, and more recently in Australia as well. During the Great Depression thousands of such earthen houses were built in America. There are still notably many earthen structures to be found in France; according to statistics, today 15% of the French population lives in house of adobe or earth (something we cannot truly believe).

Earthen walls are 45–90cm thick. They have excellent heat insulating properties and do not necessarily require additional insulation. They are fire resistant and long lasting. One can pack earth material in wood plankings or shoot it with high pressure through pumps. The architect D. Easton developed the latter technique in Napa Valley in America. Although such a structure has proven to last centuries, it is nonetheless prescribed in America with a 5–10% cement addition, causing it to become more expensive and not necessarily better.

Numerous shelter-building animals, whose dwellings have already been described above, an entire series of colonial or solitary insects such as ants and some hymenopterans, and including a vast multitude of vertebrates and invertebrates build passages, chambers, and housings in the earth. They use the relatively stable underground temperature, which cools in the summer and warms in the winter, and the conditioning earth moisture. It is a tradition of vernacular architecture, found in particular on hillsides, to build the basement level of a house horizontally into the slope so that it remains cool and moist in the summer. Architects are increasingly picking up on these local traditions again today. Figure 4.21 illustrates combined use of sun, ventilation, and underground temperature for a "house on the hill," as realized by the architect H. H. Parson in Aldrans, Austria.

"This house for an artist is divided into three different zones that are arranged according to the simplest geometric principles: A one-story exhibition room is located in the entrance area, behind which a tall glass chamber that functions as the main source of light, heat reservoir, and buffer, and finally the diverse living spaces separated into three underground levels.

The impression of the spaces rests above all on the quality of the penetrating sunlight and the visual relationships from inside and outside—but also from the

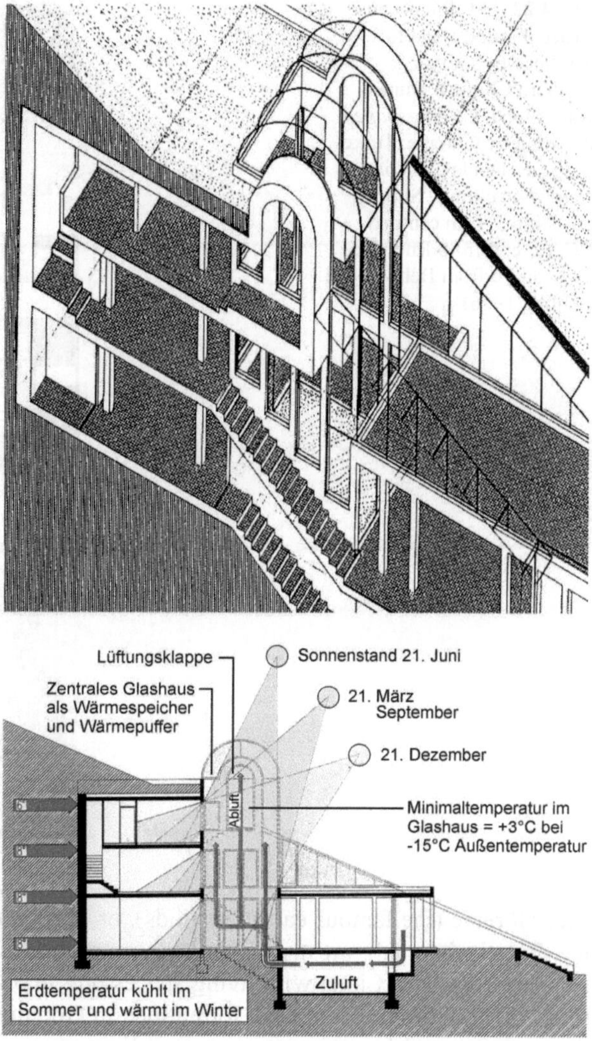

Fig. 4.21 Combined use of sunlight, natural ventilation, and cooler underground temperatures, "house on the hill," Aldrans, Austria. H. H. Parson, Innsbruck, 1984–1986. (Adapted from Herzog (Ed.) 1996)

conscious effort to produce a "cultivated cave," which satisfies the desire for reclusiveness and protection."

The "house on the hill" has a small building footprint and represents an alternative to common building forms on hillsides. The utilization of the constant underground temperature of 8 °C below 1.5 m causes in winter a decrease in heating requirements and in summer a subsequent decrease of cooling requirements for the house. The heat insulation of the walls to the soil depending on the location consists of 8 cm thick, waterproof, and closed-pore insulation panels.

Earthen dwellings in loess and tuff rock with their noted, positive structural and geological characteristics can be found in all arid and wind-eroded regions of the world or simply where loess outcroppings are, such as in northern China, in Turkey

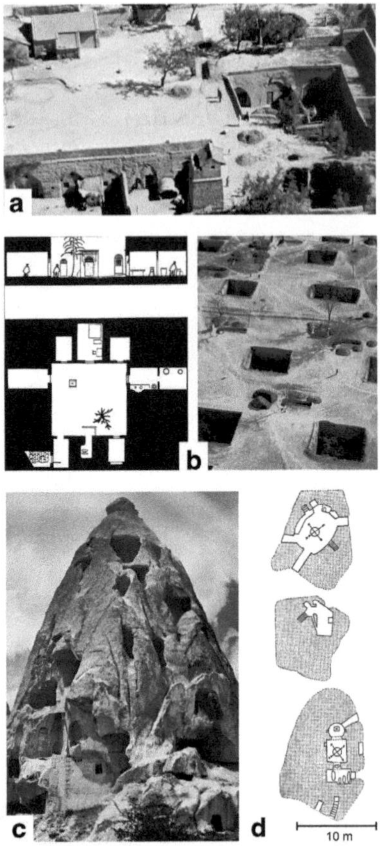

Fig. 4.22 Dwellings constructed in loess and tuff. **a** Village by Luoyang, northern China. **b** Dungkwan, China. **c** *Cone-shaped* dwelling by Göreme, Turkey. **d** Spaces inside of a *cone-shaped* dwelling, "Simeon the Stilite," fifth century BC. (**a**, **c**, **d** from Rudofski 1993 and **b** from Behling and Behling 1996)

near Göreme (the famous earth pyramids), or in southern France in the Loire Valley. Relatively small, mostly rectangular courtyards are sunk into the earth in the steppe of northern China with living spaces dug into the loess stone around them (Fig. 4.22a, b). Weathered rock formations in Göreme and elsewhere (Fig. 4.22c, d) are permeated by stairway-connected living spaces spanning several different levels.

In earlier times cave dwellings were very cheap due to an available workforce; their cost had been no more than one fifth of the cost of a common brick or wood house. Due to the good insulation ability of loess (constituted by very fine, silicate elements, baked together with lime, with a pervasive system of fine cavities) it can be 8–15 °C cooler in summer and up to 10 °C warmer in winter (without additional heating!) than the ambient temperature in the earthen dwellings of northern China. Because the sun enters the courtyards at an angle, the chambers oriented towards the south have traditionally been the most highly valued; they are formally reserved for the patriarch of the family. Dwelling in earthen and loess spaces ("troglodytism") can accordingly be very comfortable, particularly when the houses are carved completely out of the tuff stone and therefore free standing, as originally practiced in Le Beaux en Provence.

4.4 Building with Reeds and Bamboo: Rediscovered Traditions

Structures consisting of these types of materials belong to some of the oldest housing structures of humans. In East Asia bamboo is still massively used to this day due to its exceptional mechanical characteristics, and it additionally cooperates well with modern materials and techniques.

4.4.1 Ancient Reed Structures

Column-like, tightly and rigidly bound reed bundles can be variously used, such as for boat construction (Lake Titicaca) or as roof supports for houses (Mesopotamia). Figure 4.23 gives an impression of the astounding static-structural capabilities of reeds or similar representatives of the extended river vegetation family. These structural arches were fabricated out of what is known in literature as "giant reed"; also scientifically known as the *Arundo donax*, which grows over 6 m tall and possesses remarkable biomechanical characteristics, as shown by studies of biologist H. C. Spatz and botanist T. Speck of Freiburg.

4.4.2 Bamboo as Modern Building Material

Bamboo is a widespread building material, particularly in Southeast Asia; a grass, of which there are numerous varieties with very different sizes and characteristics.

There is extensive literature pertaining to bamboo as building material. A good summary can be found in the IL report of the Institut für Leichte Flächentragwerke (Institute for Lightweight Structures), University of Stuttgart, Nr. 31 from 1985. In the fascinatingly illustrated volume "Grow your own house" (Vitra Design Museum, 2000), which includes among others the bamboo architecture of Simón Vélez (Fig. 4.24a), the Zeri Pavilion of the EXPO 2000 in Hannover is extensively detailed, one of the most beautiful bamboo structures that has been built. Figs. 4.24b, c show two details of his prototype. How struts of bamboo can be integrated into modern architecture (wherever bamboo is present and wherever the bamboo building tradition has not yet died out) is shown in Fig. 4.14a in the interior view of a tall living space.

Bamboo and mud, two natural building materials, can also be easily combined. It results in a two-component material, a sort of "reinforced concrete," in which the bamboo members assume the role of tension support and the mud infill the role of

Fig. 4.23 Structure of 6 m tall "giant reeds" (probably *Arundo donax*) in the Euphrates region. **a** Beginning of construction and **b** finished building. (Adapted from Rudolfski 1993)

compression resistant matrix (Fig. 4.25). Obviously it is not suitable to northern climes, aside from experimental constructions.

As such, biomimetic consideration and thinking can be traced back to their roots, so to speak. The starting point was the ancient architecture that grew from a trial and error process, analogous to biological evolution.

4.5 Incorporation of Wind Power: Animal Structures and Ancient Building Cultures as Analogies

The orientation of animal structures toward wind or water flows is a principle of nature necessary for survival. The temperature regulation and ecological efficiency of these structures are not achieved with active metabolic energy—as the demands

Fig. 4.24 House in Arbelaez by Simón Vélez 1998. **a** Living room; X. Londonio. **b** Detail of the prototype for the EXPO pavilion (EXPO 2000), which had been constructed in Manizales, Columbia; part of the bamboo structure network. Simón Vélez 1999. **c** A node of **b**, with the use of stalk–root transition of bamboo. (Photos: **a** Xemena Londonio, **b**, **c** Zeri Foundation and Vitra Design Museum, adapted from Anonymus 2000)

of which would rapidly exceed the limits of such processes—but with passive use of forces in the environment. This is true for the orientation of ancient houses as well, relative to environmental forces such as wind. It simply appears to be the more elegant principle: instead of working, letting the work perform itself.

4.5.1 Use of the Bernoulli Principle in Animal Structures and Buildings

The Principle and Examples from Biology and Technology As generally known, negative pressure occurs in a nozzle apparatus (spray bottle principle, principle of

Fig. 4.25 Bamboo–mud house, Caldas Province, Columbia. Two stories, 60 m^2 living area, US$ 5000. **a** Overview (background removed). **b** Filling the bamboo structure with mud

the water-jet pump). According to Bernoulli the total pressure, or the sum of the static pressure and dynamic pressure, is constant in a horizontally mounted flow system: $p_{tot} = p + q = const.$ (p is the static pressure, $q = \frac{1}{2}\rho v^2$ dynamic pressure, ρ the density of the fluid, v the flow speed of the fluid).

If a tube of this type of system becomes narrower, the speed of flow and with it the dynamic pressure must increase at the narrowing point as a result of the law of continuity, and consequently the static pressure sinks in proportion to the exterior pressure and it can suck in fluid at this location from the region of outer fluid.

Steps in a continuous flow system function like "half nozzles". Lugworms of the genus *Arenicola* build their U-shaped tunnels so that one exit lies on lower level and the other on a higher plateau on a sand ripple in shallow water. Flow that runs perpendicular to the ripple line produces a negative pressure at the higher lying location and sucks fresh water through the tunnel, thereby providing the worm with its necessary oxygen. With an oblique onset of flow relative to the sand ripple the effect is reduced according to the sine law.

Familiar examples can be found among the burrowing vertebrate animals. Prairie dogs, *Cynomys ludovicianus*, fabricate their likewise principally U-shaped tunnel structures so that they always amass the excavated soil at one entrance in the form of a "conic volcano"; the opposite opening is stamped flat. The prairie winds ventilate the structure according to the Bernoulli principle; the moving fluid exits from the cone-shaped entrance. Because the cone shape is circular, the ventilating effect is independent from the direction of wind. S. Vogel and co-authors of Duke University, Durham, have measured the effect on models that simulated a prairie dog structure at one tenth of its natural size. After a build-up period the volume of flow over time is proportional to the wind speed within a broad range. Recalculated for a natural structure of about 20 m in length shows that even small wind speeds can have major effects; wind with a speed of 0.4 m s^{-1} ventilates an entire structure in 10 min; with 1.2 m s^{-1} it lasts only 5 min to completely replace the air (Fig. 4.26a, right ordinate axis).

Without induced ventilation, as explained by the Bernoulli principle, life in these types of structures would not be possible, and therefore the entire ecosystem of the North American prairies would also appear different.

As J. Olszewski and S. Skozen have shown, the branching tunnel system of moles, *Talpa europaea* (Fig. 4.26b), also uses induced ventilation. The flow speed

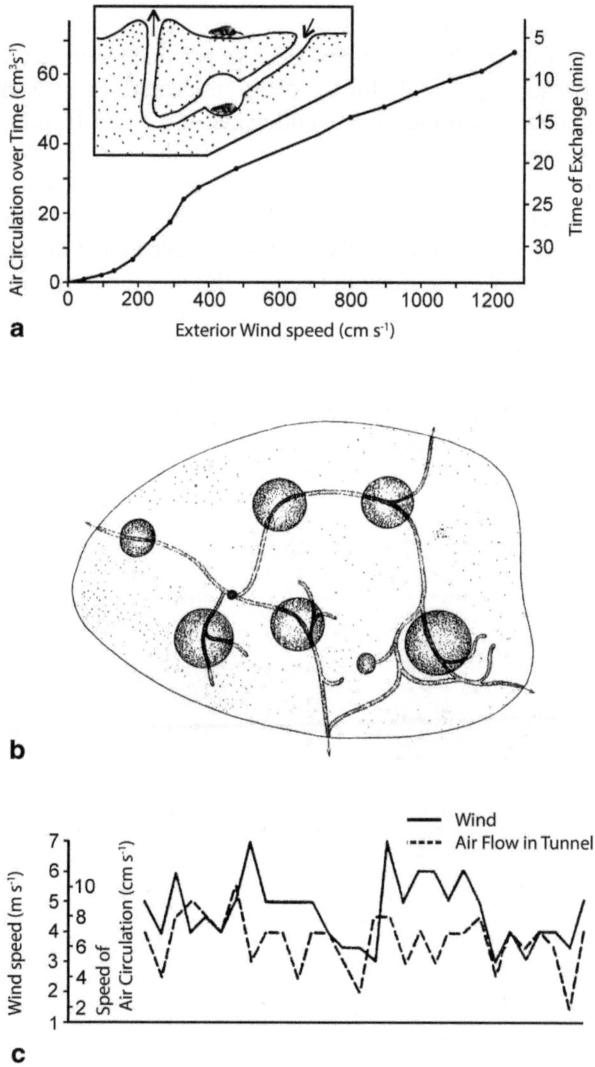

Fig. 4.26 Passive ventilation of structures. **a** Prairie dog, *Cynomys ludovicianus*. The *left* ordinate relates to a 2 m model, the *right* is recalculated for a 20-m-long structure and holds true for 10% of the wind speed abscissa. **b** Mole, *Talpa europaea*. Fragment of a field (4.5 × 2.5 m) with tunnel systems, raised hills and location of the sensor. **c** Example for the association of wind speed in 2 m distance and flow display by the sensor in b. (**a** from Vogel et al. 1973 and **b**, **c** from Olszewski and Skozen 1965)

in the tunnels follows fluctuations in the wind speed (Fig. 4.26c). How much the mole hills and the difference in elevation of the tunnel entrances play a role in relation to the local wind behaviors has not been so explicitly explained yet as in the case of prairie dog tunnels; however the moles' use of the Bernoulli principle has been verified.

The same principle was applied in ancient Iran (and still today in many North African regions) for the induced ventilation of cisternes. When the wind flows over dome structures with a hole at the highest point, air is suctioned out according to the Bernoulli principle, bringing evaporated cisterne water with it. The water is effectively cooled in this manner: 1 g of evaporated water can dissipate 2.3 kJ of heat energy at an air temperature of 20 °C. Turrets can actually strengthen the effect and therefore do not function merely as decorative elements (compare Fig. 4.28c and inset image in Fig. 4.31b).

The architect Thomas Herzog provided induced ventilation according to the Bernoulli principle for his "Design Center" in Linz, Austria (Fig. 4.27). With the contrary vaulting on the underside of a fitted lengthwise "profile" (similar to an aircraft's wing) the vaulted contour of the hall generates a nozzle effect for air ventilation. This concept is further expounded in the subsequent sections.

As D. Gruner writes, ventilation mechanisms (Botso: Dyon fu) are common for housing structures in the southern Niger inland delta. For mosque structures they have only been found in Djenné. "It takes the form of openings in small rises in the roof terrace; they function like mole hills (Fig. 4.28a). Underneath them spherical, pot-like forms are hidden, which keep a shaft to the interior open. They are consistently arranged in long rows and can be sealed with an earthen lid." Whether the likewise sealable "roof mirabs" of these mosques (Fig. 4.28b) function in the same manner is not developed in the literature.

Fig. 4.27 "Design Center" in Linz, Austria. Architect Thomas Herzog + Partner 1988–1994. (Photo: Nachtigall)

Fig. 4.28 Ventilation mechanisms in mud architecture and baths. **a** "Dyon fu" on the mosque of Djenné, Niger. **b** "Roof mirabs" on the roof terrace of a mosque on the Niger. **c** Restoration of the historic Turkish bath on Rhodos, Greece. Inset image: a dome with removed glass blocks, viewed from the interior. (**a**, **b** from Gruner 1990 and **c** from Paraskephopulu et al., Rhodos 1992–1994 (Adapted from Herzog (Ed.) 1996))

Even with the restoration of ancient buildings one is again reminded of the efficacy of Bernoulli effects with domes.

In the present reconstruction of the historic Turkish baths on the island of Rhodos, one no longer observes a humid, musty interior atmosphere. This fact is simply accounted for: The once integrated glass blocks (Fig. 4.28c, inset image) were removed from the domes. The locations now function as wind-driven ventilation holes with the Bernoulli effect.

Ventilation Using Windflow Around a Structure, Also with Termite Structures In Serengeti National Park, Tanzania, field research has shown that these structures—similar to the above depicted prairie structures—use Bernoulli and sometimes Venturi effects for ventilation air flow. Termites of this species build structures with openings. Smaller structures (Fig. 4.29a) possess only two of them, somewhat larger structures (Fig. 4.29b) possess additional side openings, and the largest structures (not pictured) have up to 12 openings.

The openings are, as a, b show, differently formed and placed at different heights. Measurements have yielded substantial ventilation flows that can be attributed to Bernoulli effects (Fig. 4.29c). Periodic effects from eddies can additionally occur, which form on the towering entrance and exit openings. A maximum of 2.7 m min^{-1} was found for the air speed in the structure, but commonly less than 1 m min^{-1}, in average only 0.12 m min^{-1}.

The open tunnels certainly regulate the air in the structure, but have no direct connection to the actual nest region. Small breaches on the exterior or to the air conveyance system are quickly repaired by the termites.

The raised openings can be further built out to attain the character of chimneys in which hot air can rise. Observations have shown that chimneys are common with structures in complete sun; with structures of the same type but in the shade, the chimneys are hardly built. Biomimetic inspirations that spring from such chimneys have been already described above. These air flows condition simultaneously the structure, as they cause water, which diffuses in from the environment, to internally evaporate, thereby yielding a more or less pronounced course over a day (Fig. 4.29d): The maximum occurs expectedly at the hottest time of day, when more powerful wind gusts also occur.

As Fig. 4.29e shows the interior temperature in the tunnel systems is more level in comparison to the ambient temperature; its daily variation exhibits none of the major fluctuations; hot and cold peaks are avoided in particular. If one blocks the tunnel system, a temperature accrual occurs at the end of the day, which can reach dangerously high values.

The utilization of the Venturi effect proves itself as surprisingly effective, doubtlessly no less pronounced than by prairie dog structures.

Even the secondary effects such as cooling and moisture enrichment are substantial; a smaller structure evaporates no less than 2 m^3 of water in a year, a larger one 25 m^3. The open structure of *Macrotermes* is likewise a system that—with governance of the inhabitants—enables a certain degree of homeostatic regulation of variables of the environment, such as humidity, air composition, temperature, and growth possibilities of microorganisms.

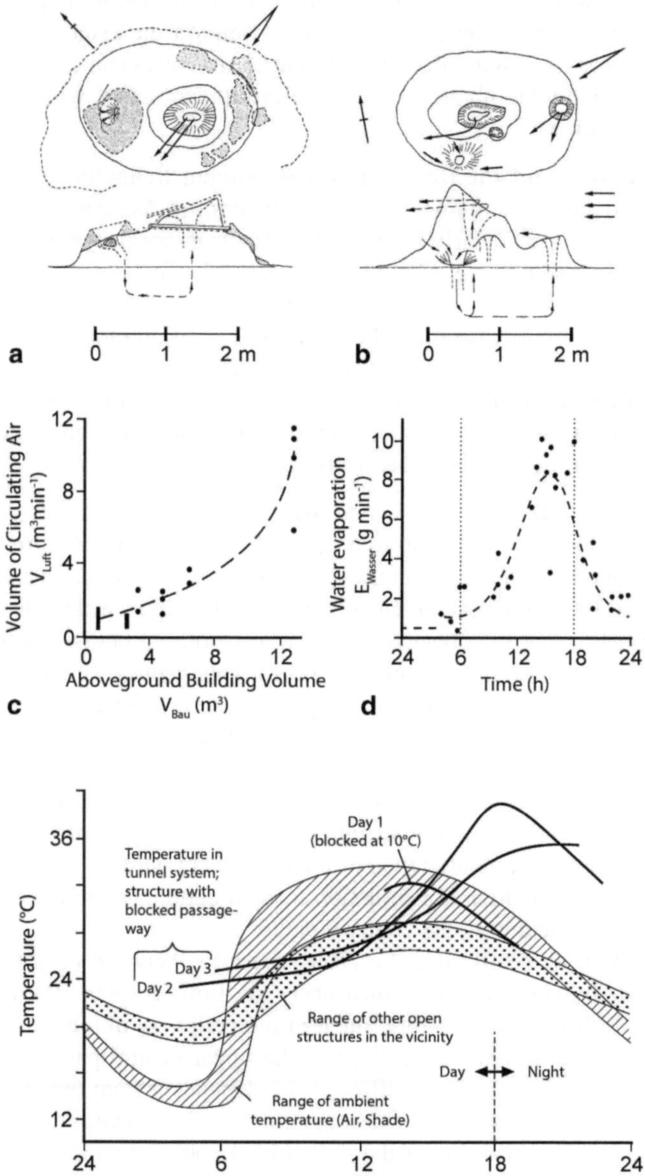

Fig. 4.29 Documentation of Bernoulli, moisture, and temperature effects in structures of the termite *Macrotermes subhyalinus* in Serengeti National Park. **a, b** Two smaller structures with two or sometimes four characteristically formed openings and indication of the wind and tunnel air flow direction. **c** Relation of the tunnel air flow volume to the size of the structure (and accordingly the number of openings). **d** Evaporated water mass, related to the time of day. **e** Temperature in a structure with a blocked tunnel system in comparison to a similar, neighboring structure (with open tunnels) and to the ambient temperature. Redrawn; measured points left out. (Adapted from Weir 1973, edited)

Because the water contains minerals, they are deposited in the passages during evaporation and accrue over a long period of time on the exterior surface due to rebuilding processes. Small, medium, and larger structures can deposit in 1 year in average 7, 26, and 92 g of calcium and magnesium carbonate, which has a noticeable ecological effect.

Air Flow Sensors and the Building and Repairing Behavior of Termites P.E. Howse studied the nest building behavior of the termite *Zootermopsis angusticollis* and *Z. nevadensis*, particularly the way they repair minor damage (openings). For this purpose they use particles of mud or wood that they cohere with a watery secretion from the anus region. They exhibit an extremely high sensitivity to the smallest movements of air, whose speed lies in area of one thousandth of the speed that occurs in normal, closed spaces (!). The sensors lie in the antennae. Hypothetically, when the earthen structures of *Macrotermes* termites are under construction, more or less random air currents occur, which the animals note and build accordingly so that these currents are maintained and strengthened.

Wind Induced Ventilation Within Ant Structures Leaf cutter ants of the species *Atta vollenweideri* build large, up to 6 m deep nests, which house up to 5 million individuals. Over the course of colony's growth about 15 m^3 of earth is accumulated and more than 1000 underground chambers are laid out. In the chambers the colony cultivates mushrooms on collected leaves, whose fruiting bodies are harvested by the ants. The ants, as well as the mushroom gardens, require a high O_2 supply and consistent CO_2 removal. Because the wingless ants, as opposed for example to the fanning honey bees, cannot actively ventilate their nest, the colony must use passive nest ventilation.

For passive air circulation, thermal effects (convection), as described for the termite structures, can be used and function with the support of wind. Due to the relatively low-temperature tolerance of the mushroom gardens (damage with temperatures over 30 °C), thermal ventilation for leaf cutter ants is limited and consequently only feasible during the winter months.

Kleineidam et al. (2001) found a clear separation between air inflow and outflow openings by simultaneously measuring the flow in some 100 nest openings (Fig. 4.30a). Inflow openings were found in the periphery of the 1 m tall nest hill, while outflow openings were localized in the upper central part of the nest. The function of an opening as in or outflow was independent from the prevailing wind direction. The air flow speed in the nest openings was, however, strongly correlated to the wind speed measured over the nest (Fig. 4.30b). A temporal analysis of the air flow behavior during variable wind behaviors (wind gusts) showed a delayed inflow of fresh air on the periphery during rising flow speeds at the outflow openings. The delayed inflow was dependent on the absolute wind speed. During high, low, and very low wind speeds measured delays were approximately 2, 7, and 12 s respectively.

These data are the first evidence for wind-induced nest ventilation in ant structures and supports the hypothesis postulated by Kleineidam et al. that the ventilation of nests of leaf cutter ants is achieved by a suction effect (Bernoulli effect) at the outflow openings. The negative pressure at the outflow openings can relate to

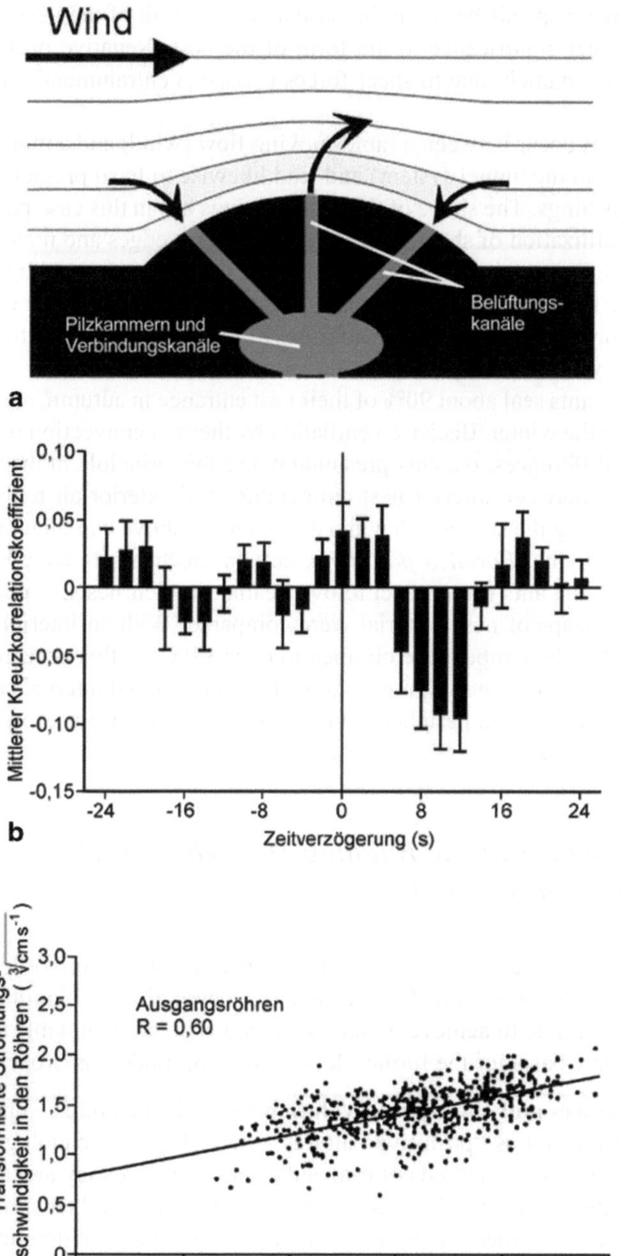

Fig. 4.30 Wind-induced air flow in nests of the leaf cutter ants *Atta vollenveideri*. **a** Scheme. **b** Time delay between in and outflow during very low wind speed (v_{Wind}=0.7 m s^{-1}, delay period 12 s). Positive ordinate values signify that a wind gust is more noticeable in the exit tunnels than in the entrance tunnels. **c** The speed in the exit tunnels (here reformulated) climbs with increasing wind speed; the increase is equal in the entrance tunnels, the distribution somewhat greater. (Adapted from Kleineidam et al. 2001)

differential wind speeds between the base and the zenith of the nest hill; therefore giving particular significance to the form of the nest. Negative pressure can also locally emerge, namely due to shear forces ("viscous entrainment") at the outflow openings.

Shear forces occur between a rapid-moving flow (wind) and a more or less stagnant fluid (air in the tunnel system) and lead likewise to local pressure differentials at the nest openings. The shape of the nest openings has in this case particular influence on the utilization of shear forces, in which sharp edges and many small openings instead of one large one increase the pressure difference, as Vogel et al. 1973 have shown. The leaf cutter ants actually form their openings in the central area of the nest accordingly; however, the addition of shear forces for ventilation of their nests has not yet been completely understood.

Leaf cutter ants seal about 90% of their nest entrance in autumn, and they remain closed during the winter. Because ventilation by thermal convection is promoted by temperature differences, the ants presumably use this principle in the winter, when the difference between interior nest temperature and exterior air temperature is at its highest; during the summer they must rely on the above explained wind effects.

From wood ants, *Formica polyctena*, certain mechanisms have become well-known, which the ants use to react to overheating in their nests. Natural structures and artificial heaps of nest material were compared. With an interior heating element (of 20 W) the temperature climbed to over 35 °C. In this occurrence the ants reduce the height of the nest dome, enlarge the openings and internal chambers, and relocate the pupae to the periphery; and the workers shift from energy production to energy consumption.

4.5.2 Climate-Suitable Building Methods in Ancient and Modern Cultures

Ancient primordial cultures developed their structures with more or less haphazard trial and error, though ultimately according to an evolutionary biomimetic strategy. What they were able to achieve with this method is certainly notable and can serve as a comparable basis for the further development of modern approaches as well.

Ancient Cultures and Biological Evolution In ancient cultures—up until not too long ago designated as "primitive cultures"—arguably all technological developments proceeded from methods of trial and error. Therefore they are, as mentioned, principally similar to natural evolution. Natural structures and technological structures of this kind are then readily comparable according to biomimetics. They have arrived at their current forms after long periods of simply "playing around" with processes of changing, discarding, and changing again.

In its current situation, building design deals with proper use of wind and underground moisture and cooling effects for the regulation of climate. The above cited case of prairie dogs' usage of the Bernoulli principle already implies such multiplicity of use. With the induced air flow through their earthen structures cool and moisture-enriched air is sucked out of the porous tunnel system and thereby moistening, for example, collected dry materials. It can with the help of evaporative cooling regulate the temperature in the structure, but also provide the prairie dog with necessary water.

An especially contrived system for passive air regulation, which still fascinates travelers interested in early technology, had evolved in ancient Iran. M. Bahadori (1978) reported about it; a series of details are collected in Fig. 4.31.

Tall wind towers of adobe material are used, whose upper window covers can be variously opened or closed and can capture the wind according to dynamic pressure and direct it downwards. There the air travels through an earthen tunnel, for example, and arrives in the basement level, where it exits through adjustable windows and doors. In the lower, cooler part of the wind tower (which keeps the cool temperatures from the night for a long period of time) the incoming warm air ((1) in Fig. 4.31a) is convectively cooled (2). In the underground ducts, moisture collects, partially accentuated by air humidity (3), partially evaporated, and thereby cooling the air (likewise 3). The characteristic temperature–moisture diagram in Fig. 4.31c corresponds to this process. During the night the flows reverse, as the air heats up from the now warm interior walls and rises, cool night air is then pulled in through the windows and doors.

Another system for wind usage combines the just mentioned effects with suctioned air, which travels for a while along an underground layer exposed to groundwater (Fig. 4.31b). The incoming (4) and convectively cooled (5) air mixes with the suctioned (7) and moisture-enriched (8) air, combining the cooling effects of convection and evaporation (9). This course is mapped in Fig. 4.31c as well. During the first half of the night (8), rising air in the wind tower lifts air that is likewise water vapor enriched.

Evaporative cooling always requires a fluid in motion that removes the boundary layer of humidity at water-leading locations. For this purpose the Bernoulli principle was also used in ancient Iran.

On the top edge of longitudinal barrel vaulted roofs a negative pressure arises that can siphon off hot air through openings located above. The system functions the best when the air current runs perpendicular to the length direction of the roof, otherwise according to a sine law, similar to the lugworm tunnels. Dome roofs function independent of wind direction according to the "volcano cone principle," similar to prairie dog structures. Elements placed at the highest point of the dome (turrets) can have, as mentioned, not only an artificial but also a thoroughly functional significance for air flow (inset image in Fig. 4.31b).

Air regulation by use of wind was important for Ancient African architecture as well. The round huts of many kraals, laid out in lines in an advantageous direction that leads the wind partially according to the Bernoulli and partially according to dynamic pressure principle. L. Ilg has compiled details for this system. The contrived structures (Fig. 4.32a) recall in their climatic, biomimetic subtlety the highly developed structures of ancient Iran. The orientation of the cities to preexisting wind direction was probably considered, as can be deduced from a site plan of the city of Khartoum, Egypt from about 2000 BC (Fig. 4.32b).

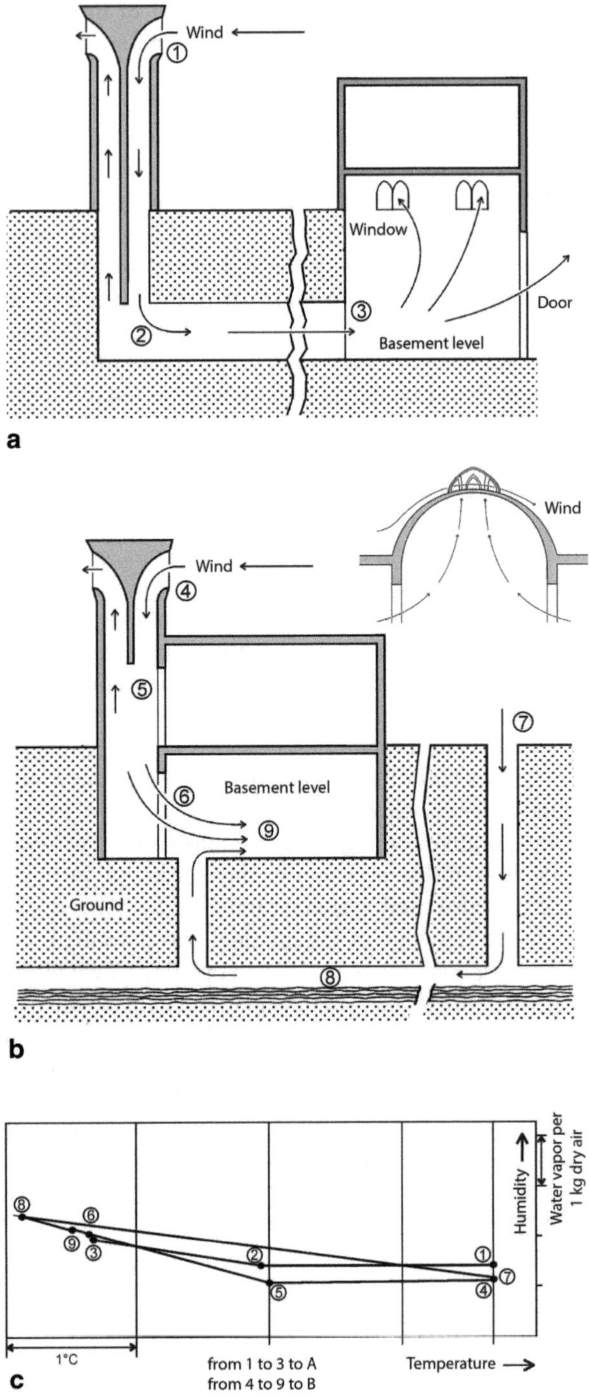

Fig. 4.31 Ventilation and cooling in ancient Iranian architecture. **a** Operation during the day, with wind through the wind tower and underground pressure duct. **b** Operation during the day, with wind tower and underground suction duct. **c** Principal temperature–moisture progressions for a and b; compare to the text. (Adapted from Bahadori 1978)

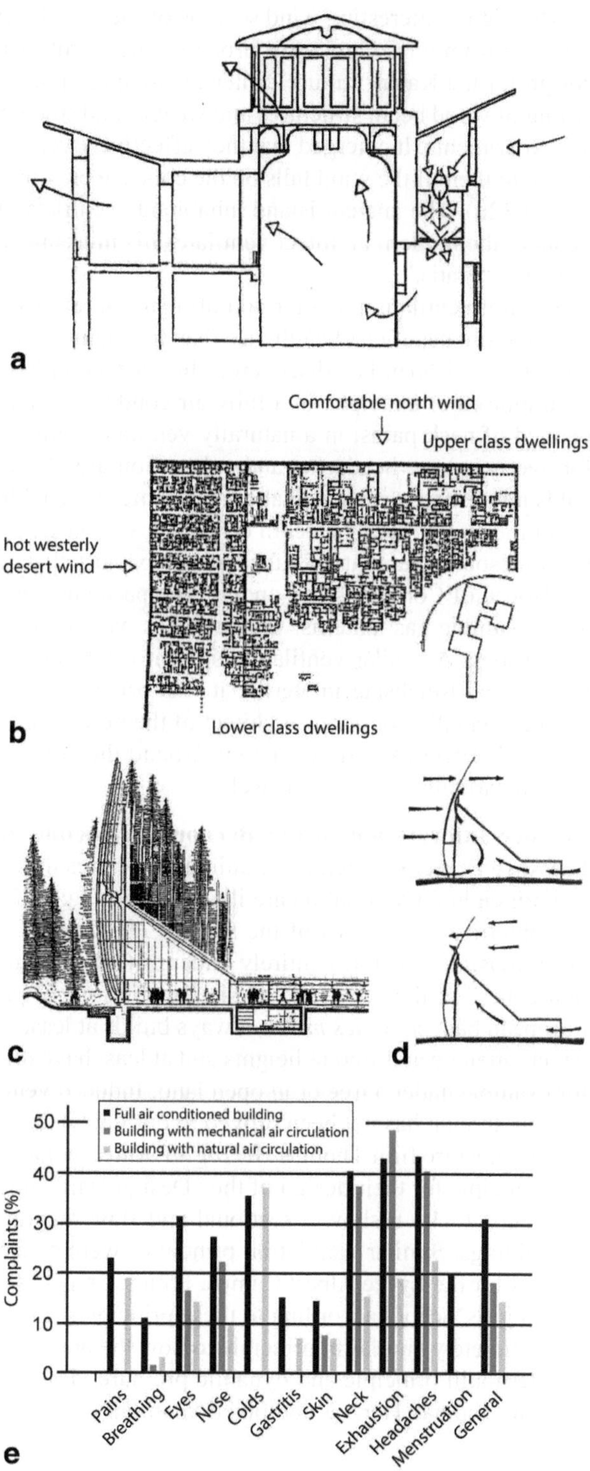

Fig. 4.32 "Natural" building ventilation. **a** North African building form. **b** City plan Khartoum, 2000 BC. **c** Wind catcher for the Kanak culture center in Nouméa, New Caledonia by Renzo Piano, 1993. **d** Building ventilation system from c functions independent from the wind direction. **e** Sick Building Syndrome study by J. Röben

The highly interesting wind screens of the Kanak settlements in New Caledonia emerged from a trial and error process. The architect Renzo Piano included in his design for the Kanak culture center in Nouméa spoon-shaped "wind screens" consisting of wood beam structures and weaves and studied their functions in wind tunnel experiments. It emerged that they effectively ventilate length-oriented rooms attached to them if the wind falls on the concave or convex side of the "wind catcher" (Fig. 4.32d). The ancient island inhabitants ventilated their large gathering houses as such; the modern architect ventilates his museum spaces according to the same principle "gratis."

Natural ventilation and air conditioning of an office building, for example, are clearly more beneficial healthwise than mechanical ventilation or even full air conditioning. J. Röben has discovered this fact in an SBS study (SBS: sick building syndrome). For example, in a fully air conditioned building 40% of workers complained of neck pains; in a naturally ventilated building only 15%. Similar results for eye irritation, headache, and exhaustion are also notable, while the values for colds interestingly remained about the same (Fig. 4.32e).

At the end of the nineteenth century, architect D. Boswelt had already provided natural "source ventilation" (through an exhaust channel, hidden in a tall tower) for the House of Commons in London. The parliamentary hall was illuminated by 46 air-consuming gas lanterns, whose fumes were removed by means of this ventilation system. A similar ventilation system was also already provided by Wallot for the original Reichstag in Berlin; it was principally adopted and improved by the environmentally conscious architect of the new Reichstag, Sir Norman Foster. The Bernoulli effect caused by air flow around the dome of the Reichstag incidentally plays a particular role here as well.

Further Information to the Bernoulli Principle and Transition to Dynamic Pressure Principle Two more animal structures that almost certainly use the Venturi principle for ventilation are illustrated in Fig. 4.33.

Subterranean termites of the species *Hodotermes mosambicus* live in their almost invisible structures entirely underground, although they do maintain "ventilation cones" at the ground level similar to those of prairie dogs (Fig. 4.33a). The European badger, *Meles meles*, always builds at least two exits to its tunnel system, which often lie at different heights and at least have differently structured openings, for example under a tree or in open land. Induced ventilation is to be assumed here as well, though has not been proved to our knowledge (Fig. 4.33b).

Architecture firm Thomas Herzog + Partner included an effective use of the Venturi principle for their design of the "Design Hall" in Linz, Austria (Fig. 4.27).

Figure 4.34a, b show a sectional and flow diagram for this system as applied to buildings. Similar ventilation principles were conceptualized in France as well, such as for the Lycée Albert Camus, Fréjus, France, where Foster & Partnerts have built a girls' school according to these principles (Fig. 4.34c).

Completely similar in principle to how the ancient Iranians combined the use of the Bernoulli principle and dynamic pressure, described in more detail in the next section, Thomas Herzog combined "naturally ventilated portions" for his Hall 26

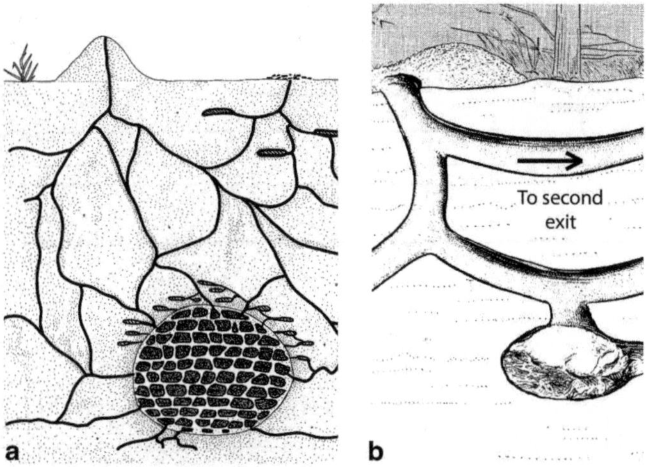

Fig. 4.33 Additional animal structures with possible induced ventilation. **a** Subterranean termite, *Hodotermes mosambicus*. **b** European badger, *Meles meles*. (Adapted from v. Frisch 1974)

for the German Convention Center AG (Industry Convention in Hannover and later EXPO 2000). Figure 4.35 clarifies the details. As already shown with the "Design Hall" in Linz, Venturi wings were lengthwise applied at the highest points to siphon off the air; the perforated, oppositely positioned slanted walls now function additionally as wind pressure catchers according to the dynamic pressure principle.

Altogether it results in very effective induced ventilation that is necessary for the large glass surfaces.

4.5.3 Usage of the Dynamic Pressure Principle in Animal Structures and Man-made Buildings

The contained kinetic energy in 1 m^3 of mass m that flows with velocity v, is equal to ½ ($m_{1\,m^3\,air}$) v^2. One can also state, the energy related to unit of volume V is equal to ½ ρ v^2 (ρ=m/V=air density). If the observed air volume is flowing against a perpendicular wall and decelerated to v=0, its kinetic energy manifests itself in the occurrence of dynamic pressure $|q|=|\frac{1}{2}\rho v^2|$. This can have different effects, for example setting a trapped air volume in motion. The same principle applies to flows of water.

South American caddis fly larvae (*Hydropsychidae*) have developed a system with fluid flow according to the dynamic pressure principle, which astoundingly resembles the badgir system (compare Fig. 4.36a with 4.39a).

These larvae build a vaulted tunnel system with protruding "dynamic pressure capturers" and an extremely fine mesh network (mesh width only about 3–20 μm) attached inside of their lower, U-shaped tunnel. Below this mesh lies the tube-shaped dwelling chamber for the 2-cm long larva. Bernoulli effects could also play

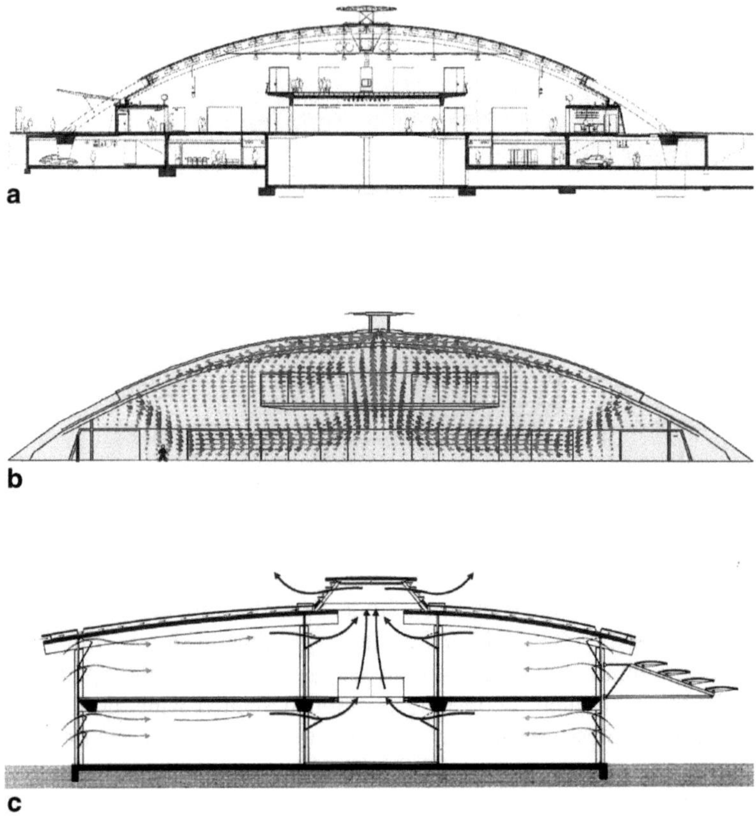

Fig. 4.34 Induced ventilation by the Venturi effect on the upper side of barrel vaulted structures. **a** "Design Center" (Exhibition and conference building), Linz, Austria. Thomas Herzog+Partner 1986–1994. **b** Air currents in a with "central updraft"; outflow and Venturi-induced flow not depicted. **c** Lycée Albert Camus, Fréjus, France. Foster & Partners, London, 1991–1993. (Adapted from Herzog (Ed.) 1996)

a meaningful role for this current-driven water circulation, though it has not been empirically proven.

"Chimney mussels" of the genus *Clavagella* carve out a living chamber in limestone and extend their parallel inflow and outflow tubes far into the open water.

They are sheathed with limestone which extends with the growth of the mussel. It can be assumed that the roof-like protrusion resulting from this growth has a flow-mechanical function, particularly in the region of the opening (Fig. 4.36b), but this has not been proven in detail. The same can be accepted for the "windcatcher-like" entrances of the structures of the stingless bee *Trigona testacea* (Fig. 4.36c) and the wasp *Angiopolybia pallens* (Fig. 4.36d). With (b) to (d) structural porosity could also assume the role of exit openings.

As a rule, ventilation and air cooling principles are implemented in combination, for example, the utilization of cool underground temperatures and moisture coupled

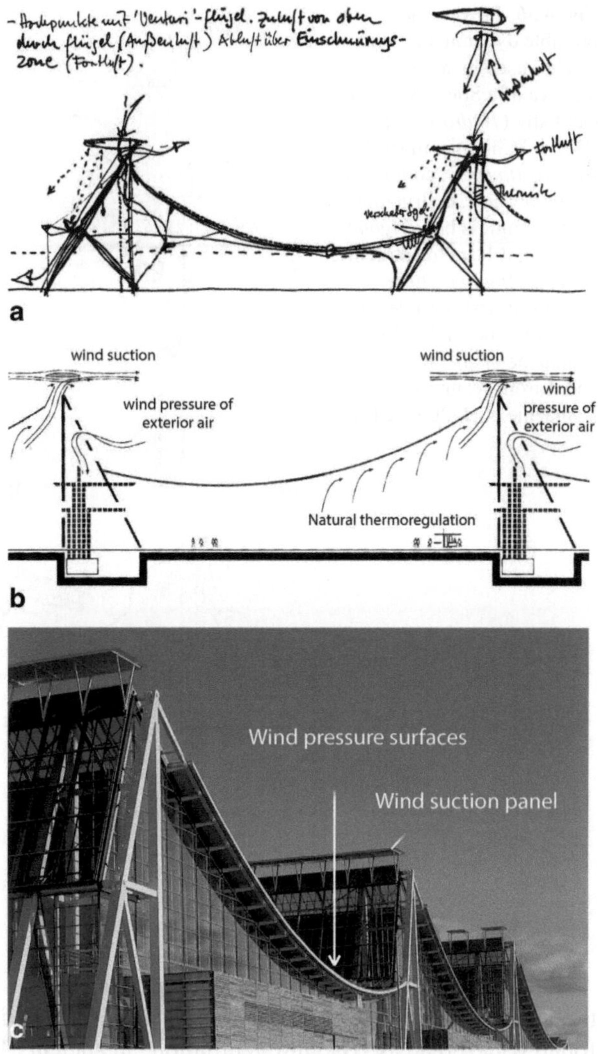

Fig. 4.35 "Natural" ventilation portions of the Hall 26 for the German Convention Center AG, Industry Convention 1996, later EXPO 2000, Thomas Herzog. **a** Concept sketch, June 1994. **b** Function schematic, August 1994. Ventilation portions from **a** drawn in, other functions (natural and mechanical intake and exhaust) removed. **c** View of the finished structure, April 1996. (Adapted from Herzog (Ed.) 1996, edited)

with the Venturi principle. This combination becomes clear from the construction of a garden pavilion in Isfahan, Iran, from the second half of the seventeenth century (Fig. 4.37a) and likewise from Italian villas of Palladio. The Italian Renaissance master had similarly conceptualized his design for a rotunda near Vicenza in 1566 (Fig. 4.37b). He was fascinated by the use of cool and moist air from underground grottos and drew on this principle for the Costozza Villas near Vicenza, for example (Fig. 4.37c).

The lengthwise-protruding roof structures of the Toraja in the tropical rain forest of South Sulawesi are presumably air catchers built according to dynamic pressure principles and function similar to Renzo Piano's "windcatchers" in Nouméa as well (see Fig. 4.32c, d); however, no exact measurements are known to us.

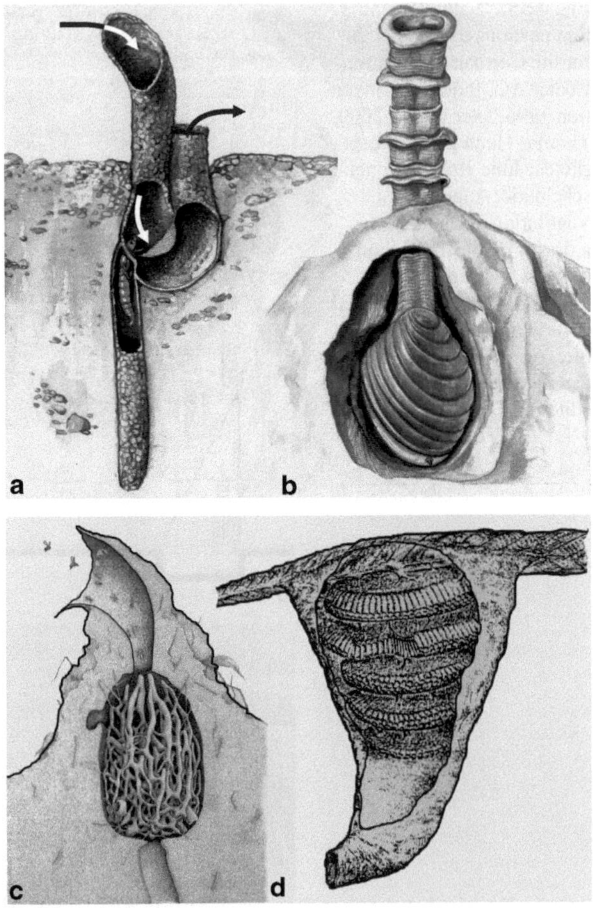

Fig. 4.36 Proven and possible dynamic pressure catcher used by animals. **a** Larva of a South American caddisfly (*Hydropsychidae*). **b** "Chimney mussel," *Clavagella spec.*, combined in and outflow. **c** Structure of the stingless bee *Trigona testacea*. **d** Hanging structure of wasp *Angiopolybia pallens*. (**a**, **b** from Freude 1982, **c** from Michener 1974 from Camazine et al. 2001, edited, and **d** from Jeanne 1975 from Camazine et al. 2001, edited)

Extensive use of the dynamic pressure principle, as it has been described with the ancient Persian windcatcher towers and illustrated in Fig. 4.31a, b, has been found widespread as "Badghir ventilation" in ancient Mesopotamia or Pakistan, for example (Fig. 4.38).

Behling and Behling (1996) write:

"Devised technologies for ventilation can often be found in the buildings. In Hyderabad, Pakistan, the cool winds come mostly from the same direction. Therefore, the buildings are outfitted with immense wind catchers that direct the air flow into the rooms (Fig. 4.39a). Many traditional houses in Baghdad are equipped with a badgir that receives the air current out of the Northwest. A badgir is a sort of chimney in the wall of the house that extends to the highest point of the roof parapet. A badgir is therefore particularly effective when it is installed with wide openings diagonal to the prevailing wind direction. As soon as the air is captured, it gains moisture in the cooling passage, lowers in temperature, and sinks. An example for an air conditioning system in its most polished and energy-efficient form. In order for the technology to function, however, large temperature fluctuations are necessary.

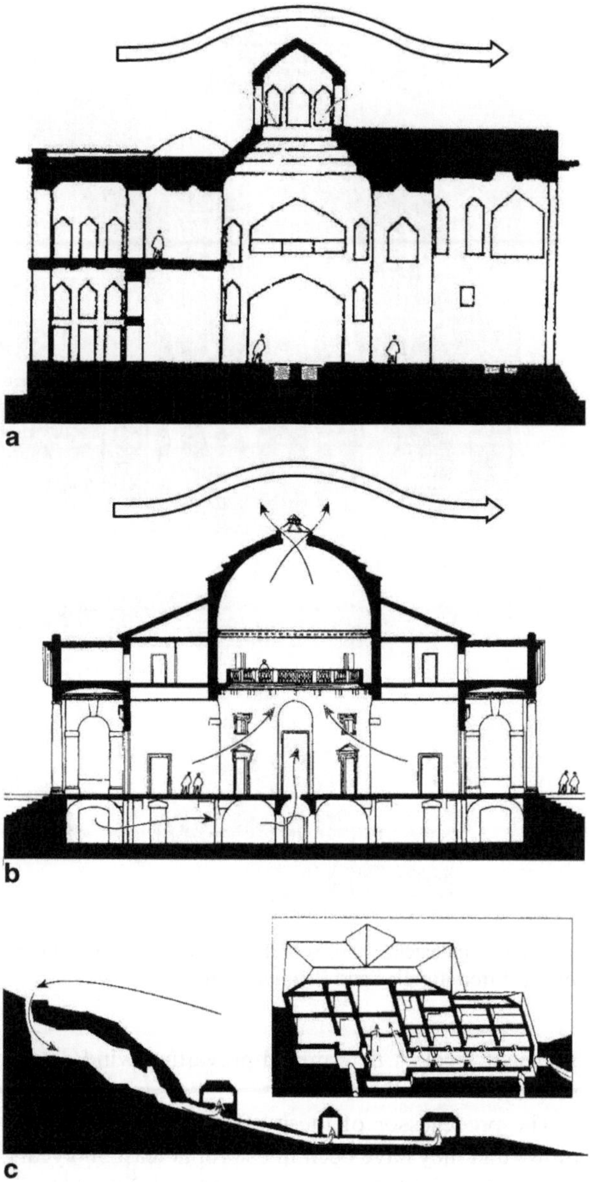

Fig. 4.37 Use of the earth's cool temperatures and moisture underground, partially due to ventilation according to the Venturi principle. **a** Garden pavilion Hashtbihishd, Isfahan, 1669–1670. **b** Villa Rotunda near Vicenza, Italy. Palladio 1566. **c** Costozza villas near Vicenza with "cave cooling," Palladio, sixteenth Century. (Adapted from Behling and Behling 1996)

The air rises again in a central, tower-like structure, warms up and exits the building through dormer windows (Fig. 4.39b). Shaft-like inner courtyards function in principle in the same manner as an air passage.

As illustrated in Fig. 4.39c, the temperature difference between the basement level and the roof surface can reach 20 °C (!).

In Hyderabad Sindh, western Pakistan, this "air conditioning" system defined the roof landscape. The chimneys ventilate only one room and extend down to the

Fig. 4.38 Windcatcher houses of the Toraja in the tropical rain forest of Palawa, South Sulawesi. (Adapted from Behling and Behling 1996)

basement level. A determined prevailing wind direction is a prerequisite for their efficacy.

The predecessor of these effective badgir structures is unknown, though one knows that they have been in use for at least 500 years." (These passages also serve as an "internal telephone" between the different stories)

4.5.4 Example for Ventilation and Air Conditioning: Incorporation of Biomimetic Inspirations in the Structural–Architectural Planning Process

The possibilities for the use of environmental forces in technology are immense; a variety of uses in animal structures and in classical and modern architecture were

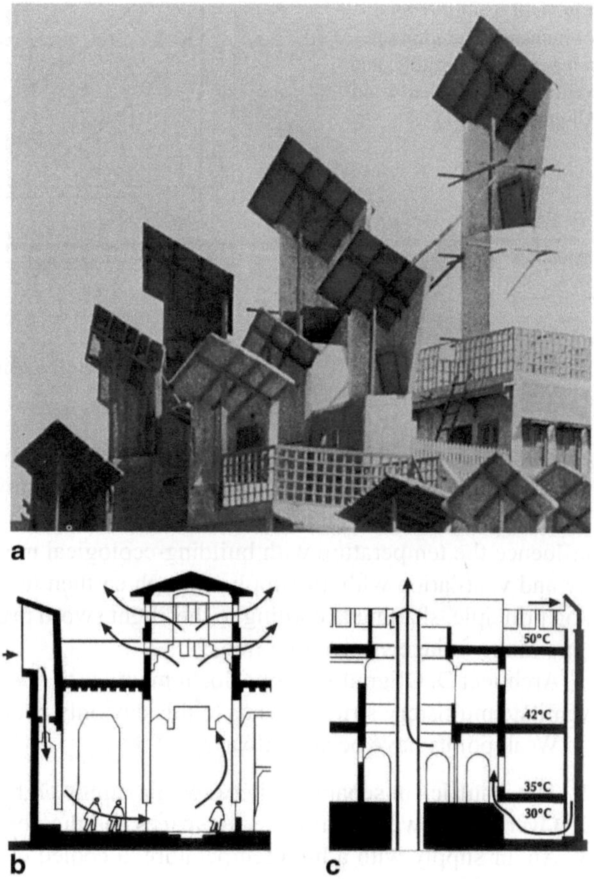

Fig. 4.39 (48) The principle of badgir ventilation. **a** Windcatchers ("badgir") on the buildings of old Hyderabad, Pakistan. **b** Buildings without a courtyard. **c** Buildings with a courtyard. (Adapted from Behling and Behling 1996, edited)

demonstrated as examples for ventilation and air cooling. What should be considered when biomimetic inspirations are allowed to influence the design process of architects and engineers? Such observations of the living world surely cannot lead to direct copies; ultimately they disappear from the construction chain; a finished building does not necessarily look "natural" only because natural prototypes are considered during a building's conception. Issues of this sort are illustrated with many completed and planned structures found in the second half of this book. However, subject-related aspects will be included here as well as examples.

4.5.4.1 The Further Development of Double Facades in Relation to Ventilation and Light Distribution Systems

In a Cartesian graph of temperature in relation to the relative air humidity one can designate the middle region as the "human comfort zone" (see Fig. 4.40). In the region adjoining this zone the situation by moisture supply, ventilation and shade

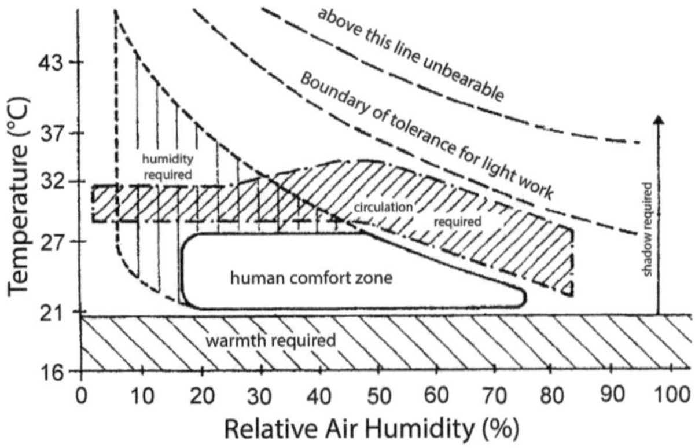

Fig. 4.40 Comfort zone of humans in relation to temperature, humidity, and ventilation. (Adapted from Oligmüller 2001)

approximates the ideal zone. According to topography or climate region one will more strongly consider one or other facet for the planning of a building.

Transparent heat insulation offers itself as a multifunctioning system. One can influence the temperature with building-ecological measures; for example, humidity and ventilation with air supply through earthen tunnels according to the prairie dog principle, shading according to the "light sword principle," which has its prototype in the foliage system of a tree.

Architect D. Oligmüller from Bochum writes to this end: "The desire to naturally ventilate multistory structures has to this day only been partially fulfilled."

Weak points have been as always:

1. The insufficient separation between air supply and removal
2. Overheating with continuous air spaces and thereby necessitating
3. An air supply with a high temperature, a cooled exhaust, and draft effects with ventilation in winter
4. Elaborate partition systems for the fulfillment of sound-technical, and fire prevention demands

For facade construction two building methods can be named that offer an excellent starting point for multifunctional systems.

1. A façade structure that allows the interstitial space to be usable (essentially a multistory greenhouse or veranda structure)
2. An updraft façade, that is closeable and then stores warmth in the buffer space, or in an opened state positively influences the balance of warmth in the building and prevents overheating in the summer

Both solutions can be combined so that their air supply through the earth is always either pre-cooled or pre-warmed according to the season. Additional pre-warming in the winter could be provided by the passive use of sun energy in form of thermohydraulic regulation that directs the air current through a buffer during heavy sun infiltration.

For their workspace the students at the Knobelsdorffschule in Berlin constructed an earthen canal that essentially improved the climatic conditions in this space with exterior pre-warming or cooling of air. As such the interior temperature is reduced for example from ca. 29 to 24 °C in the summer.

The office building for the regional administration of Bad Segeburg was to be redesigned, as the work conditions in the individual rooms were no longer tolerable due to the high amount of noise and exhaust from the directly bordering B 206 road. Due to structural reasons the incorporation of full air conditioning was ruled out and the abandonment of the building was already being considered. Architects F. and W. Lichtblau from Munich suggested a ventilation system through earthen passages. Due to the bad air quality in the vicinity of the building, the passage leads from a nearby park, supported by a motor for the constant supply of consistent air flow. The hung façade functions simultaneously as sound insulation and as space for the distribution of the air supplied by the underground passage.

American architects B. Yanda and R. Fisher can be counted as well among the pioneers of the ventilation-heat transfer system discussed here. In 1980 they provided a general design for a house suitable for summer and winter utilizing this type of device.

In the 1990s Pohl Architects designed similar transfer systems for a technology center in Erfurt with functions as well for ventilation and solar gain, completed in 2002. The building became a component of the pilot research commission within the frame of an EU-supported research project "SOLARBAU-Monitor." The biomimetics-inspired systems of thermally reactive building and earth masses, as well as buffer and updraft facades, were investigated based on their practicability and user friendliness. As such it is one of the earliest projects in Europe that had tested the efficacy of the complex ventilation systems according to biological prototypes with the use of earthen masses and building parts as hot or cold reservoirs. The effectiveness of the measures was confirmed during a monitoring phase; at the same time the planners were able to gain important insights for later projects. A brief overview can be found in section "Complex climate systems 1: new buildings". The building method received press from the related subject literature, particularly in climate-efficient building technology (compare Voss, Löhnert, 2006; Bürogebäude mit Zukunft, FIZ Karlsruhe. Bohne, D., 2004; Ökologische Gebäudetechnik, Kohlhammer) (Fig. 4.41).

Biological functionality for temperature control can also provide inspiration for the highly current topic of building reuse. In section "complex climate system 2: building reuse," one can find a brief illustration of an older brick building, which was completely restored and reused. The team at Pohl Architects integrated the lecture hall and studios for the department of media and the university datacenter of Bauhaus University in Weimar into a former brewery building. The building masses are activated throughout the entire structure by a devised ventilation system and combined with an offset, updraft, and buffer façade. The concept functions with the intelligent use of the massive, preexisting walls in the basement level, a glass-sheathed climate buffer mounted in front of the massive, existing façade, and the ability to switch from air supply to air removal openings of the either windward or leeward oriented ventilation openings (Fig. 4.42).

Fig. 4.41 Terminal EF—Technology center in Erfurt. Pohl Architects (Fig. G. Pohl)

Fig. 4.42 Climate buffer for the Konrad Zuse Media Center in Weimar, Arch.: Pohl Architects (Fig. M. Miltzow)

The application of biology-inspired technologies for existing structures is one of the biggest challenges of the future.

4.5.4.2 The Transparent Light Sword

Transparent shading plays a meaningful role today with the use of daylight (Fig. 4.43).

The effect of transparent light swords rests on the staggering of shading elements so that they do not form a complete surface. The foliage of a tree serves here as precedent.

The reflection inside of the leaflike lamellas leads to an essentially transparent shade, which does not cause too much shading in the proximity of windows underneath the light sword, even with diffuse light, at least 60% of the yearly condition in northern latitudes. "Light swords" have played a recent role in the use of light and shade in the designs of Le Corbusier, who designed elements of the roof of his Chapel Notre Dame du Haut in Ronchamp, France (1950–1955) as "light swords" and thereby creating unique interior light conditions.

4.6 Principles of Self-Organization

Principles of self-organization play a role in the living world in the construction of membrane structures or path networks. These principles are also used for technology, for example in the material sciences. In the area of architecture the example is often called self-evolving path networks.

4.6.1 Self-Organization in Nature

With extensive use of self-organization nature avoids two dead-ends that would have crucially compromised the development of life, if not made it impossible.

1. *The problem of complex, active control and regulation*: Instead of actively inserting modules of membranes or cellular subsystems "in the exact, right place" and making them functional—which would already require a complex regulation system—it formulates conditions to which each building element organizes itself.
2. *The problem of energy allocation*: Instead of "actively" applying metabolic energy—which would rapidly overtake the capable production of metabolic energy by an organism—nature uses, where possible, "passive" energy available in the environment to provide for its self-organizing structures and systems.

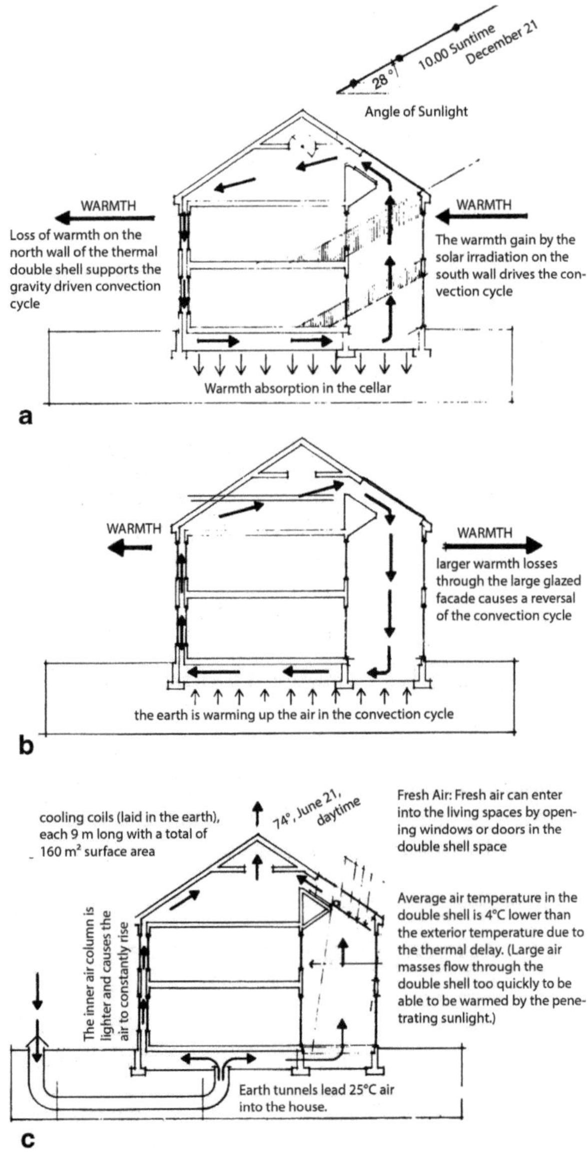

Fig. 4.43 a, b, c General design of a house for 38°N with exogenic temperature regulation. (Adapted from Yanda and Fisher 1983)

Examples for these principles can be found in practically all life forms and life functions, from the microscopic scale of molecular mechanisms up to macroscopic global ecology. The structuring of biological unit membranes (cell membranes, plasma membranes) can be taken as an example.

This membrane consists of lipid and protein molecules and exhibits a basic structure uniform to the entire living world and therefore also a particular thickness ("double membrane" with a thickness of about 45 Å). It consists of lipid molecules in elongated form, mostly phospholipids, with a hydrophilic (water-soluble "headgroup") and a hydrophobic end (fat-soluble "tailgroup") and protein molecules incorporated in the lipid layers.

If the lipid molecules assemble in watery environment, they align themselves into a mono-molecular layer, with one side hydrophilic and the other hydrophobic. If additional lipid molecules are available, they form "by themselves" a double layer on the cell boundary; the lipophilic portions of these layers lie in the middle on top of each other. In this second layer the elongated, clustered membrane proteins "swim," which protrude from both sides of the lipid bilayer, can form pores, and are responsible for the transport of ions through the membrane and as such responsible for the construction of a membrane voltage—the basic functions of life.

Figure 4.44 shows a block diagram of the biological membrane. Through these membranes cellular structures are isolated from one other, so that compartments emerge, in which—in close vicinity—different life processes can operate. This basic principle of the living world essentially rest upon "foreign"-energized self-organization.

4.6.2 *Self-Organization in Urban Planning*

Aerial images of "naturally grown" cities in developing countries often give the impression of chaos or at least randomness. However, one should not overlook the fact that the in between structure of narrow alleys—the accommodative street network—is the central framework, around which and to which the individual buildings are developed, expanded, changed and adapted.

Because no recognizable urban planning or architectural guidelines existed for ancient cities, they evolved over the centuries with complex self-organization prin-

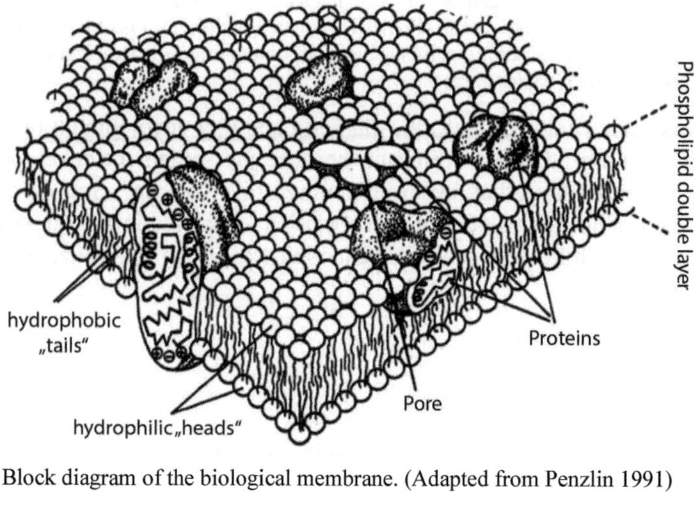

Fig. 4.44 Block diagram of the biological membrane. (Adapted from Penzlin 1991)

ciples (Fig. 4.45). The dimensions, angles, the arrangement of small and larger open spaces (plazas), the positions of the courtyards with respect to the street, etc. are anything but dysfunctional. In retrospect, this kind of development represents for a given space the most effective arrangement of more or less standardized yet individualized and distinct living units, which are of varying quality but ultimately accessible and developable.

These street networks, which emerge from gradual self-organization processes, are analogous to corresponding networks in nature (Fig. 4.46). For example, the complex branching vein networks of plant leaves, which form through coincidental criteria but are ultimately highly functional, represent the most efficient and equal connectivity to each individual cell body. Computer simulations have unearthed astounding details in these network systems. Beyond that only a few generalities can be briefly mentioned at this point.

Figure 4.45 shows "random" building arrangements in settlement structures of ancient Africa, from Ethopia, Niger region, and the cities of Zanzibar and Marrakesh. For comparison, Fig. 4.46 also displays more or less "natural" network and connectivity structures that require self-formation processes. S. Becker et al. (1994) have given thought to these structures. They found that one can compare "path networks" of nature and humans from the viewpoint of fractal geometry. They write:

> Hierarchical construction is characteristic for these fractals. The basic structure always reappears in various scales and layers. This spatial hierarchy results from a strict internal order, a hyperbolic subdivision of mass, which finds its expression in the fractal dimension. Fractal characteristics do not conform to consistent rules of generation. Inkblot-like structures are generated with the integration of random processes. Numerical algorithms allow the investigation of the fractal characteristics of such structures and the determination of the fractal's dimensions. Using these methods more than 20 urban centers were studied.
> In all cases the settlement area exhibited fractal-like characteristics. These settlement bodies accordingly follow a spatial organizational principle despite their irregular morphology. Surprisingly, settlement bodies follow a structural law that is observed in the organic and inorganic world; approximated by sedimentation on material surfaces or leaf vein systems. The analysis of the temporal development of urban spaces permits the grasping of information about the urban dynamics. Fractal dimension values in equilibrium, as they are observed in some cases, indicate an allometric growth principle, as it is observed in biological systems. The results additionally suggest socio-economic self-organization processes that promote fractal city structures.

The authors derive consequences for planning and city development but reject the idea "that planning becomes irrelevant, because the settlement development runs according to certain laws that deprive planning of its purpose": Self-organizational processes can rather be influenced by regulation structures in a large scale. There follows then a multitude of more current topics of urban planning, of which the most important are:

1. Relationship with the edge
2. Creation of identity
3. Necessity of forming connections

The understanding of self-organizing processes for early, "unplanned" urban architecture quickly leads to the formulation of solutions for urban development in the

Fig. 4.45 "Random" settlement structures of ancient Africa: **a** Soqota, Ethiopia; **b** a village in Ethiopia; **c** Hara, Ethiopia; **d** Labbezanga, Niger; **e** Old Zanzibar; **f** Old Marrakesh. **a–d** from Schaur, IL Report 39, 1991 and **e**, **f** from Rudofski 1993)

future. For more detailed results of these studies performed at the Urban Planning Institute of the University of Stuttgart, refer to the summary work of Becker et al.

4.7 Solar Effects: Multitude of Possibilities in Nature and Technology

Animals and plants have developed a variety of possibilities for the use of solar irradiation. They are foundational for the functioning of the living world. Despite substantial improvements only in the last few years, technical utilization of the sun's energy—whether in large, commercial scale or small scale—is only at the very beginning in comparison to nature. A thorough investigation of analogies found in

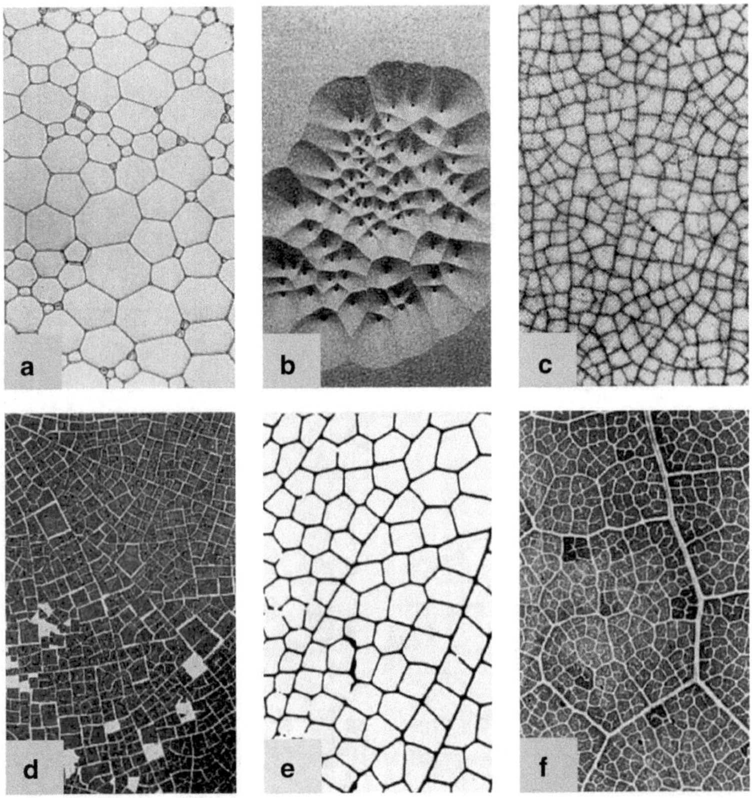

Fig. 4.46 Network and connection structures involving self-formation processes. **a** Soap bubble layer between two glass plates. **b** Leaking sand model based on an Ethiopian settlement (Soqota). **c** Crack structure in porcelain. **d** Dried-out crack structure of a gelatin layer. **e** Dragonfly wing. **f** Maple leaf. (Adapted from Becker et al. 1994)

the biological sciences, of actually existing and evolving, highly developed biological structures and functions, will certainly be of use. One crucial example of such an analog would be the development of leaf-like "organic solar cells" following the precedent of the natural photosynthesizing green leaf. One could then clad large facades with such solar cells or construct a "green window."

4.7.1 The Sun as a Source of Energy

Aside from geothermal energy and moon-induced tidal forces, the Sun is the only source of energy that is available to living organisms on Earth—be it directly, utilizing solar radiation, or indirectly, as with photosynthesizing plants or even wind (ultimately an effect of the sun). Norbert Kaiser compiled the different flows of solar energy in his paper "Maximen für solares Bauen" (" Axioms for Solar Building").

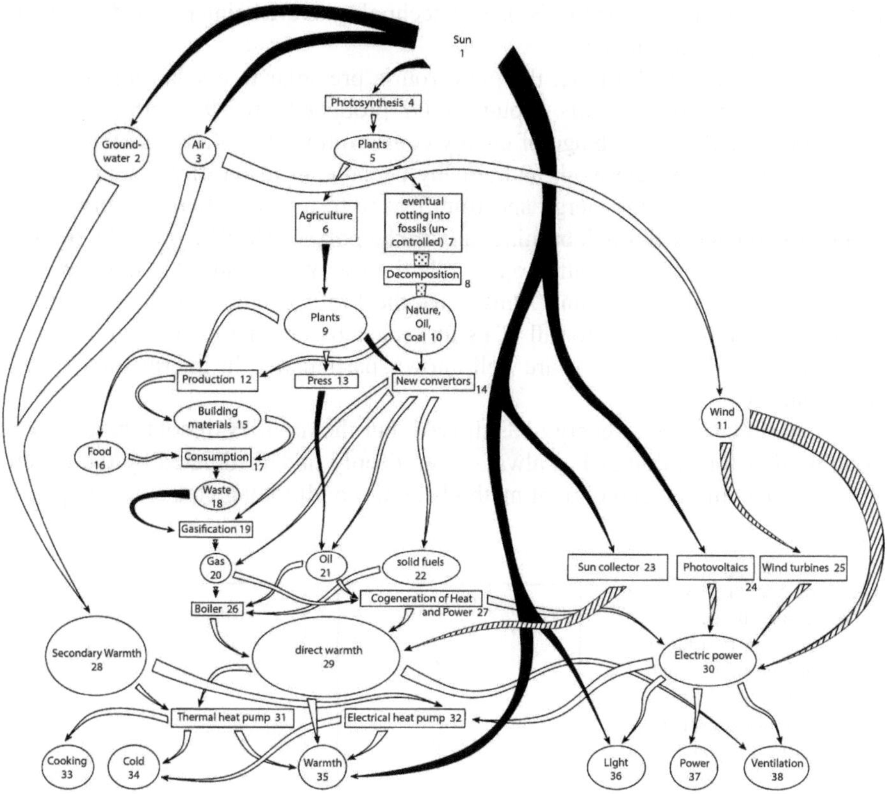

Fig. 4.47 The sun is—based on different, direct or indirect, single or multi-step routes of shorter or longer duration—ultimately the only energy source that is also available to humans. *Large black arrows*: direct use (emissions-free). *Hatched arrows*: simple (physical) transformed use (emissions-free). *Small black arrows*: multiply (biological/physical) transformed and uses renewable resources (emissions from burning; CO_2 neutral due to short-term absorption cycles). *Dotted arrow*: multiply (biological/physical) transformed, use of non-renewable resources (emissions from burning; greenhouse gases). (Adapted from Kaiser from Herzog (Ed.) 1996)

He sorts the different applications and forms of solar energy accordingly (Fig. 4.47):

1. Direct, emissions free
2. Simple transformed, emissions free
3. Multiply transformed, producing emissions but CO_2 neutral
4. Multiply transformed, producing emissions but not CO_2 neutral

Except for the last form, in which fossil fuels are burned and CO_2 gases are released, all types of usage are ecologically unproblematic. Some technologies for solar energy have been available for some time; others undeveloped but due to their principal simplicity could be quickly developed, though practical applications for some remain at the moment only in fantasy (above all solar–hydrogen technology on the basis of "artificial photosynthesis"). New technologies are being pursued, even with insufficient political support.

As Figs. 4.48a, b illustrate, the path from a pre-solar to a solar era means the subsequent abandonment of—though solar produced—fossil intermediate storage of energy. The assemblage of energy consumption (Fig. 4.48c) of the last 150 years shows that muscle work had already become an insignificant source of energy relative to the total energy spectrum by the turn of twentieth century; use of wood-stored energy through burning only in the post-World War II era. In 2000 the burning of coal begins to flatten out, while the use of oil, natural gas, and nuclear energy is drastically increasing. Only since the 1980s can one speak of noticeable solar energy usage, which for all of its forms has barely more than a 10% portion. The problems of its utilization are well-known, particularly the intermediate storage of the energy.

In certain cases solar energy has thrived, but the abilities of solar-regulated air conditioning for buildings has always been deeply underestimated by the general public. The common and current methods of air conditioning consume a large part

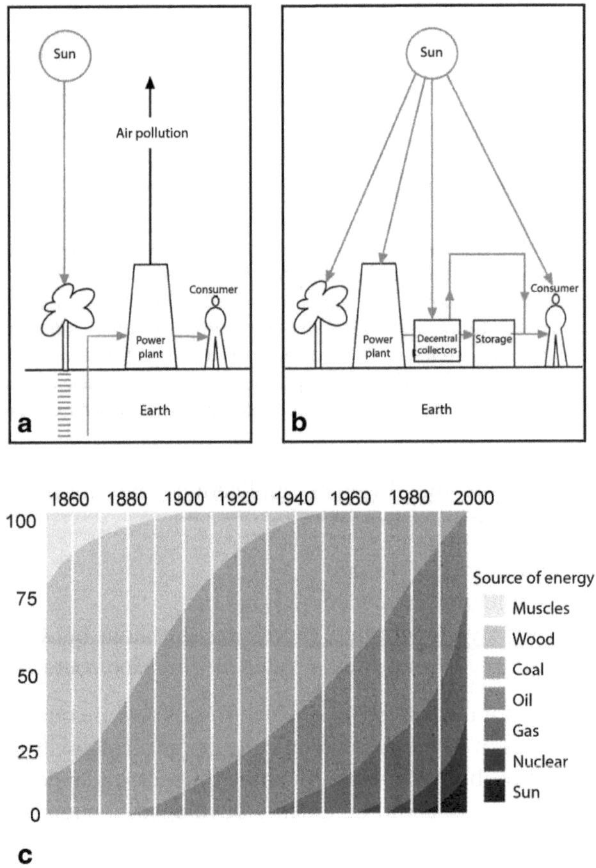

Fig. 4.48 Sun energy and energy consumptions. **a** Pre-solar era. **b** Solar era. **c** Assemblage of energy needs of the last 150 years. (Adapted from Kaiser from Behling and Behling 1996)

of the energy available for households. This portion could be dramatically reduced first and foremost using relatively uncostly methods—which would only acquire their significance however with general acceptance and application.

4.7.2 Biological Adaptations to Solar Radiation

H. Tributsch, radiation-biophysicist and physical chemist, has contemplated how preindustrial architecture and living organisms have developed and still develop methods for the use of solar radiation. He observed that Scandinavian houses, which are often colorfully painted, are lighter in color in the South and are even completely white in the "hottest" southern regions; apparently to better reflect the more powerful solar radiation and avoid overheating. Houses are "immobile"; however, mobile animals similarly use the effect of "adaptable color formation." Many lizards can accordingly adapt their skin color to the amount of sunlight.

"An iguana from the Fiji Islands for example becomes ever so lighter the hotter the sun shines and the warmer its body becomes. During intense heat, black beetles in the Namib Desert coat themselves with a wax film, which they can secrete from glands. Using this wax, they can reflect 40% more of the sun's rays and significantly cool themselves."

Similar capabilities are observed in the bird world, though there are exceptions. Black-colored birds in hot regions survive as such, for example the Black Stork in Yemen or the Oystercatcher in West Africa. Due to the condition that a feather coat functions as heat insulation, the black feathers do indeed heat up, but the warmth can be convectively dissipated, such as during flight.

Advantages then emerge when a black bird rests in the shade. Due to the reversibility of radiation absorption and emission, a bird with a black plumage can better dissipate warmth than a white bird. Perhaps it is also the reason why Bedouins and other desert peoples are often clothed in black.

As to how much heat radiation can play a role, it has been observed that in the summer deer prefer to rest at the edges of forests, in particular the sun-shaded side. Because the air there is relatively cooler, the edge of forest emits much heat, and is in average 2–3 ° cooler than in the interior of the forest.

"Traditional architects from warmer regions have used this effect. They built verandas so that during the mid-day heat one could likewise observe sunless, unobstructed sky from them. Thus one needed a tree-less lawn in front of the veranda to obtain its temperate effects."

Why can seagulls leave their eggs in full sunlight without them overheating? In the shells of the seagull *Larus heermanii* a pigment was verified that reflects 42% of heat radiation (near and mid-wavelength infrared). In full sun the eggs only heat up to 30 °C; without the protective pigment they would heat to 45–50 °C and thereby damaging the embryo. Why not design white façade colors with similar reflective capacity? Black buildings in the desert according to Tributsch would also be imaginable; they must only be well insulated. "A sheep on the meadow loses as much warmth due to heat emission as it does due to deflection. To consider the laws of radiation would be an advantage for our architecture, and traditional builders have learned to use these laws based only on pure experience."

This statement is correct and references the wealth of experience in working with sunlight and wind, geothermal warmth and moisture, light and shade, water abundance and lack, and so on. One can speak of "cumulative effects." It becomes important then to not single out solar effects alone and study only their details, but to suitably combine them with other important building-ecological effects as well. H. Tributsch envisions here a broad subject field, what the viewpoint of solar energy alone concerns already:

> Even building engineers of future solar-powered homes will have to apply numerous energy technologies to reach near perfect and optimal living conditions. It is not an uncommon development in technology. One must only consider how many individual pieces of technology are contained in a car or airplane alone. For the use of solar energy we must also allow different technologies to develop parallel to each other and synergetically cooperate with one another.

Modern architects are slowly considering again the (simple) physical fundamentals and on the other hand the wealth of traditional experience embodied in the so-called "primitive structures," which developed—in analog-to-natural evolution—in a trial-and-error process without the influence of specialists.

4.7.3 Macroscopic, Solar-Driven Energy Systems

In its balanced simplicity of sum formulae, photosynthesis of green plants is doubtless the most fascinating solar energy system, perhaps with the most technology potential for an energy economy of the future as well. However, there are other macroscopic—the engineer would say "direct"—possibilities for solar energy usage.

H. Tributsch included these possibilities in his assembled considerations. Generation of hot and cold belong here as well, including light collection, interaction between light and exterior surfaces. The principle of solar-driven evaporation used by plants can also be mentioned in this context. The most important principles are collected in Fig. 4.49.

Warmth, Cold Because the regulation of warmth makes a major portion of the total energy required for functionality in biology as well as in technology, a skillful handling of warmth and cold should be strongly considered. In biological development there exists more selection pressure. Technology has often—due to by far too low-energy costs—ignored this aspect, but in a future era of energy scarcity will have to use all conceivable mechanisms.

Both small and large animals have developed various strategies for protection against heat and/or cold, understood simply by the well-known Bergmann rule: the (heat producing) volume of an animal is proportional to the length of body cubed, the (heat exchanging) skin only squared. Therefore, arctic animals should be large and thick and additionally developed a relatively thick insulating fur layer; desert animals in contrast have developed forms with more heat-exchanging surface area

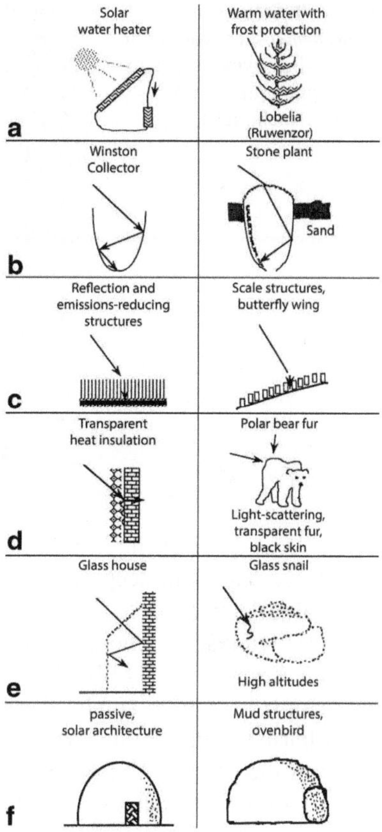

Fig. 4.49 Six examples for macroscopic, solar-driven energy systems in nature and technology. **a** Solar-heated water system. **b** Light collection. **c** Optimization of light reflection and scattering. **d** Transparent heat insulation. **e** Greenhouse effect. **f** Temperature regulation with *thick* mud walls. (Adapted from Tributsch 1995, edited)

in proportion to volume. The ears of the desert fox are therefore relatively larger than those of the arctic fox. The heat insulating capabilities of pelts and plumages and their encapsulated air pockets are ad hoc regulable and can vary with the season as well (by fur replacement or molting). There are numerous heat exchange functions by animals in water or land.

Devices for solar heat gain warm water in radiation-absorbing collectors. One can fill an intermediary store with the warm water and only let it circulate during the cooler night. In the mountains of Ruvenzori, Uganda the lobelia plant works according to analogous principle. Rain water collects in the seams and is enriched with a secreted antifreeze substance, thereby preventing frost—the nights can reach approximately – 15 °C—from clinging onto the lobelias (Fig. 4.46a).

Light Collection, Systems for Sunlight Maximal light yield (and additionally yield from longer IR waves) cannot be captured with lenses due to their limited aperture, but rather with internally reflected, parabolic funnels (Winston collectors, Fig. 4.49b). Their light yield is proportional to the square of the refraction index of the material comprising the funnel. A special advantage: the reflecting surfaces need not be positioned towards the sun. Analogous biological fiber optic systems

Fig. 4.50 Stone plant *Fenestraria* spec. with light penetrating "windows" and fiber optic transmission. (Photo: Nachtigall)

can be found for example in the ommatidia of crabs' eyes or also in plant buds, which can lead light deep into the finest root tips according to the principle of total internal reflection.

In H. Tributsch's laboratory the small South African stone plant *Frithia pulchra* was investigated for its fiber optics. It is built, similar to a light bulb, as a Winston reflector (Fig. 4.49b), and bears a transparent window directly on top and normally lives buried underground; only the window bulges above the earth's surface. In its form the plant is almost identical to the complex calculated Winston collector (Fig. 4.49a). The species from the genus *Fenestraria* (Fig. 4.50) is built similar. The dome-shaped window provides for the collection of light independent from the position of the sun. The deep directed light is used by the photosynthesizing cells on the edge. The warmth only heats up the large volumes of the water-retaining window cells and is quickly re-emitted ("water-warmth filter"). By weakening the intensity through the fiber optics, the light reaches the deep cells at the proper intensity for photosynthesis. Because the photosynthesizing cells sit on the periphery of the plant column, they benefit from an additional cooling principle (the soil's cool temperatures).

While our modern desert architecture is only little adapted to the climate and must work with large cooling mechanisms, the precedent of the stone plant shows how much better one could design climatically independent, desert-adapted architecture.

A suggestion by the author for a desert-located building is sketched in Fig. 4.51b. The dwellings would be accordingly built deep into the earth, "covered by a glass dome with a heat-absorbing water filter. Similar to the stone plant, the light is scattered deep within the funnel-shaped structure, where it can reach living spaces or even gardens, which would benefit from the cool temperatures of the earth."

"Intelligent" Skin Structures The iguana *Brachylophus vitiensis* from the Fiji Islands appears dark with lower temperatures of the early morning and therefore absorbs the most possible warmth from the sun. As sun travels higher in the sky

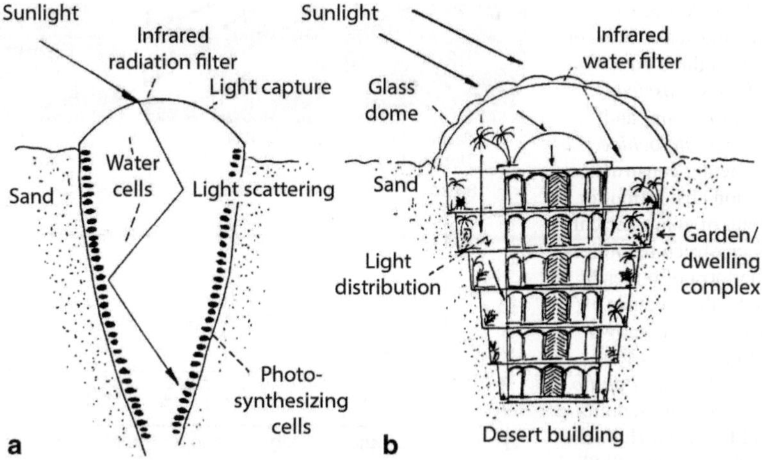

Fig. 4.51 Suggestion for a desert building **b**, abstracted from the build-concept of the stone plant *Fritia pulchra* **a**. (Adapted from Tributsch 1995)

the iguana becomes an ever lighter shade of green; the "intelligent" skin protects it from too intense sunlight. Plants turn their leaves towards the direction of the sun or they orient them with their broad side in the North–South direction (use of the early morning sunlight: "compass plants," represented in northern climates for example by the milk thistle, *Lactuca serriola*). Humans' skin contains sweat glands that cool the skin's surface with the evaporation of moisture during threat of overheating.

Many butterflies collect solar energy in their wings (Fig. 4.52c). As such the large butterfly *Ornithoptera priamus poseidon* from New Guinea can reach temperatures in its body of up to 61 °C with intense sun irradiation, as experiments with dummies have shown (it will correspondingly end the quickly occurring heating process early). Once the wings are removed the induced experimental body temperature sinks to about 50 °C. Such wings are then, as explained more thoroughly below, solar energy collectors.

The problem of transparent heat insulation in technology and biology has been already handled above. Extensive literature exists for greenhouse effects (even glass snails, which live in high altitudes, use these). Massive adobe structures, as they are built by the Pueblos with their thick mud structures and even by the ovenbird and potter wasp (Fig. 4.49f), consistently maintain their interior temperatures through extremely varying day and night temperatures (heat reservoir capabilities, delayed heat emission).

4.7.4 Butterfly Wing as a Solar Panel

The very delicate, submicroscopic scales found in butterfly wings (Fig. 4.52c) are multifunctioning. They increase the aerodynamic lift by about 10% by reducing the fluid friction and enable fluorescent colorations, which are based on the principle

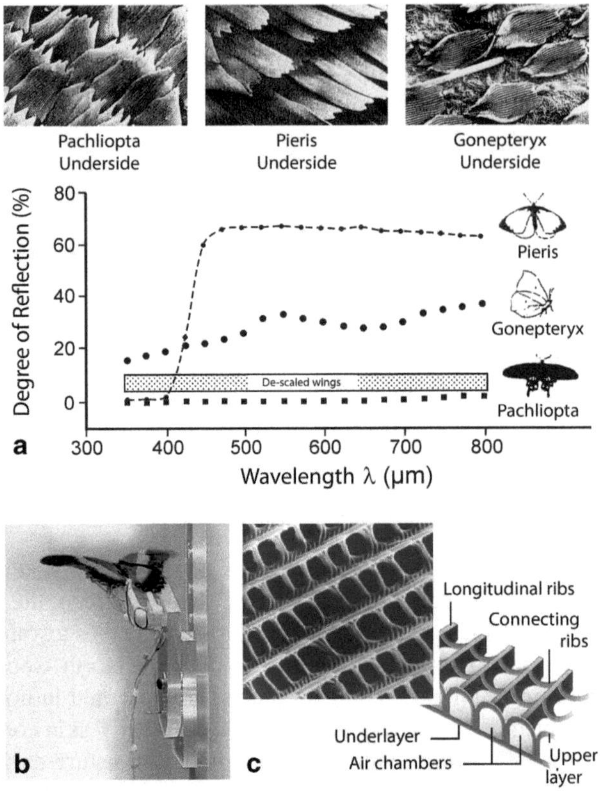

Fig. 4.52. **a** Degree of reflection and scale form for species from three butterfly species *Pieris brassicae*, *Gonepteryx rhamni*, and *Pachliopta aristolochiae*. **b** Experimental construction for radiation absorption for the cooling of computer chips using microstructures following to the form of butterfly scales. **c** Scale structure of a butterfly of an unknown species (REM) and of the tropical swallowtail *Papilio palinurus*. (**a** from Schmitz 1994 and **b**, **c** from Helmcke O.J. and from Wong 1998)

of colors from thin lamellas and other physical principles; here they play a role in the formation of gender and thermoregulation. The surface structures move in the micro- to nano-scale, are formed in the shape of the individual scales by "digitalized self-formation processes," and can, with their multifunctionality, give inspiration for corresponding technical skins.

Butterflies need a temperature of about 40 °C in their thorax in order for their muscle motors to function properly. Many species rest with their wings in a slant position with respect to the sun so that the wing surface reflects the sun at the ideal angle to their thorax. Special, cushion-like scales on the thorax hinder the escape of absorbed heat. Hemolymphs flow in certain veins in the wings. The hemolymph is directly heated up in the exposed wings, then flows back to the body, and warms it supplementarily.

Different species are differently outfitted with various reflection and absorptions capabilities in their wings, which one can correlate to their different lifestyles. Figure 4.52a shows the reflection and absorption spectrum for the members of the genera *Pieris* (albino butterflies, white), *Gonepteryx* (brimstone butterfly, yellow), and *Pachliopta* (southern festoon, dark), as well as the difference between descaled wings and normal wings. The forms of the scales are illustrated in the small figures. The reflecting ability is understandably the highest with the white *Pieris* and the lowest with the dark *Pachliopta*. Wings without scales accordingly have a very low capability to reflect light, of only a few percent. Conversely, the spectral absorption (the largest with the intact *Pachliopta*, intact wings always have a larger spectrum than descaled) and the spectral penetration (the smallest with the dark Pachliopta, intact wings always have a smaller spectrum than descaled) of the wings are the largest with darker colors.

Pachliopta folds its wings together and absorbs warmth with the black scales on the underside of the wings; the air between the folded wings is also warmed. In contrast, *Pieris* and especially *Gonepteryx*, which is already active in springtime, spread their wings to a slanted angle and reflect sunlight to the thorax with the topside.

Radiation effects are reversible. Warmth could then also be released at too high body temperatures using the same structures. The temperature balance of the tropical butterfly *Papilio palinurus*, which can absorb about 80% of sunlight, was measured by T.J. Wong (Fig. 4.52b) of the Institute for Machine Engineering at Tufts University, Medford, USA. A surface structure following the prototype of gridded scales with longitudinal and cross ribs of this butterfly could be imprinted on the surface of electronic chips that easily overheat during use. It would also be assumed that the heat emission could be improved with a corresponding microprint of aluminum sheathing plates. They would then be self-cooling—in a reverse of the butterfly principle. The subtleties of the submicroscopic structures appear to be essential. The tropical butterflies *Papilio palinarus* and *Urania fulgens* only differentiate themselves by the morphology of their scales with small, geometrical peculiarities. The "chitin fences" of the former are approximately one quarter of the wavelength of typical color apart from one another, which leads to interferences and ultimately "cancels out" the radiation inside of these layers, resulting in the layers being warmed up. With the latter the distances are somewhat larger causing much more radiation to be reflected and losing the direct warming qualities. These aspects could be reversed for the cooling of computer chips and panels.

As Fig. 4.53 illustrates, the theoretical trendline for the degree of reflection as a function of wavelength is dependent on refraction index n for butterfly scales. Experimental data stay in compliance with the calculations for wavelengths up to about 800 nm and a common refraction index of about 1.6. Within certain boundaries one can predict then the reflection capabilities of thin-layered structures with mathematical programs.

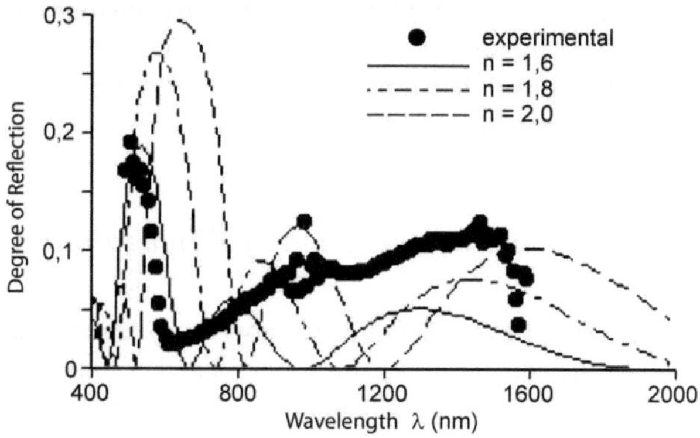

Fig. 4.53 Spectral reflectivity of scaled wings of the butterfly *Papilio blumei*, a tropical butterfly with fluorescent green wing stripes. (Comp. Biomech. Lab., Tufts University, Medford MA 1999)

Schmitz and Tributsch have shown how effective these systems function on stretched and dried butterflies that had been exposed to a radiation emitter with 0.1 W cm^{-2}. Particularly interesting are the Phoebus Apollo butterflies, which appear in the high-altitude biotopes. The wings of the Phoebus Apollo, *Parnassius phoebus*, are strong absorbers: 87% with 350 nm (ultraviolet range) and still 28% with 800 nm (infrared region). If the scales are removed, the percentages reduce to 38% with 350 nm and 3% with 800 nm, with the thorax of the prepared specimen reaching 59 °C and the base of the wings reaching 56 °C. If one removes the wings entirely, the body of the specimen only reaches a temperature around 10–11 ° lower. The wings serve then as warming devices, and their capability as a solar absorber is three to seven times better (!) when they possess scales. The dried butterflies were clamped in natural "sunning position".

4.7.5 Adaptive Solar Usage

The utilization of solar energy is also possible, next to stationary forms in nature, for example in the foliage of trees, within contexts of adaptive systems. The example of sunflower, which turns itself over the course of a day in relation to the sun's position, is probably the most well-known. Lesser known natural systems, like the sunlight capture performed by some sea sponges, represent a form of solar adaptation using light-directing elements. In the meantime, it has been attempted to technically apply different natural systems, for which researchers of the AIT, Austrian Institute of Technology have played a pioneering role. They have analyzed examples of natural functions using the methodology of Pool Research to make them available for the second phase of development—abstraction for applications in facade developments. Some examples are extensively discussed in Chap. 6.

4.8 Photovoltaik: Solar-Contingent Electricity Generation in Nature and Technology

There are analogies from the living world even to this technically self-sufficient and already much implemented technology, although these analogies are still disputed, for example with hornets. However, the photo-electric basis of this technology will be briefly recapitulated first.

4.8.1 Principal Function of Photovoltaic Cells

These cells consist of semiconductors (e.g., silicon). Intentional impurities (doping) change the conductibility due to facilitated or impeded release of electrons. Doping with phosphorous, for example, results in the formation of n-silicon (n negative), which contains more free electrons and is better conductor. Doping with boron results in p-silicon (p positive), which contains more gaps (less electrons) and is thereby positively charged. If an n- or a p-silicon are layered on top each other, they generate an electric field (Fig. 4.54a) that builds barriers of electrons on the boundary surface and only enables a one-directional flow of electrons (in the + direction), until it reaches equilibrium.

An absorbed quantum of light can normally strike and release an electron and thereby forming a gap; these have the tendency to wander to the oppositely charged side.

If one connects the both sides to an external resistor, electrons can flow to the p-side, where they bind with the relocated gaps (Fig. 4.54b).

The product of generated voltage and flowing current corresponds to the electrical power of the photovoltaic cell. Its efficiency can amount at the most to 25%; in reality it is lower. The principal construction of such cells is sketched in Fig. 4.54c. These types of cells can be built from monocrystalline silicon, but also polycrystalline silicon, amorphous silicon (lower effectiveness, but also lower price), gallium arsenide, copper indium diselenide, cadmium telluride, and others. In each case the construction is complex, relatively expensive, energy intensive, and generally requires a highly purified working environment.

These technological difficulties drive the price high; the high-energy consumption during production particularly makes these types of photovoltaic cells ecologically problematic, as it lasts a certain amount of time before they have delivered enough energy to offset the energy lost for their fabrication alone.

Other alternatives are understandably being sought. These alternatives are designated as "organic solar cells," and they are currently being developed, as shown in the examples of subsequent sections, for various locations according to botanical prototypes. There is however, as mentioned, an example for "animal solar cells," whose potential certainly has not yet been fully grasped.

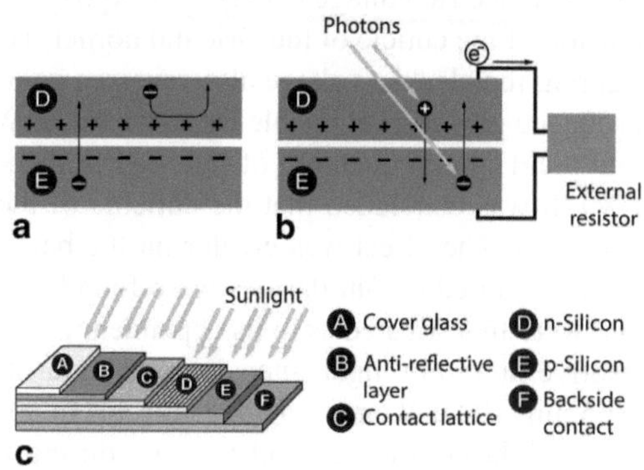

Fig. 4.54 Principle function and construction of photovoltaic solar cells. **a** Effect of an electric field (voltage generation). **b** Flow of current through an external resistor. **c** Construction (new drawing, based on Aldous 2001)

4.8.2 Problems of Photovoltaics on Basis of Silicon

The path from quartz sand to a finished silicon wafer requires complex, chemical, and mechanical (silane production and monocrystal formation) processes with high-energy costs above all (reduction at 1100 °C). The use of diamond saws to the cut the monocrystals into wafers also generates waste. Despite these production losses the "energy return" is today generally positive: Calculated over its life span, silicon-based photovoltaic elements provide more energy than it costed for their production. However, the amortization period lasts at least several years, and it is not clear whether these calculations include all additional costs (i.e., cost of transport). In any case, the silicon technology is still currently irreplaceable, but due to its high-energy cost and complex technology it should be replaced in an intermediate timespan.

The photovoltaic industry already needs a new source for silicon; silicon scrap from the semiconductor industry is no longer sufficient, as Bernreuter has calculated. If the demands for silicon for solar energy had been limited to 2300 t in 1998, then one calculates 8000 t for 2010. The purest silicon with only one foreign atom per 109 silicon atoms is already valued today at 100 € per kg. A series of companies have submitted concepts for innovative production methods for cheap, pure silicon, for example Bayer/Leverkusen, Wackerchemie/Burghausen, Kawasaki Steel Corporation (who wants to reduce the price to a few dozens euro per kilogram), and others. The good intentions of the "100,000 roofs program," which together with the renewable energy law trigger a photovoltaic boom, would be sunk if photovoltaic panels were delivered either in too small scope or only overpriced; according to the predictions both problems will probably combine at some time or another, another reason to look for alternatives in the area of organic solar cell technologies. Prototypes for these technologies can be found in a multitude of plants. The only known "animal use" of a solar cell to this day is described in the following section.

4.8.3 Photovoltaic and Thermoelectric Effects of Hornets

At the beginning of 1990s J.S. Ishay of Tel Aviv University observed the occurrence of electric voltage between an exposed portion and a neighboring, unexposed portion of the cuticle of the Oriental hornet, *Vespa orientalis*. With a reversal of the lighting the electric voltage also reverses polarity. A small, site-specific, radiation-produced power from visible light of a few mW cm^{-2} was already found. The maximal yield from a quantum of light lay in the spectral region of 360–380 nm (near UV). It was concluded that the cuticule of these hornets functions as a biological solar cell. The effect was greater on the back edge of the abdominal tergites than on the front edge. Similarities were found in the pupa cocoon of the same species and were investigated as to its dependency on edge conditions such as temperature, relative humidity, light intensity, and time of exposure. Each 2 min illumination (365 nm; 100 μW cm^{-2}) yielded currents of a few nanoamperes with time constants $t_1 = 18$ s during the rise and $t_2 = 30$ s during the decrease in current (Fig. 4.55a).

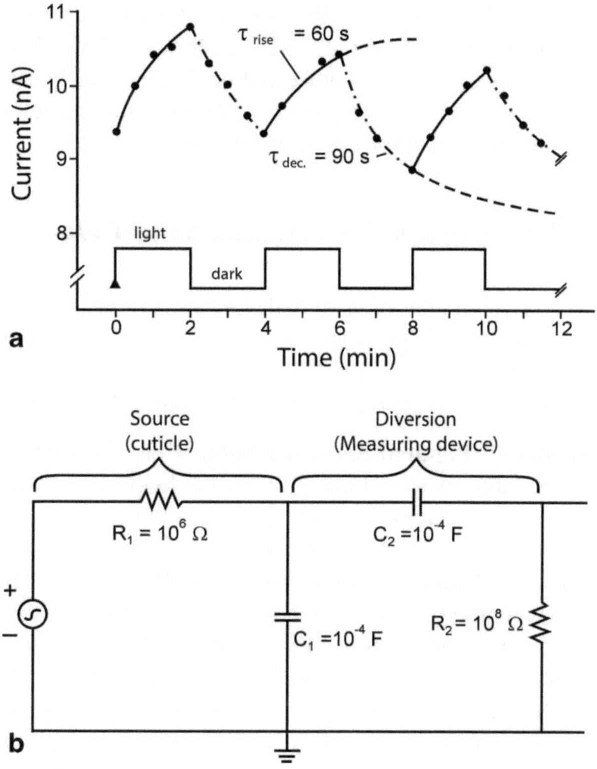

Fig. 4.55 "Bio-solar cells" (?) in the cuticule of the Oriental hornet, *Vespa orientalis*. **a** Measurements on the front lid of the pupa shell (I=365 nm, P_{rel}=100 μW cm^{-2}). **b** Equivalent electrical circuit diagram, compare to the text. (Adapted from Ishay et al. 1992, supplemented)

These measurements were consistent with earlier findings, according to which the cuticules of hornets behave like an organic semiconductor, the measuring area like a diode. The entire process was interpreted as a combination of photovoltaic and warmth effects; both are caused by the absorption of radiation.

As the equivalent electrical circuit diagram (Fig. 4.55b) illustrates, the internal resistance of the diverting branch is orders of magnitude higher than that of the production branch; with measurements of current the decrease in voltage on the instrument is orders of magnitude smaller than on the cuticule, both are to be empirically claimed. Simulation experiments with the use of the equivalent electrical circuit led to principally similar results. It was concluded that the cuticule experiences changes in polarity under illumination or heat, as they are known among photosynthesizing membranes and exhibit an electret effect. It is well-known that electrets, such as illuminated beewax, form under high-voltage gradients from about 10 kV cm^{-1}; there are suggestions as well that this could occur with very low gradients (a few dozen mV cm^{-1}), as it appears in the cuticules.

The reaction of the cuticule of hornets to light can be depicted as "extra-retinal photo-perception." The differentiation between thermoelectric effects in darkness and photoelectric effects in light and their reduction to fine-morphological and submicroscopic mechanisms is not yet completely clear.

Therefore, the study focused primarily the ascertaining of effects. One should, however, not undervalue the discovery of "effects themselves." On the one hand, it is already good even if one can model them "partly causal" at the beginning. On the other hand, not yet completely explainable effects belong to the strongest stimulants that are known among natural science researchers.

4.8.4 Organic Photovoltaic Solar Cells

The development of such cells has already lasted two decades and has always led to products of impractical applicability. However, a practical result would still be of great significance. Therefore it is still being further researched in several locations. Three classical approaches as well as more recent developments are described here.

Grätzel's Pigment-Sensitive Solar Cell Conventional solar cells convert light energy into electrical energy with the help of photovoltaic effects on the interface of a semiconductor. The semiconductors must be highly pure and defect free, which makes their production complicated, energy consuming, and expensive. In the work group of M. Grätzel in the Laboratory of Photonics and Interfaces at the Swiss Institute for Technology, Lausanne, a pigment-sensitive solar cell was developed. While the semiconductors in conventional, silicon-based photovoltaic panels simultaneously absorb light and provide for the separation of electrical charges (into "electrons" and "holes"), a monomolecular pigment layer assumes the task of light absorption and the semiconductor boundary layer the task of charge separation: Each task is divided into a separate element. The method, which is simple in principle but is controlled by several critical parameters, indicates the way to "simply" constructed, environmentally compatible, low-energy cost, and clearly more economic solar cells. Kalyanasundaram and Grätzel describe the principle as follows (Fig. 4.56a) :

> The light absorption occurs through a monomolecular pigment layer (S) that is coupled by a chemical bond onto a semiconductor surface. After excitation by a photon (S*) the pigment layer is shifted into the position for the transfer of an electron to a semiconductor (TiO_2; "injection process"). Due to the generated electric field an electron can be removed from the semiconductor material. Formally speaking, a positive charge is therefore transferred from the pigment (S+) to a redox mediator (A) (process of "interception"), which contains the solution between both electrodes. From there the positive charge reaches a counter-electrode. As soon as the mediator returns to its reduced state, the circuit is closed, and current can flow through an external resistor. The theoretical maximum voltage of such a device corresponds to the difference between the redox potentials of the mediator and the Fermi state of the semiconductor.

Alongside developments from D. Wöhrle of Bremen, this Swiss development, which dates back to the beginning of the 1990s, led to the first experimental "bio-solar cells." It uses nanocrystaline films of TiO_2.

> The solar cell consists of two conducting glass electrodes in a sandwich configuration with a redox electrolyte in between. A TiO_2 layer of a few μm in thickness is struck from a colloidal solution of monodispersed particles of TiO_2. This layer is porous and exhibits a large surface area, to which the pigment molecules in monomolecular division can cling. After appropriate treatment with heat, which should reduce the resistance of the film, the

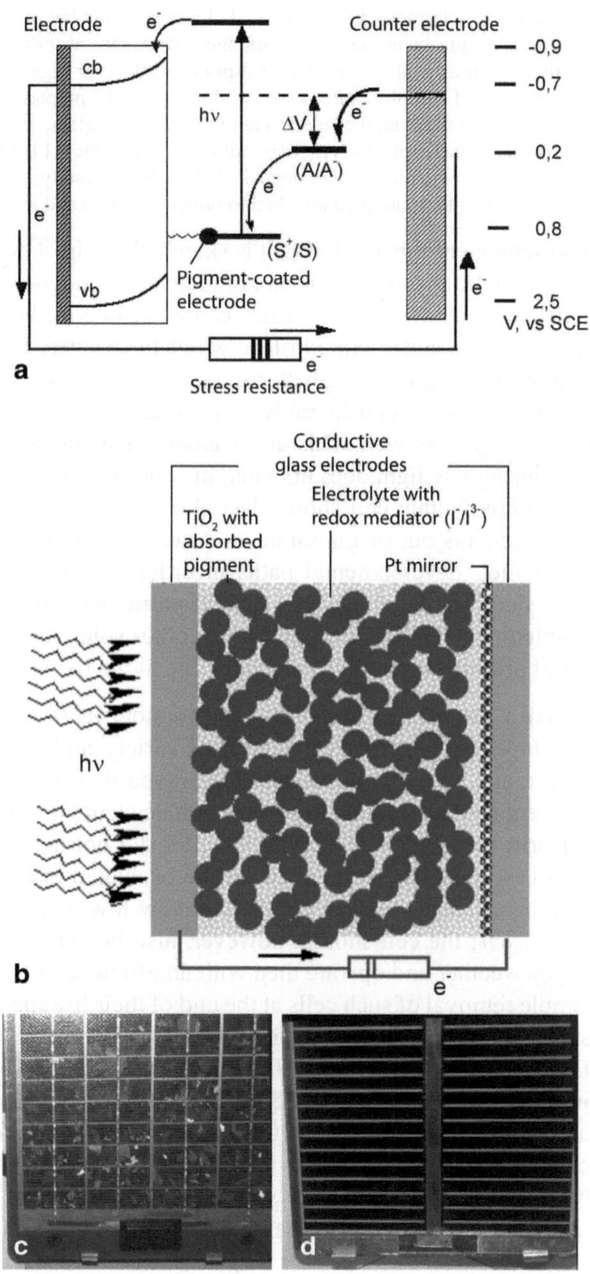

Fig. 4.56 Principle of the pigment-sensitive solar cell according to Grätzel. **a** Scheme of a solar cell. **b** Nanocrystalline solar cell, compare to text. **c** Photo of a silicon-compound cell. **d** Photo of a Kurth-compound cell. **c, d** Parts each with the same size. (Adapted from Kalyanasundaram, Grätzel 1999, **c, d** photos: Nachtigall)

> electrode is immersed with the oxide layer in a suitable pigment solution for an hour. The porous oxide layer acts like a sponge, effectively absorbing the pigment molecules and their color as well. Molecular absorptions of three or higher are easily obtained using RU-polypyridyl complexes inside of the thin layer. The prepared electrode is then brought into connection with another electrode of conductible glass, and the interstitial space is filled with an organic electrolyte (commonly with a nitrile) ($I^{-}E/EI^{---}$). On the counter-electrode a thin layer of platinum is deposited that should catalyze the reduction of triiodide into iodide. Contacts are then attached to both electrodes and the entire circuit is closed.

The scheme of this kind of cell is sketched in Fig. 4.56b.

The light absorption on monomolecular pigment layers ensues with low yield. Sufficient photovoltaic efficiency cannot be obtained with smooth surfaces, but with sponge-like nanostructured films, which have a very high internal surface area. Penetrating light is therefore better scattered and crosses hundreds of absorbing monomolecular layers, considerably increasing the possibility of absorption and with it the light gain as well. This arrangement also ensures that the effectiveness of the cell during low light does not sink, in contrast to traditional silicon-based systems.

A photovoltaic cell should last about 20 years and thereby pay for itself. The developments out of Lausanne, and not only these, must overcome many hurdles along their developmental paths in order to obtain practicality and resistance to corrosion. The latter problem area is particularly weighty. Due to these durability problems the Grätzel cell, similar to other solar cells, has yet to be developed for serial production, more specifically due to the oxidation of conducting layers.

Kurth Cell M. Kurth and his collaborators R. Monard and F. Flury have further developed cells of above-mentioned variety and, particularly by the application of ceramic corrosion protection, improved their durability. By its exterior form a conventional arrangement of photovoltaic elements hardly differentiates itself from a Kurth cell (Fig. 4.56d). The inventor did not answer inquiries into the concept details so this development cannot be evaluated in comparison to the others. With sun the efficiency of this cell is relatively low in comparison to silicon solar cells with 7.8%; the cell should, however, also be active by diffuse light, even during foggy weather and operate then with an efficiency of 5.5%. Also emphasized is the simple removal of such cells at the end of their life span. This concept celebrated by science journalists was considered for the entrepreneurial prize of the Swiss W.A. De Vigier Foundation in 2000, although even today nothing closer is known about its application in praxis. If it eventually had come to permanent application of these and other similar cells in 2010, the development time for the first comparable solar cells with 1% yield, given by C.W. Tang of the Kodak Corporation in 1986, would have amounted to almost 25 years. That would not be a lot in comparison to other technologies, though more development is yet to be awaited.

4.8.5 The Plastic Solar Cell

The Sariciftci Cell (Plastic Solar Cell) The physical chemist S Sariciftci, who researches in Linz, Austria, presides over one of several worldwide active research groups that would like to develop organic solar cells on the basis of artificial photosynthesis. The development goal is a plastic solar cell that can be fabricated automatically and impervious to mechanical stress (Fig. 4.57a). Figure 4.57b shows a possible structural variant, in this case still utilizing a glass carrier. As recognized in Fig. 4.57c, polyester films or glasses (resistance between 10 and 100 Ω cm^{-2}) coated with indium tin oxide (ITO) are used. 3,7-Dimethyl-octyloxymethyloxy-PPV is applied as a donor (generally: Alkaloxy PPV). A fullerene is used as an acceptor, specifically 1-(3-methoxycarbonyl) propyl-1-phenyl [6,6] C61 (abbreviated PCBM). Figure 4.57d shows a measurement sample.

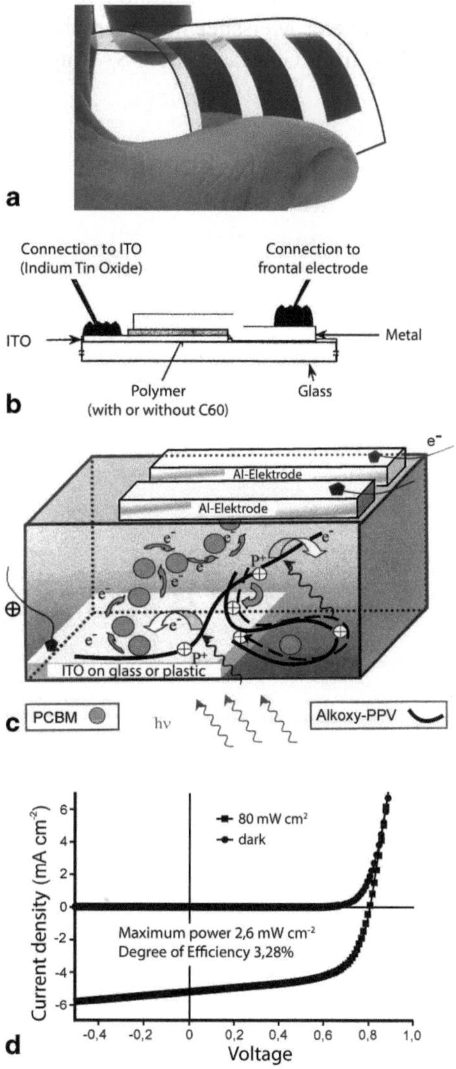

Fig. 4.57 Plastic solar cell according to Sariciftci. **a** Design example. **b** Example of organization of the layers. **c** Concept principle. **d** Measurement sample. (**a**, **b** from collection of papers by Sariciftci and **c**, **d** from Doppler Laboratory for Plastic Solar Cells and Quantum Solar Energy, Linz 2001)

A great future can be predicted for concepts like this one. Certainly, as already mentioned, there are many practical aspects yet to be solved with this concept, for example the durability during its life span, resistance to breaking, the translation onto installations of large area, and a reduction of the now too high cost of production.

The vision that physical chemists are now developing rests upon the use of the countless facades and windows on buildings. A lightly tinted window pane, through which one can let an electrical current flow when exposed to solar radiation—there are countless window panes!

The research into organic cells has found a broad basis today. However, intensive research had begun too late, only taking place for the last 15 years or so, since the holding of the first international conference for this subject field under the leadership of Dieter Meissner in Cadarache, France in 1998. It united the researchers at the time from Germany, the USA, Japan, and other countries.

Chapter 5
Biological Support and Envelope Structures and their Counterparts in Buildings

In this chapter, biology stands in the foreground. The examples from biology are, however, categorized according to their structural characteristics, and in each case technical prototypes will be indicated that are analogous to the biological structures. In a few of the examples, the study of natural forms and structures influenced and inspired the development of certain technologies. With others the influence can be assumed but unable to be verified by sources; the connection can be found then by the juxtaposition of the analogous structures. Some aspects have already been covered in earlier sections, namely tensegrity and tensairity structures, panel structures, fold structures, and aspects of bee honeycombs. With the exception of the latter, these will not be touched upon again. Some has likewise already been spoken about the earthen and "ceramic" nests of certain animals.

The subsequent Chap. 6 proceeds in the reverse direction. In that chapter, structural and architectural ideas as well as completed structures stand in the foreground. In both of these chapters the goal is to illustrate how a technological product was developed from a biological precedent. Beyond that it will be shown which cross-connections "back to biology" exist.

5.1 Lightweight Structures

Biomorphic?—Not Biomorphic? At this point considerations for the meaning of "biomorphic" should be reiterated.

In the end, it always remains problematic comparing structures of technological basis and those from nature. Even today, though we are practiced in these formal comparisons, the danger persists that they only remain superficial. If swelling forms of vertebrate organs and postmodern "biomorphic architecture" are juxtaposed, it can be determined that they both have no right angles or sharp edges and both somehow "seem organic." One can also attempt the psychology behind the formulation of criteria for well-being, but the physical–functional basis for the comparison is missing.

G. Pohl, W. Nachtigall, *Biomimetics for Architecture & Design*,
DOI 10.1007/978-3-319-19120-1_5

When one compares a technological construction to one of the living world, such as a television tower to a stalk of grass, the physical basis for comparison is, in contrast, already given; during still wind for example, purely central compression forces are acting parallel to the vertical axis. On the other hand, the danger remains that one draws the wrong structural conclusions, as laws of similarity can be easily violated in comparisons, such as overlooking the nonlinear dependencies of structural parameters of length. In addition, the materials for structural stability and rigidity in biology and technology have entirely different E-moduli and Poisson ratios.

Then when is a building "biomorphic" and when is it not?

One can search for commonalities in the manner of how the criteria for lightweight structures are used, for example. For the Eiffel Tower, measureable organic forms such as bones gave substantial inspiration to the structural engineer Koechlin; the consistent use of a framework of beams, whose alignments are based extensively on force trajectories, corresponds to the lightweight-structural principle of a bone. One could ask for example: Could the femoral bone itself be built even lighter for its given, average structural stress? Could the Eiffel Tower also have been built lighter according to these principles? Are they or are they not both structural systems, optimized through principles of lightness? Such questions would not be meaningless. One could pursue each of them with the use of completely similar structural analyses. In contrast, each comparison of a biological structure to a structure like the Great Pyramid of Giza would be nonsense; the "structural intentions" are obviously different.

There are, then, entirely meaningful possibilities for comparison to be made in the space between structures of nature and of humans. They can quickly lead to tangible results, if both are clearly worked out from fundamental physical laws. The comparisons can also remain qualitative, if they are oriented on the functionality of specific parts, which can offer inspirations within their respective boundaries of observation, whether it is for biology purposes and a better understanding of natural structures ("technical biology"), or as inspirations from nature for architectural design ("biomimetics").

Ultimately comparisons are always possible for the development of a building or product, even when their results do not correspond to the representative criteria defined here. It would then be that the comparison leads to insights only in a relative early stage; after this point it would either become meaningless, because no further basis for comparison is definable, or not be sensible, because recognizable "structural intentions" are diametrically opposed. One should then end the comparison there. Although it would be completely wrong though to not use this method at all.

5.1.1 Diatoms → Geodesic Domes

Diatoms, defined in botanical language also as bacillariophyceae ("rod-shaped plants"), are algae of sea and freshwater plankton, which fabricate finely porous silicate skeletons. Figures 5.1 and 5.2 illustrate some forms and details captured by a classical transmission electron microscope (TEM). The petri dish-shaped forms usually have a diameter under a tenth of a millimeter and float freely in the plankton of the ocean or freshwater. They can additionally attach themselves to each other to form chains that can be anchored to the ocean floor. Such chain formations or other "Aufwuchs" (organisms that grow on open surfaces in aquatic environments) of diatoms form the brownish, slimy coating on stones in slowly flowing creeks in springtime. Under the microscope the structure reveals itself as a lacy, mesh-like skeleton (Fig. 5.1a). More intense magnification under the electron microscope shows that the apparently open pores are actually covered with another mesh layer, whose pores in turn can be covered in a sieve-like man-

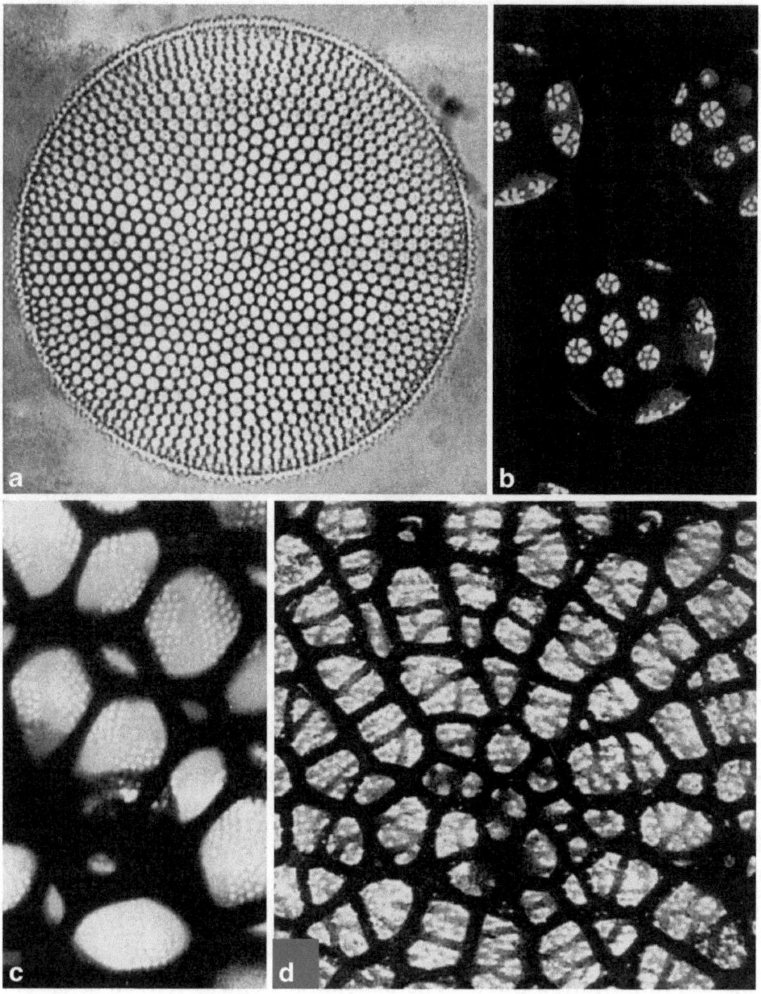

Fig. 5.1 a–d TEM images of diatoms. (Adapted from Roland 1965, edited)

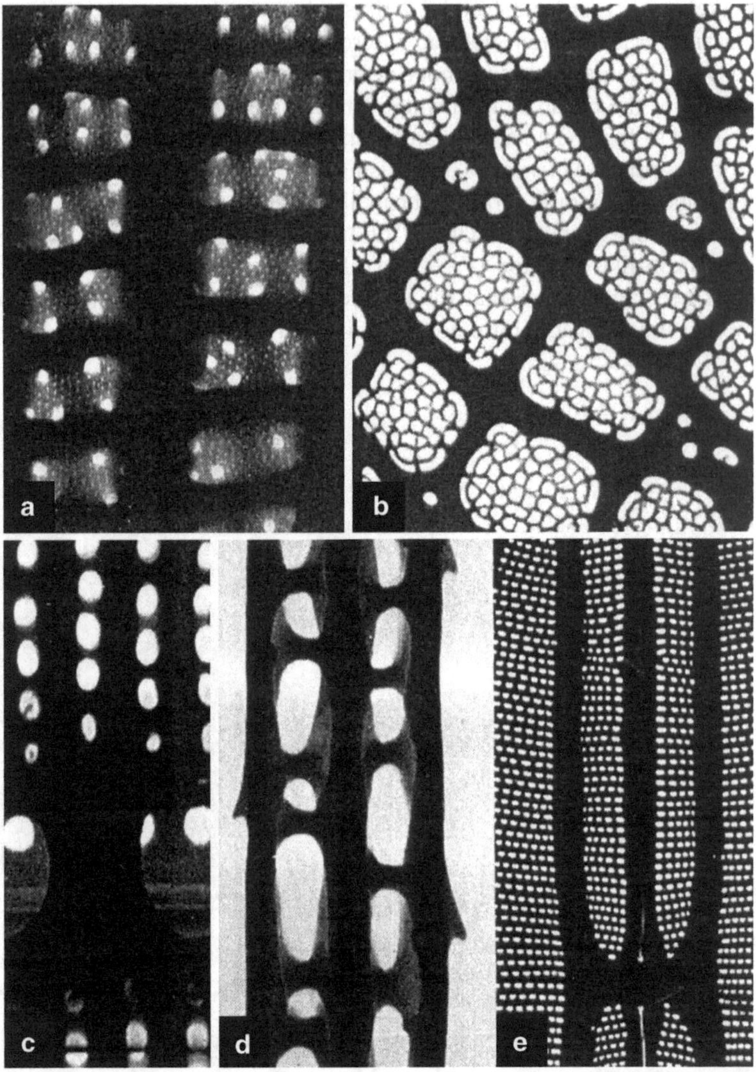

Fig. 5.2 a–e TEM images of diatoms. (Adapted from Roland 1965, edited)

ner (Fig. 5.1b, 5.1c, 5.1d, Fig. 5.2). One ultimately finds a system of up to three fine layers nested inside of one another. The finest pores are actually open, but they are however already so small that they do not allow multi-molecular proteins to penetrate.

Closer observation shows a typical irregularity that suggests a strong role of random processes in the micromorphological formation.

Diatom → Train Station Shed The diatom *Surirella* is formed as an elongated ellipse (Fig. 5.3a). A central beam supports spanning arched ribs on both sides, onto

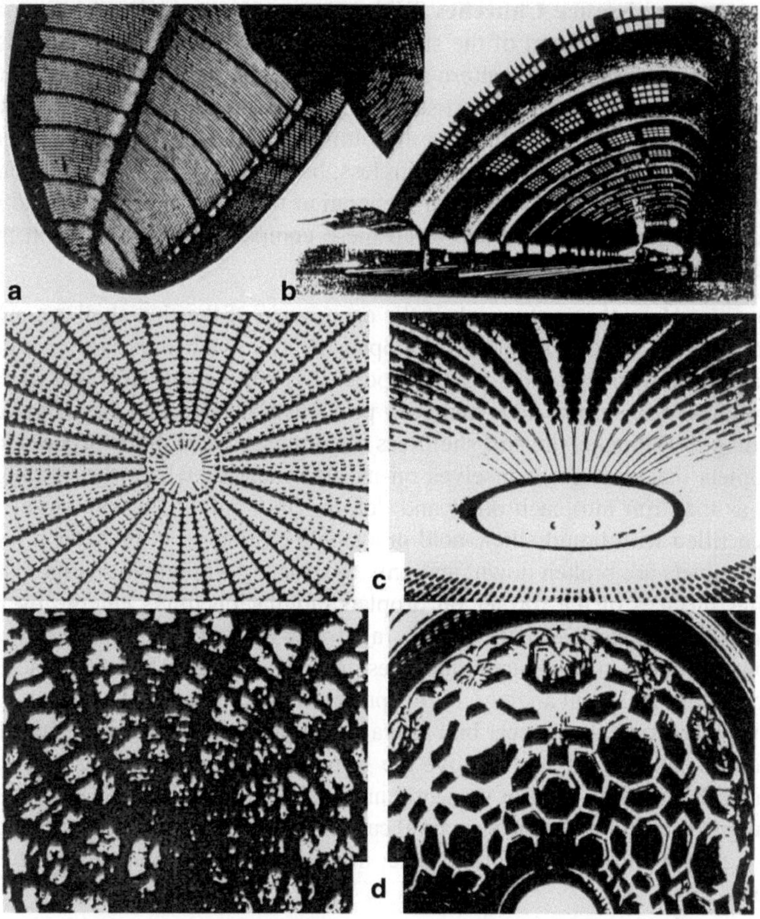

Fig. 5.3 Similarities of form, diatom structures: **a**, **b** diatom *Surirella* and train station (E. Torroja), **c** diatom *Arachnoidiscus* and Palazzetto dello Sport, Rome (P.C. Nervi), and **d** diatom *Thalassiosira* and Renaissance church, Rome. (Adapted from various authors from Nachtigall 1974)

which a finely porous perforation system spreads itself. Its basic form is remarkably similar to a train station canopy designed by E. Torroja in the 1940s. It is not known whether the similarity of form is simply coincidental or whether the architect had actually been informed about diatom structures.

Diatom → Stadium In contrast to the previously mentioned diatom *Surirella*, which exhibits an elongated oval shape, the diatom *Arachnoidiscus* is circular in plan view; it belongs to the subgroup of "Centrales." Its structure is correspondingly radially symmetric; radial ribs run from the center outwards supporting the fine mesh layer in between. The famous roof structure designed by P.C. Nervi for the Palazzetto dello Sport in Rome (Fig. 5.3c) can be viewed as an analogy for this case, although it is again uncertain whether the similarity is intentional or not.

Diatom → Renaissance Churches With stronger electron microscopic magnification the structural system of the shell of the diatom *Thalassiosira* reveals itself as a highly sculptured brace pattern that is equally unimportant for its structure as are the heavy articulations and artistic cornices on the ceilings of many Renaissance churches (Fig. 5.3d). Although the similarity is purely superficial, even such comparisons are not completely meaningless, because in design and architecture it eventually comes down to pure form generation that need not always and everywhere have a functional purpose. Even these comparative exercises can provide inspiration for creative activities.

Fat Droplet Hypothesis To explain the occurrence of the fine shell structures of diatoms (Fig. 5.7a) G. Helmcke developed the following hypothesis. When the plasma globule slips out from its auxospore, it is still naked. However, the shells rapidly form at this point. Fats had been previously detected on the outer surface. According to Helmcke, the still shell-less diatom uses a metabolic process to form fat droplets that arrange themselves on its outer surface where they struggle for space, as they run into each other and deform (Fig. 5.4a). The interstitial spaces are then filled with liquid silicic acid in a single casting process. After it hardens, the fat droplets are broken down, and only the casted form remains (Fig. 5.4b). On touching surfaces of the former fat droplets circular openings emerge by mutual flattening of the spherical shape (Fig. 5.7a). Today this hypothesis is no longer supportable, though it still inspires much design work. Delicate building bricks were developed as a technical analogy to this principle (Fig. 5.7b).

If one were to press a soccer ball into a hexagonal box, cast the interstitial space with a hardening liquid, and remove the form, a hexagonal building block would remain with an empty void in its center and a circular opening on each side (eight in total) yet maintain a certain structural capacity. With these blocks one could build lightweight partition walls, for example; in the 1960s this was actually attempted (compare Fig. 5.4b).

Cast Concrete Shells For his doctoral thesis, architect T. Noser, whom G. Helmcke had mentored during his time in Berlin, attempted shell forms according to the "Helmcke Principle." This principle, as mentioned, was later proven as not entirely accurate; in detail the diatoms construct their shell differently. Though the fat droplet hypothesis still developed as an entirely separate but important heuristic principle and provided many inspirations for unconventional building designs.

T. Noser pressed soccer balls between pressure plates and casted the form with plaster or polyester (Fig. 5.5a). It resulted in the previously stated hollowed hexagonal frame form. Shells with sectional forms that follow a kind of catenary curve were also attempted (Fig. 5.5b). After evaluation and turning them upside-down, they were self-supporting. One could produce spans of up to several meters, which are preeminently suitable for roofs over swimming halls, for example. Covered with a film that can hold for a few years, they can defy surface stress due to wind or snow as well. Details can be found in the Figs. 5.6a, 5.6b, 5.6c, 5.6d, 5.6e; the legend provides information. Interestingly, one practically cannot distinguish between the image of a diatom shell and its replication in casted forms when they are photographically compared at similar sizes.

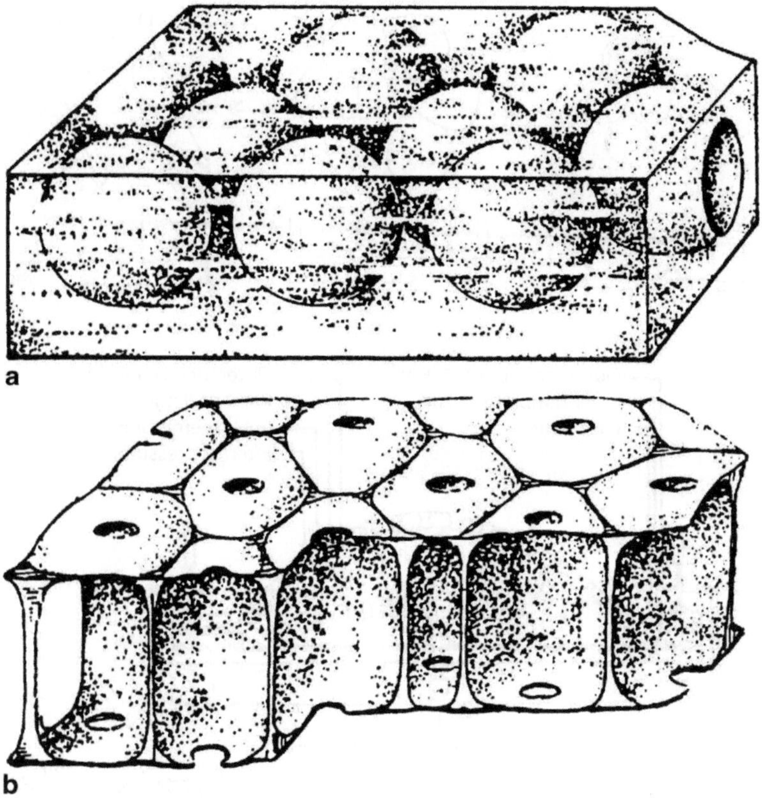

Fig. 5.4 Diatom concrete cast forms, **a** fat droplet hypothesis of G. Helmcke, ca. 1956 and **b** result from **a**: diatom shell cavity. (adapted from Nachtigall 1987)

Lightweight Structures: Bell Towers An economical system was sought for the reinforcement of large surfaces, in which individual elements can be brought together that can sustain heavy wind forces (e.g., for the reinforcement of screens for open air theaters). Hexagons built of glass fiber-reinforced polyester, which correspond to the diatom-cast forms, have proved themselves useful for this task (Fig. 5.7c). As an analogy, F. Otto and M. Mahnleitner, who in Berlin in 1960 worked in continuous exchange with botanist G. Helmcke under the keywords "biology and building," welded tetragonal cube structures consisting of steel panels together (Fig. 5.7d), whose basic form has a similar biomimetic background. The cubes became components of a bell tower for a church in Berlin (Fig. 5.7e), which have several advantages: relatively lightweight construction, easy to erect as one piece, enough resistance against torsion to withstand the vibrations from the bell and wind forces, and ultimately simple to breakdown and completely renewable: the disassembled structure can be melted down. The Mühlau-Mahnleitner minimal support structure of hexagonal honeycombs (Fig. 5.7c) proved itself in practice for a 500 m² movie theater screen for the Waldbühne in Berlin and the diatom-inspired bell tower of Otto-Mahnleitner for a church in Berlin.

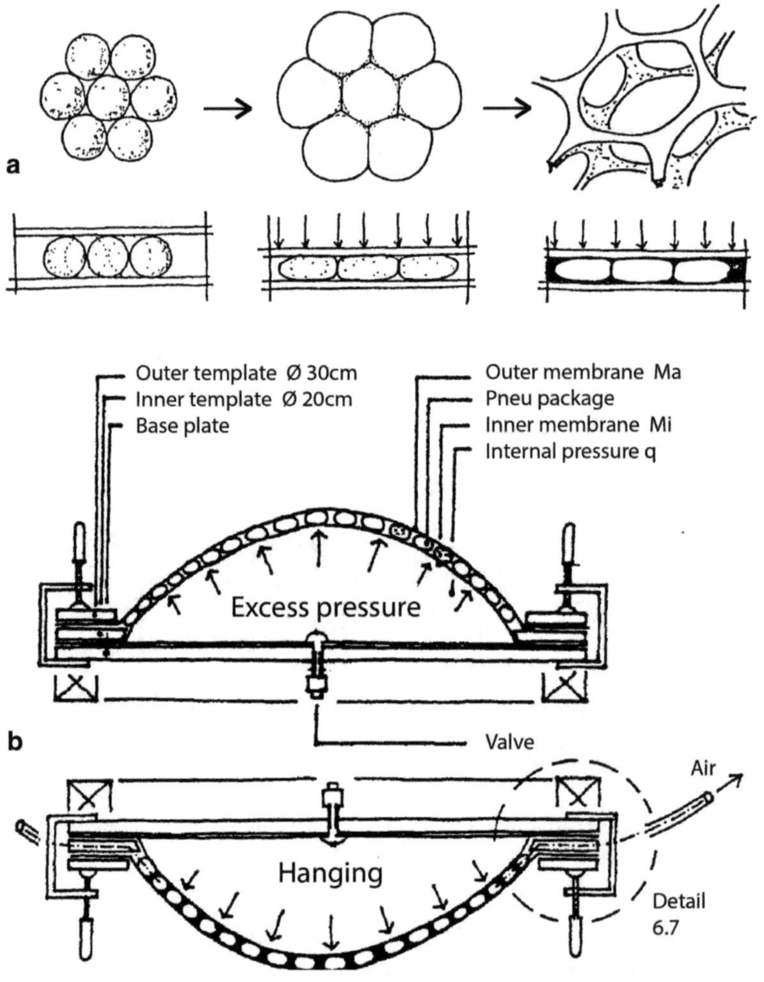

Fig. 5.5 Diatom-like, lightweight panels and shells: **a** principle of panel formation and **b** principle of shell formation. (adapted from Noser 1983)

Steel-Reinforced Concrete Shells Diatoms with the form of an equilateral triangle (approximately *Triceratium alternans*, Fig. 5.8a) provided inspiration to the engineer and architect E. Torroja for the construction of large, triangular, reinforced concrete shells (Fig. 5.8b), which were to cover water reservoirs. If one structures them so that the vaults from the edges outwards are formed as sloped surfaces, they are self-supporting. Triangle configurations of this kind can be combined to form hexagonal structures, as J. Joedecke has shown with his diatom-like, experimental reinforced concrete shell (Fig. 5.8c).

Geodesic Domes Since the 1950s B. Fuller has become known for his geodesic domes, for example the canopy of a greenhouse ("Climatron") in the Botanical

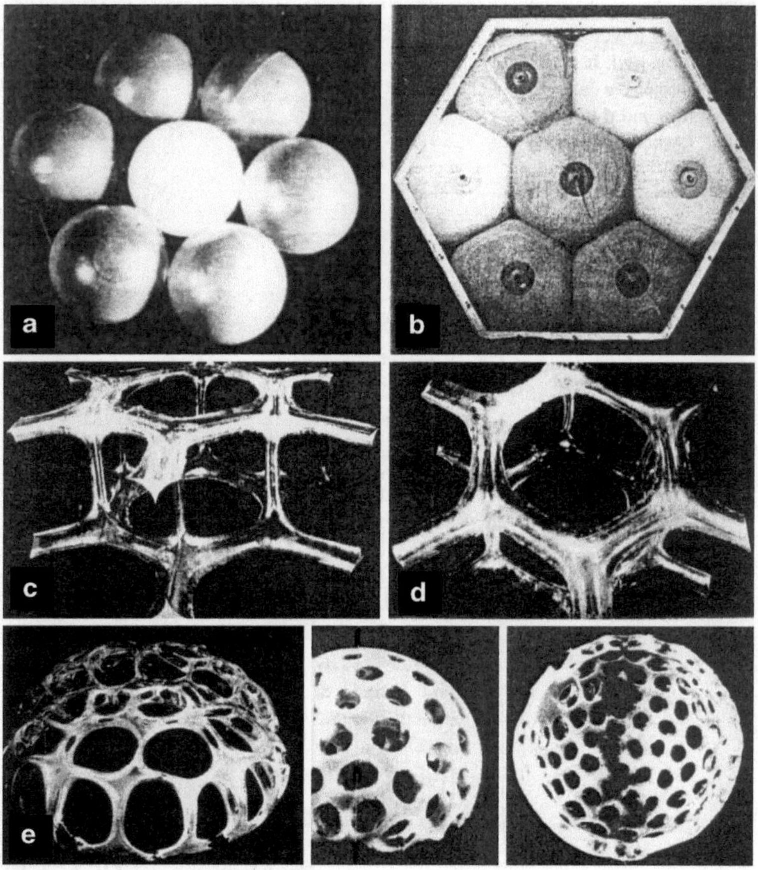

Fig. 5.6 Formation according to the diatom principle: **a** soccer balls arranged around a center, **b** soccer balls deforming to a hexagonal form, **c**, **d** "interstitial form" of diatoms and from **b**, and **e** hemisphere forms. (Adapted from Noser 1983)

Gardens in St. Louis from 1953 (Fig. 5.9). The structure consists of a double shell with hexagonal grids that mutually support each other and are covered with triangular plexiglass panels.

There are diatoms, which are structured principally similar and also represent a doublelayered, self-supporting shell framework, whose cavities are lined with finely punctured silicic acid membranes. B. Fuller denied that he received his inspiration for his structure from nature; the similarities have only been discovered a posteriori. It seems strange, as it is well known that Fuller had occupied himself with small and microscopic life forms, though it may also be apparent that general influence resulting from these studies is completely unavoidable. In honor of the architect, soccer ball-like molecular cages of carbon were named "Fullerenes".

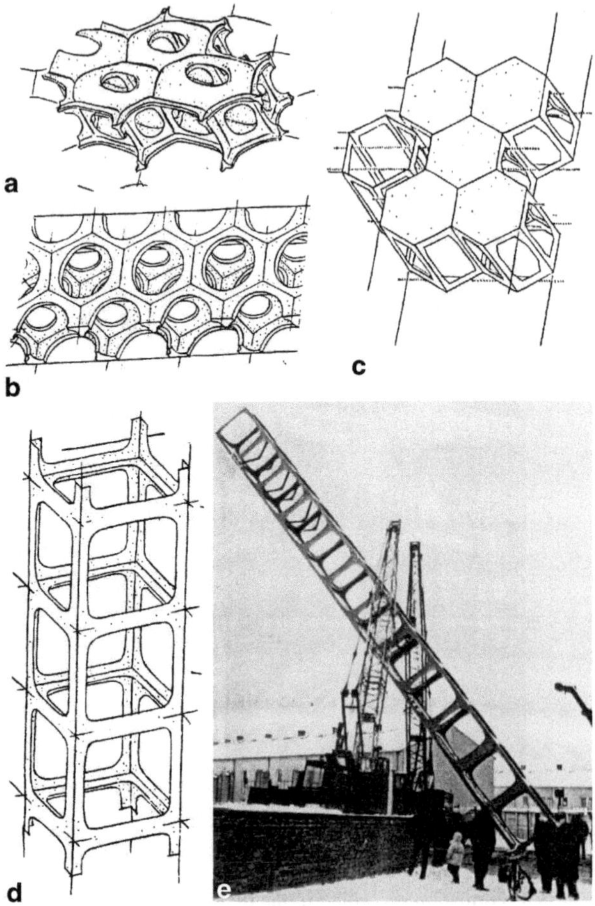

Fig. 5.7 Diatom-inspired cast forms and bell tower concept: **a** diatom shell, **b** building blocks (cast), **c** cardboard box analog, and **d** bell tower analog. (Adapted from Mahnleitner from Otto et al. 1982, partly redrawn)

5.1.2 Radiolaria → Radiolaria-Inspired Structures

While the structures of diatoms (algae of freshwater and of the oceans) are limited to round, cylindrical, or somewhat elongated, boat-like forms and build a variety of shapes within these basic frameworks, the similarly sized radiolaria vary in their forms much more drastically. There are spherical base forms (Fig. 5.10) among the radiolarians as well, including those with long needle-like offshoots and "spheres within a sphere" (Fig. 5.11b, 5.11c, 5.11d). Beyond that, however, there are still various, complexly symmetrical (Fig. 5.11a) or frame-like entities (Fig. 5.12a, 5.12b).

Radiolarians have also provided influence in the decorative arts and decoration of buildings, such as belt buckles of the Jugendstil movement and the famous, consistently fashioned entrances of the Paris subway system.

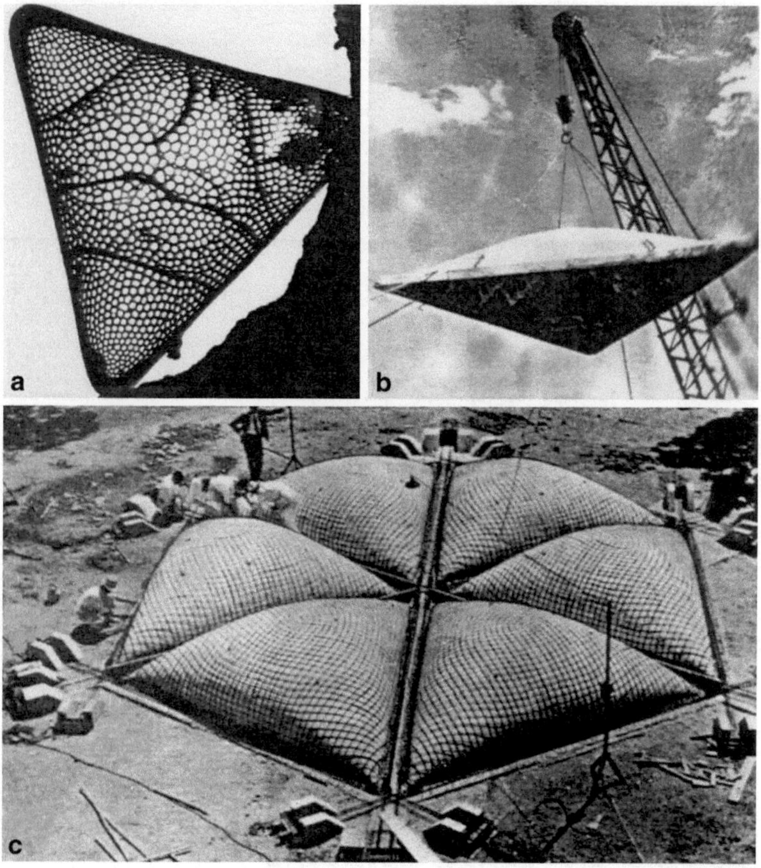

Fig. 5.8 Diatom Triceratium and diatom-like shells, a TEM image of the diatom *Triceratium alternans*, b steel-reinforced concrete shell for a water reservoir covering, E. Torroja, 1950, and c experimental, diatom-like, six-piece-reinforced concrete shell under construction, J. Joedecke. (Adapted from various authors from Nachtigall 1987, Coineau, Kresling 1987)

5.1.3 Radiolaria → Radiolaria-Analogous Spatial Structures

Radiolaria of genus *Callimitra* (Fig. 5.13a) are formed yet more differently than the previously mentioned genera. They have provided inspiration for tetrahedral spatial constuctions (Fig. 5.13b, 5.13c.) and are briefly described in Sect. 4.5.2.

In the 1940s the French architect Le Ricolais became famous for consistently imagining new structural systems following the principles of rod and node frameworks (Fig. 5.14a). These sometimes resemble B. Fuller's geodesic domes.

Le Ricolais formulated structures consisting of compression-resistant members and tensile connections that sustained symmetrical deformations during pressure

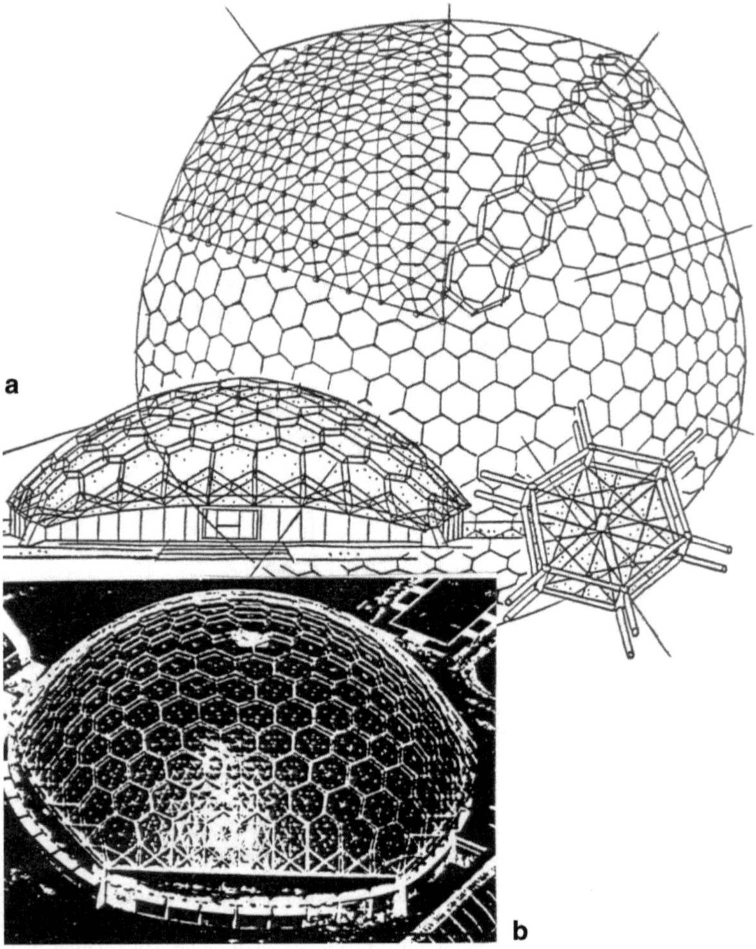

Fig. 5.9 Geodesic dome: "Climatron," St. Louis, B. Fuller 1953. (Adapted from Nachtigall 1987, Coineau, Kresling 1987)

tests without local structural failures (Fig. 5.14b, 5.14c, 5.14d). One can apply them as reinforcement elements in a smaller scale or possibly also as shock-absorbing material.

The length–width proportion of man-made tower structures in comparison to stalks of grasses must be reduced due to structural laws of similarity; such structures not only appear plump, they must have a broad base to absorb the torsional forces at the foundation. In technology it has often been attempted to dissolve these structures to stilt-like elements to reduce the overall mass, as with the television tower in Libereć by H. Hubaćek and in Ostankino by C. Nikitie and colleagues (Fig. 5.15a, 5.15c). In a similar manner the bases of the spines of radiolaria and diatoms are

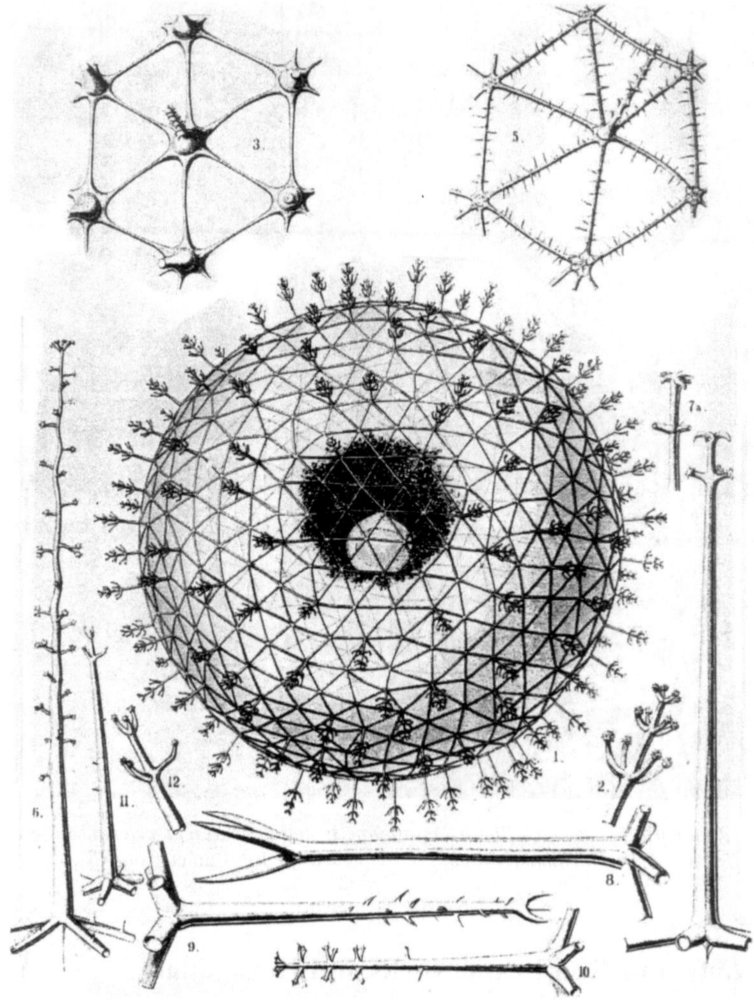

Fig. 5.10 Radiolarian *Aulosphaera* spec. and details. (Adapted from E. Haeckel's famous monograph, 1878)

also "dissolved" in a lightweight constructional principle, as illustrated by the long spines of the genus *Chaetoceras* (Fig. 5.15b).

Diatoms and radiolaria in particular had inspired building engineers during World War II and in the period afterwards (once the use of transmission electron microscopes had flourished) to start working with extreme biomorphic lightweight structures. The structures that actually resulted were unique and anomalous (Fig. 5.16a, 5.16b, 5.16c).

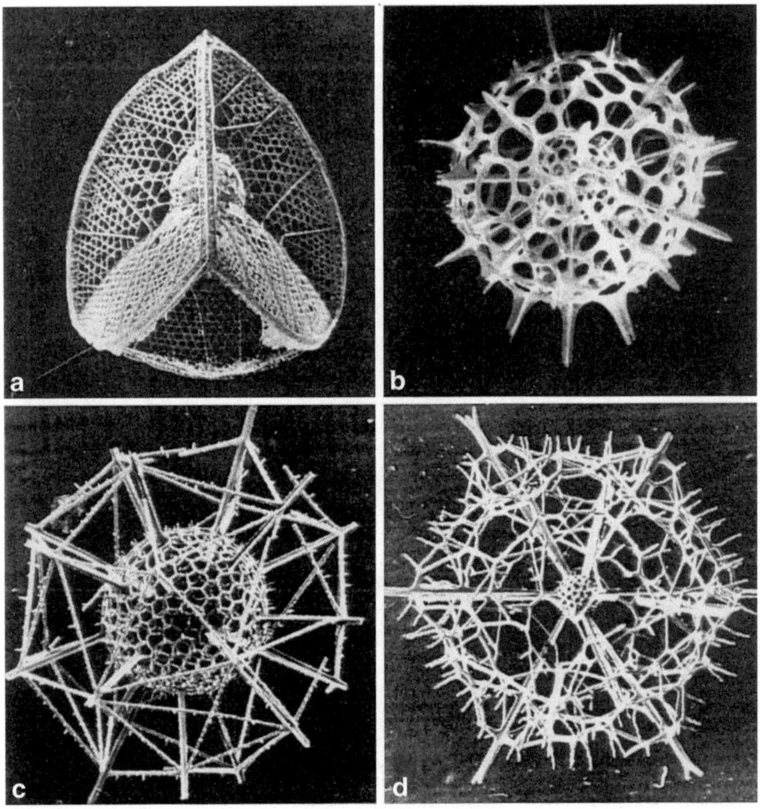

Fig. 5.11 Radiolaria forms: **a** *Callimitra carolotae,* **b** *Actinommia trinacricum,* **c** *Astrosphaera hexagonalis,* and **d** *Lychnosphaera regina.* (Adapted from J.P. Caulet from Coineau, Kresling 1987)

5.2 Node-and-Rod Frameworks and Hexagonal Structures

Stable node-and-rod structural networks are composed of triangular meshes that can be combined to form hexagonal structures. There are many examples of these kinds of structures in biology and technology.

5.2.1 Pith of the Juncus Plant → Unbendable System

The pith of the rushes of the genus *Juncus* consists of star-shaped cells that accrue onto each other (Fig. 5.17a). At the points where they touch the cells become somewhat wider and meld into each other with crossing walls (Fig. 5.17b). In the middle emerge hexagonal structures composed of equilateral triangles, whose edges consist

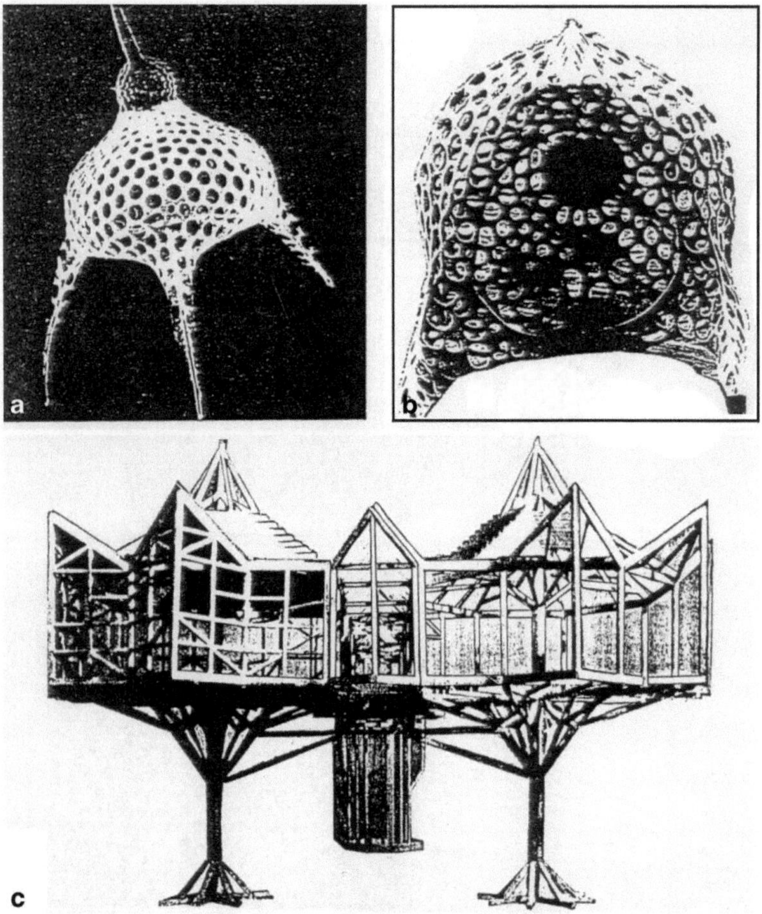

Fig. 5.12 Radiolarians and radiolaria-inspired structures: **a** radiolarian *Pterocamium* spec. and **b** radiolarian *Dictyoceras* spec. **a**, **b** from Helmcke, 1960 and later, **c** after a concept from D. Oligmüller

of offshoots of two different cells meeting halfway. This arrangement spans the entire elongated, cylindrical interior space of the rush "stalk" and generally contributes to the stabilization and rigidity of this self-supporting system. In contrast to bee honeycombs, the system must not be constructed strictly symmetrical. Along with the (commonly formed) hexagonal lattices others exist with fewer or sometimes more edges, that is, "cells" with seven spokes (Fig. 5.17b).

Bending tests were performed on fresh "stalks" of this variety, in natural state and with the pith removed (Nachtigall, unpubl.). Although the pith hardly accounts for any of the weight, it is responsible for almost 50%of the resistance to bending and sturdiness: an "intelligent" material, which obtains major effects with small, yet "shrewdly distributed" masses.

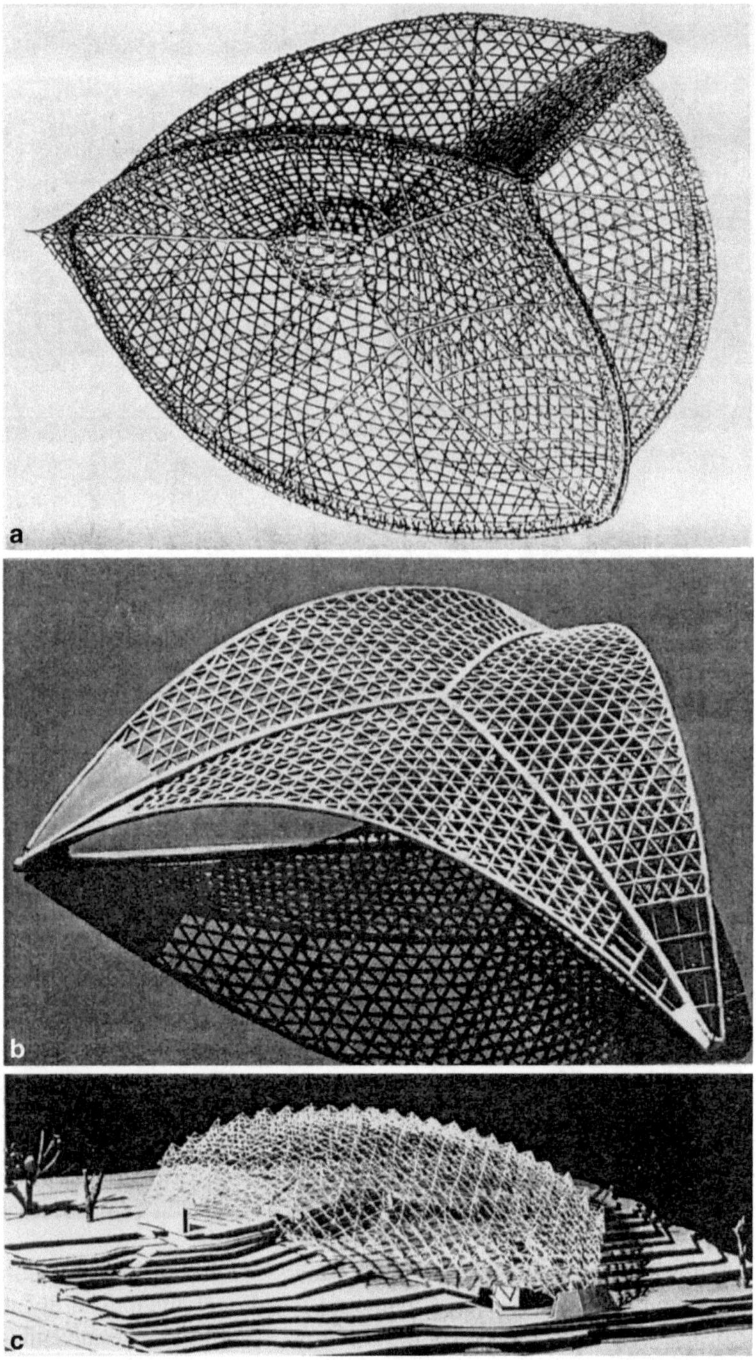

Fig. 5.13 Radiolaria and structural systems: **a** radiolaria *Callimitra* spec., **b** tetrahedral node-and-rod system on cylinder barrel surface, P.C. Nervi, Paris, **c** model of a long spanning frame system, G. Wujina, St. Petersburg. **a** from E. Haeckel 1878, **b**, **c** from Lebedev 1983)

Fig. 5.14 Radiolaria-analogous spatial structures of Le Ricolais, ca. 1940: **a** Le Ricolais with spatial framework, **b** pipe structure, **c** pressure test of **b**, and **d** pressure deformation of **b**. (Adapted from Coineau, Kresling 1989)

5.2.2 Panel Bracing → Experimental Structures

Hexagonal frameworks can obtain stability against spatial buckling like the pith of the *Juncus*. This stability can be further increased with thin panels that brace the structure between the struts. G. Pavlov and others have introduced experimental works for the study of this structural system (Fig. 5.17c, 5.17d).

5.2.3 Bee Honeycombs → Hexagonal Systems

Much has been reported about the bee honeycomb as an extremely material-efficient lightweight structure with rhombic dodecahedral-dovetailed connections (Fig. 5.18a,

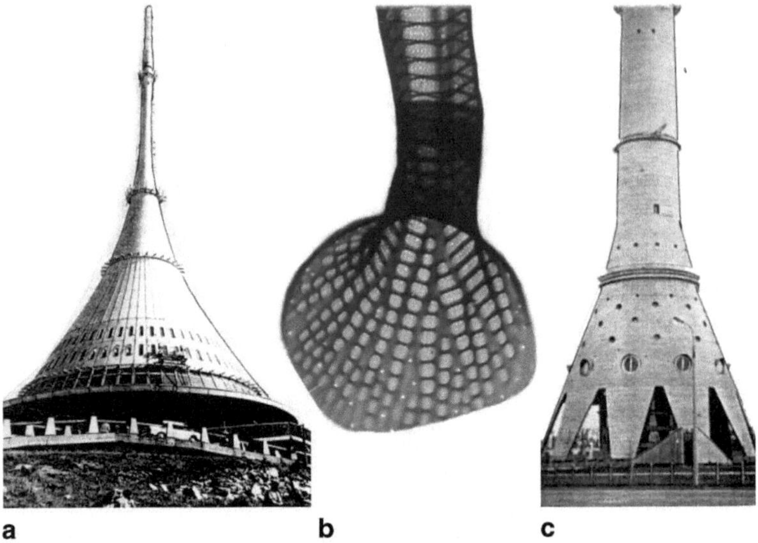

Fig. 5.15 Broadened "tower" bases using lightweight principles: **a** television tower in Libereć, H. Hubaćek, **b** base of a spine of the diatom *Chaetoceras* spec., and **c** television tower Ostankino, Nikitie et al. (**a**, **b** from Lebedev 1983, **c** from Helmcke from Nactigall 1974)

5.18b). Interesting for historical reasons is the concept of the Trelement house, which had become famous in the early 1970s. The base form of these houses, which are based on the hexagonal grid arrangement of bee honeycombs, allows any shape, provided that it can be composed of hexagons or from the triangle-shaped enclosements (Fig. 5.18c, 5.18d). The concept was at the time very progressive, and there was once talk of recapturing this idea once again in an improved form. For its construction it requires only a few lightweight concrete anchors in the earth, which can be removed relatively easy if the structure were to be broken down. A forest of aluminum columns with star-like radiations is erected, between which the connecting struts are bolted together. Arbitrarily large areas can be spanned with different compositions; the building forms had ultimately proven themselves suitable for kindergartens and houses, which need not be built to last forever, but intended rather to be inhabited by only a few generations. The structure was easy to dismantle, and because the support structure only consisted of aluminum, it was also completely recyclable.

In this sense the concept corresponded to biomimetic points of view. Disadvantageous, for example, was the thermal bridge generated through the beams bordering directly on the exterior, which led to formation of condensation. A modernized concept that would avoid this drawback and better consider the recyclability of the connecting elements and their assembly and disassembly could provide a good basis for modern structures, which can last possibly two to three decades and eventually pay for themselves.

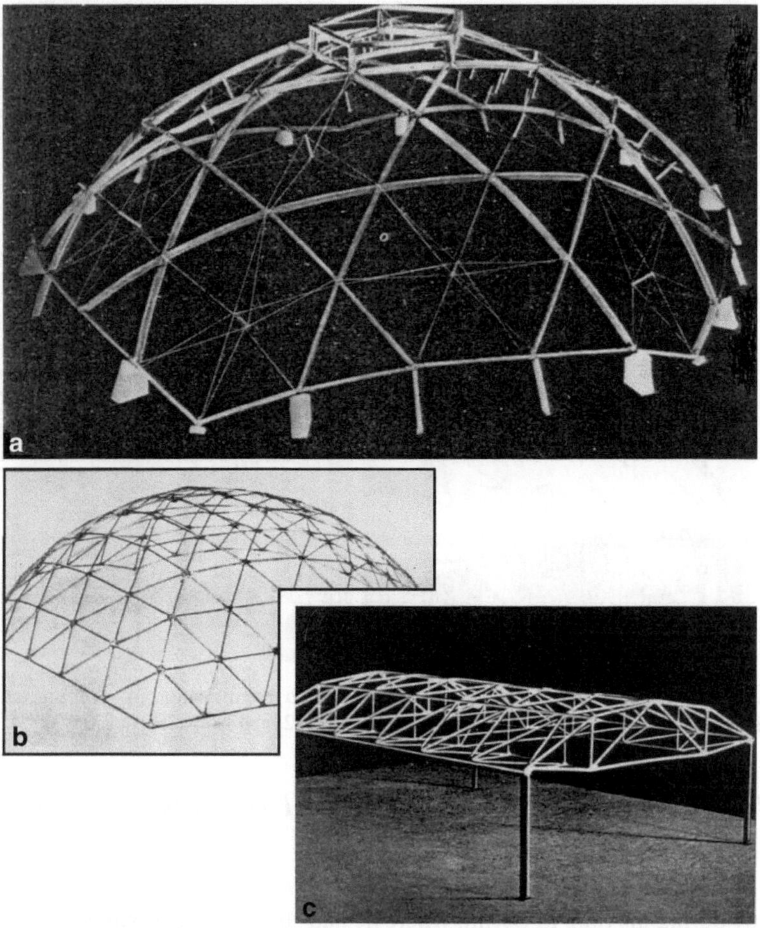

Fig. 5.16 Design for a dome framework, partially inspired by radiolarians: **a** geodesic dome with compression-resistant members and tension cables. Tupolev, ca. 1940, **b**, **c** icosahedral node-and-rod dome structures, Le Ricolais ca. 1942. **a** from Lebedev 1983, **b** from Patzelt 1972)

5.3 Rigid Nodes and Tubes

In technology, nodal points in a system are often built massive and their connecting pipes decidedly thick walled due to safety reasons. Nature also completes such systems, but because of the need for material efficiency and low mass (which also means lower energy for their formation) they mostly appear in the form of lightweight structures.

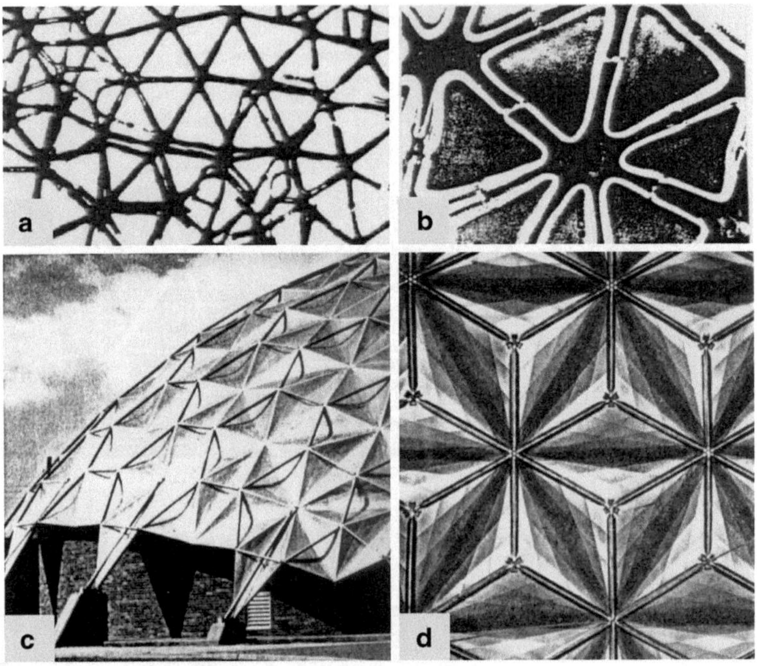

Fig. 5.17 Hexagonal structures: **a** pith of *Juncus* spec., **b** magnification of **a**, reprint with phase contrast, **c**, **d** Kiosk, G. Pavlov. (**a**, **b** from Nachtigall 1972, **c**, **d** from Lebedev 1983)

5.3.1 Nodes with the Lowest Material Expenditure → Analogous Nodal Structures in Technology

Already during his time in Berlin, where he had strong contact with biologists from the Technical University, Frei Otto had concerned himself with particularly lightweight structures of all kinds. One aspect of this research pertains to the construction of a very light yet torsion-resistant node system as a basis for larger, long-spanning, lightweight structures. The comparison of such a nodal system in a bend-resistant support system with the lowest material cost from 1960 (Fig. 5.19a) to preserved skeletal elements of a fossilized sea sponge (Fig. 5.19b, 5.19c) illustrates the conceptual correlation.

In the structure of siliceous sponges the connectors between nodal points are more dissolved and form a network of branching struts. Specific locations of minimal structural stress are handled as if there were none and therefore left vacant of material. The entire structure is oriented on trajectories of forces. They can grow with the accumulation of corresponding "standardized" elements when under intense compression but also can be strengethened with incorporation of another "intermediate floor."

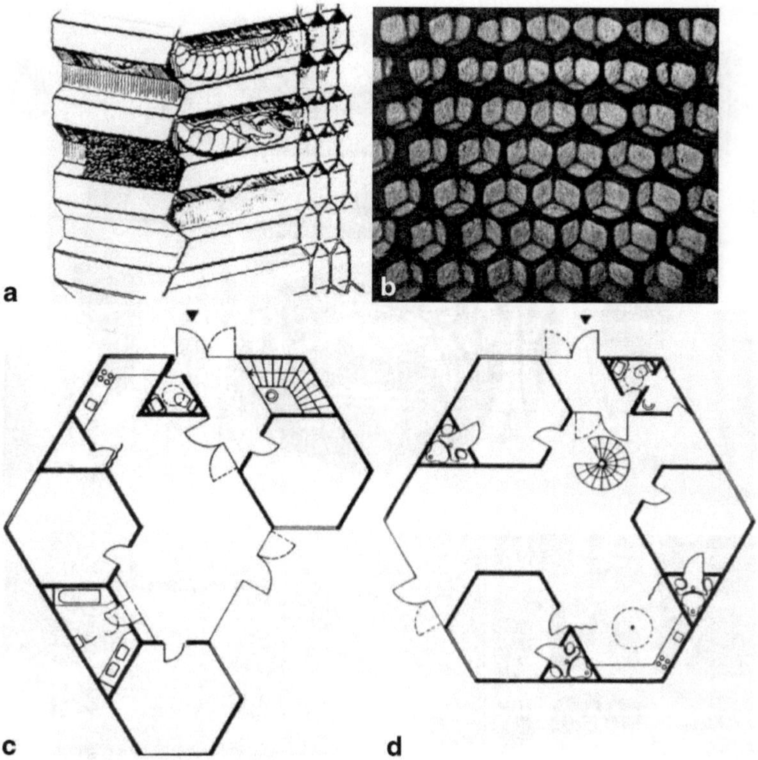

Fig. 5.18 Bee honeycomb and hexagonal spatial arrangement: **a**, **b** bee honeycomb and **c**, **d** two possibilities for spatial arrangement in Trelement houses. (**a**, **b** from Nachtigall 1974, **c**, **d** from Trelement broschure ca. 1972)

5.3.2 *Tetrahedral Node Networks → Long-Spanning Structural Systems*

An interesting geometric form is exhibited by radiolorians of the genus *Callimitra* (Fig. 5.13a) and other similar genera that have been variously investigated and experimentally re-created as spatial structures. P.C. Nervi, for example, developed a tetrahedral node network on a cylindrical surface (Fig. 5.13b) and the Russian architect G. Wujina a long-span spatial framework (Fig. 5.13c), whose concept had been influenced by structural principles of radiolarians.

5.3.3 *Plant Rigidity → Tubes of High Rigidity*

In this subject area, the historical incorporation of "technical biolgy" can be well demonstrated. The biologists learn to better classify and understand natural

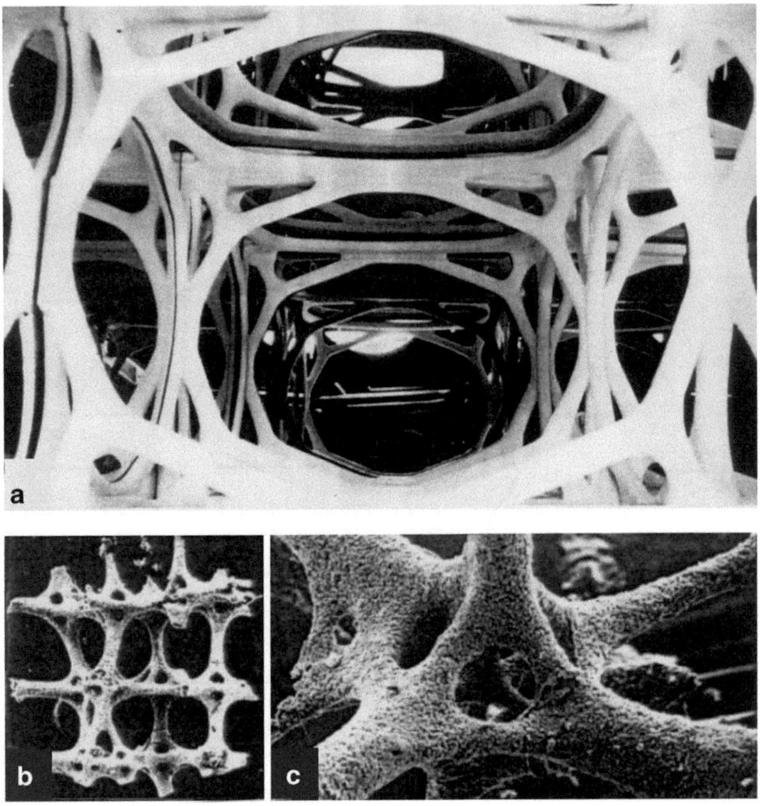

Fig. 5.19 Spatial node–and-rod framework: **a** nodes of a rigid panel system with the lowest material expenditure and **b**, **c** skeleton parts of a fossilized siliceous sponge. (Adapted from Otto et al. 1992)

structures by using comparisons to technology. Technical–physical insights have already been used from early on as principles for explaining the way biological entities function.

Sections through plant stems or stalks, such as those of the sedge plant *Cladium mariscus* (Fig. 5.20a), are normally denoted by their ring-shaped, coalesced sclerenchyma structures. These often form circumferential ridges. Such structures often exhibit sectional forms that resemble architectural I-beams. They have therefore also been described as "biological I-beams".

It has been known of these types of architectural support beams since their introduction in concrete and rail construction of the nineteenth century that they exhibit a particularly high area moment of inertia and is therefore relatively bend- and torsion-resistant, above all when they are merged into a radial complex (Fig. 5.20b). S. Schwendener, botanist and biomechanic of the late nineteenth century, was inspired "by the observation of iron bridges and train station sheds and their numerous I-beams" to interpret the rigid stalks of plants as systems of these beams. In 1888

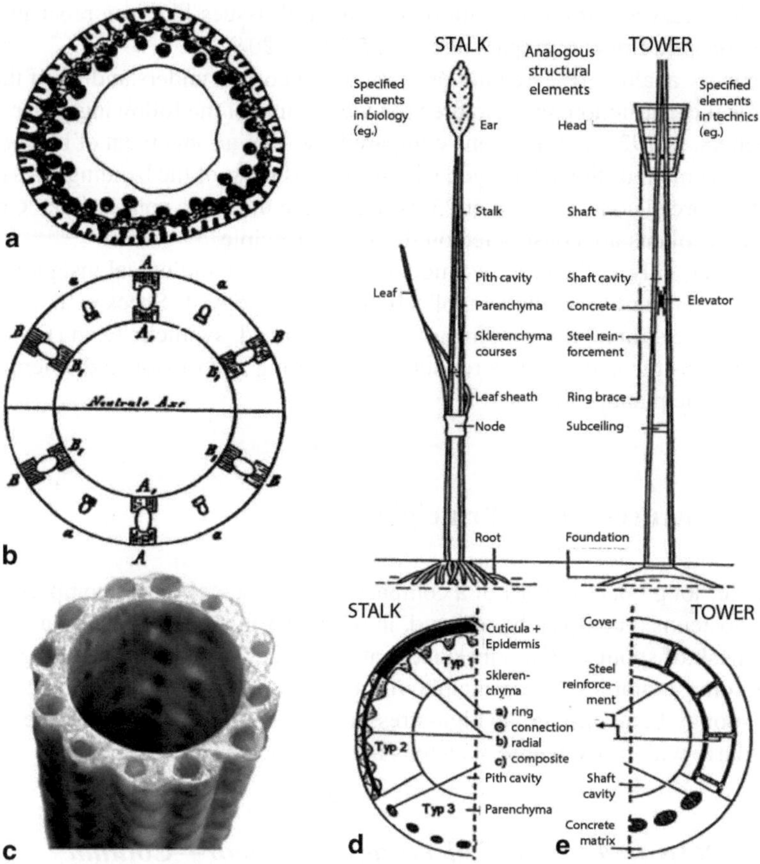

Fig. 5.20 Structure of grass stalks and their technical interpretation: **a** section through *Cladium mariscus*, **b** Schwendener's imagining of the stalk as an "I-beam system," **c** technical interpretation of a stalk principle by Speck et al., and **d**, **e** plant structure and building structure analogies. (**a**,**b** from Schwendener 1888, **c** from Speck et al 2004, **d**, **e** from Nachtigall 2002)

he wrote in a treatise, "The plant doubtlessly structures itself according to the same rules as the engineer, only that its technology is much finer and more perfected," which is formally an observation of the biology–technology analogy as well.

The importance of researching analogies in this subject area was initially emphasized by W. Rasdorsky with the example of reinforced concrete. "By observing lectures in 1906 and 1907 about reinforced concrete construction" he arrived at the concept that the plant can be interpreted as a composite construction, "in which the sclerenchyma strands correspond to the steel reinforcement and the parenchyma tissue to the concrete matrix" (he probably meant the cement matrix): the correct way towards a functional understanding of these plant structures.

With these words the important role of analogy research had already (1911) been revealed. "*Between technical composite structures and plant organs accordingly exists an extensive analogy in their principle structures.*" K. Giesenhagen (1912) noted that "leaves form grids with their structural tissues like iron rebar in a reinforced concrete floor" (compare also Fig. 5.20d, 5.20e).

The early analogy research not only led to the correct understanding of the morphological construction, but also bred the viewpoints of the following generation of researchers. In 1922, F. Bachmann compared the fiber arrangement of bamboo with "a reinforcement of the outer layer, which sustains most of the bending stress (similar to reinforced concrete)." In 1923, Bower made the same connection: "Ordinary herbaceous plants are constructed on the same principle."

However, only in the present time has this technical–biological insight played a role in the sense of biomimetic application in technology. T. Speck et al. (2004) introduced tube structures consisting of fiber-reinforced, synthetic resin (Fig. 5.20c), which obtains the considerable resistance to bending and torsion and other positive aspects of the plant prototype.

5.4 Structures on the Principles of Bone

The larger long bones consist of a compact bone substance that is composed of coalesced osteons ("compact bone") and, in particular in the joint regions at the ends, a fine mesh of spongy bone substance that permeates the cavity space. Figure 5.21 shows some examples. The principle of tubular bones (thin walls, structural spans in the spongy bone only in the high-stressed joint regions) is driven to the extreme particularly with birds, whose skeletons must be especially light.

5.4.1 "Ossified Force Trajectories" → Floor—Column Structures

Spongy bone is, provided it consists of larger mesh network, generally structured so that its spans align with the major stresses acting on the bone.

Correspondingly there are courses of spongy bone, some of which correspond to the stress trajectories of compression and others which correspond to force trajectories of tension. If the network of joists is constructed correctly, the trajectories will always meet at a right angle, and that not only functions in one dimension. The joists arrange themselves in space to "surfaces of equal tension," which likewise run through the bone and always intersect perpendicular to each other (Fig. 5.22a; example for the proximal femur area of a human). The orthopedist F. Pauwels and later his student B. Kummer have hinted at this particular characteristic; the latter had been concerned not only with medicinal–orthopedical issues, but also with the structural principles of the skeleton of mammals.

The knowledge that spongy bone is constructed on stress trajectories goes back to as early as the nineteenth century. It originates from researchers such as physician P. Wolff and engineer K. Cullmann, who recognized in 1870 the principle of

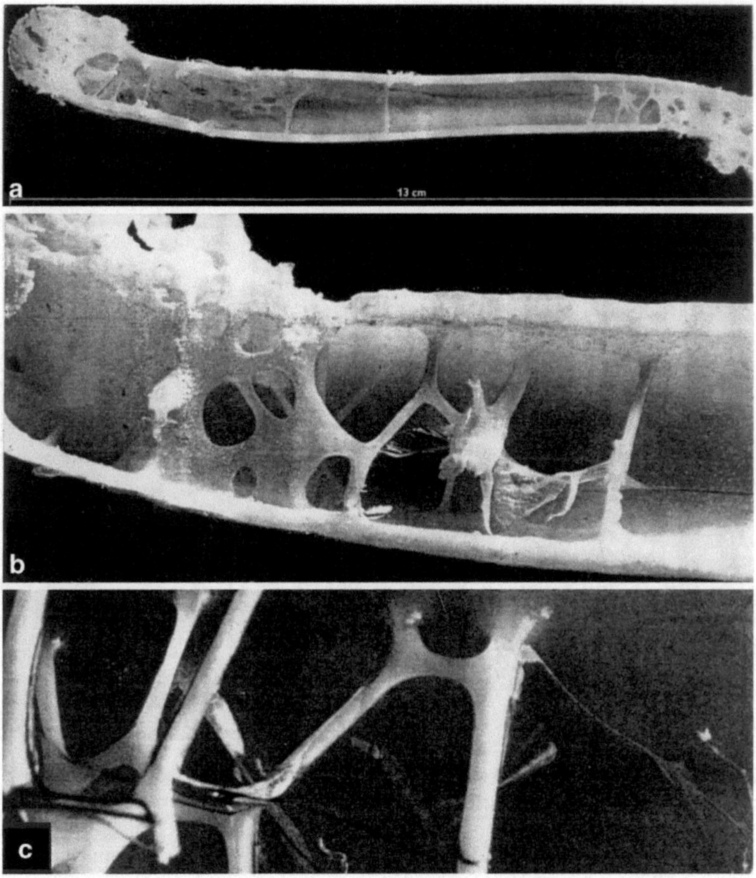

Fig. 5.21 **a–c** Brace network in various long bones. (Adapted from IL report 35)

"ossified stress trajectories" and afterwards conceptualized a highly resilient crane head (Fig. 5.22b) according to this principle.

5.4.2 Isostatic Ribs

One can design the beams for concrete flooring or roof systems particularly light, if one aligns them with stress trajectories. P.C. Nervi demonstrated this principle in 1951 with his Gatti Factory in Rome (Fig. 5.23). Nervi was one of the few famous architects who were not afraid to hint at the inspired forms of natural precedents, as the citation in Fig. 5.23 shows. He actually did receive his idea for his stress trajectory—following prefabricated concrete elements from the studies of bones.

On the cover page of the festschrift published by the architects H.D. Hecker, L. Degerloh, and B. Krupp in 1967 the classic illustration of the force trajectory-

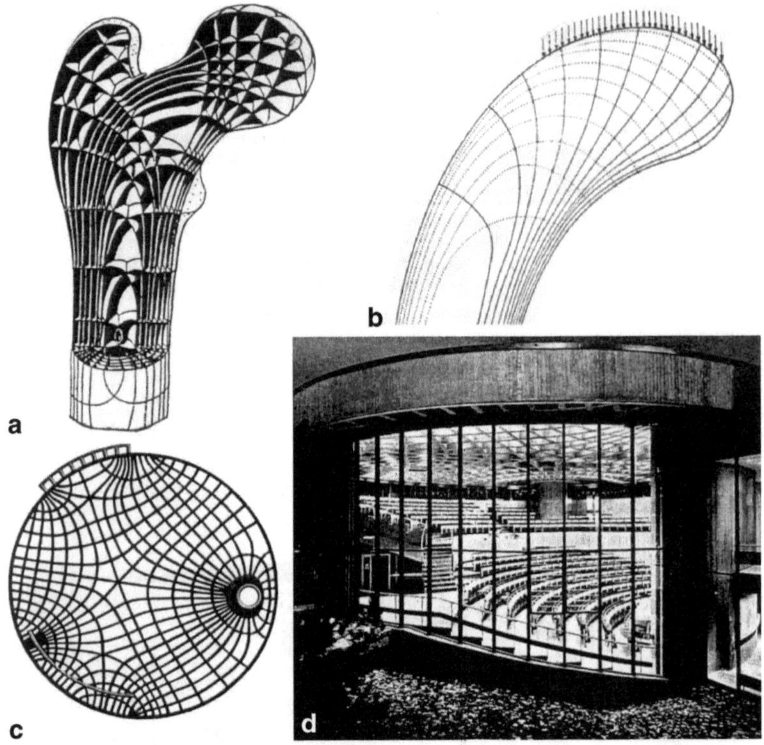

Fig. 5.22 (90) Bone and bone-like structures: **a** spatially curved surfaces of equal forces in the femur of humans, **b** classical representation by Cullmann of the crane cantilever principle based on **a**, **c** scheme for the joist and support system of the ceiling of the old biology lecture hall at the University of Freiburg, **d** lecture hall from **c** during the building process. **a** from Kummer 1985, **b** from Cullmann 1870, **c**, **d** from Hecker at al 1967)

Fig. 5.23 Isostatic ribs, Gatti Factory, P.C. Nervi, Rome 1951. (Adapted from Lebedev 1983)

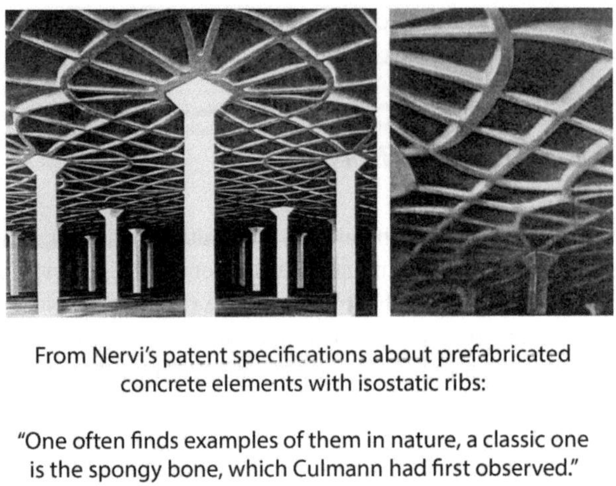

aligned spongy bone structure from two centuries ago. It probably made sense to display something from this structural principle from nature for a future biology lecture hall. The ceiling of the circular lecture hall was supported on one column and a few circumferential elements systematically provided with force trajectory-aligned beams, which makes it very light and also—obviously—interesting, not to mention "biological." In Fig. 5.23c these compression and tension absorbing beams are illustrated; Fig. 5.23d shows the lecture hall as built in 1967.

5.4.3 Bone Braces

The architect and engineer S. Calatrava has become famous for his "biomorphic"-appearing structures, particularly visible on numerous structures in his home city of Valencia. Figure 5.24 shows his concept for the new façade for the train station in Lucerne. The design of the form does not imitate any particular biological principle, although one can see the resemblance to rib arrangements, for example the ribcage of a whale or bird. It breathes in a nature-like rhythm and appears more elegant and interesting than a conventional wall design, is of course notably more expensive as well.

If one understands architecture as an esthetic environmental factor that is capable of dramatically influencing the mood and mental well-being of people, who unavoidably have something to do with buildings from day to day, then one perceives this kind of structural reduction and lightness, the swoops and the forms differently than if one "only" considers them in functional–analytical manner. As accordingly: Calatrava is not judged wholy uncritically by certain architects and architectural historians, depending on the canon to which they prescribe.

Fig. 5.24 Train station Lucerne, Model, S. Calatrava. (Adapted from Blaser (Ed) 1989)

5.5 Shell Structures

Shells are characteristic structure forms in nature. The most well-known are snail or mussel shells. The shells of mussels (examples in Fig. 5.25) in particular have inspired building forms, and not only as decorative elements. Even in antiquity one had attempted to realize their wide-stretched, often thin-walled forms, but analogous forms were only enabled with the building materials of modern era, pre-stressed concrete above all.

Architects such as Le Ricolais attempted early on to observe, understand, and abstract such shells as building–structural entities.

5.5.1 Mussel Shells → "Isoflex"

In adoption of the rib structure and other structural idiosyncrasies of the shells of the large scallop, *Pecten jacobaeus*, Le Ricolais conceptualized a structural system

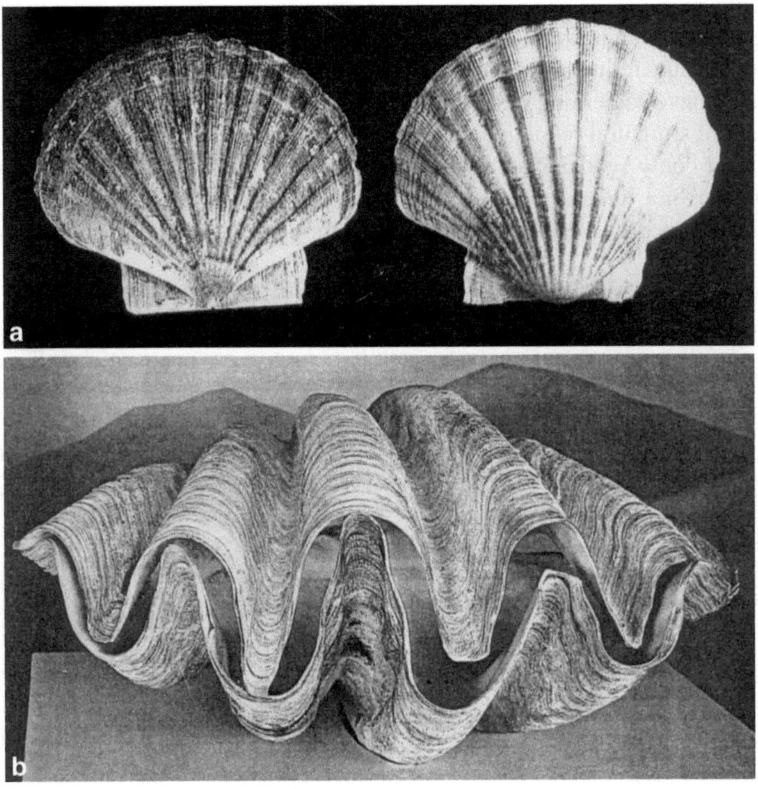

Fig. 5.25 Mussel shells: **a** the pilgrim's scallop *Pecten jacobaeus*, width 10 cm and **b** giant clam, *Tridacna spec*, width 120 cm. (Adapted from Coineau, Kresling 1987)

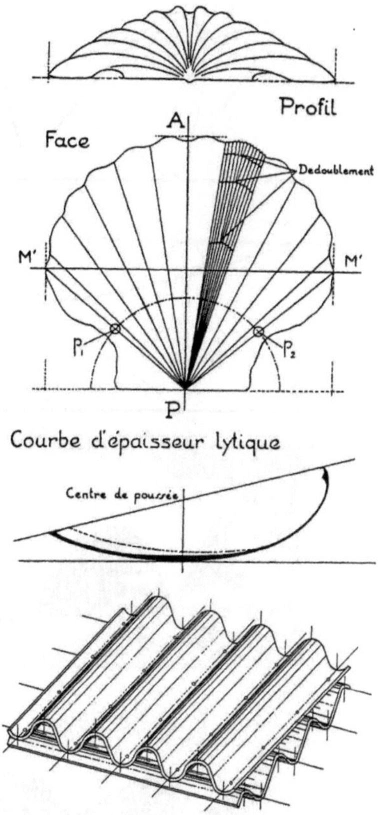

Fig. 5.26 Pilgrim's scallop and "isoflex" abstraction. (Adapted from Bull. Ing. Civ. de France, bottom drawing; Kresling from Nachtigall 1987, presentation par A. Bougrain-Dubourg)

consisting of perpendicularly crossing corrugated panels, which he described as "isoflex" (Fig. 5.26).

One can bend a layer of this structure into a tube form as well and anchor it in a given round tube, thereby producing lightweight structural tubes with significant strength and rigidity.

5.5.2 Shells Similar to Tridacna → Shell Structures

Figure 5.27 shows some examples as to how the analyses of the shells of the giant clam have influced the conception of shell-like, long-spanning structures. A restaurant in Xochimilco, Mexico (Fig. 5.27a), represents a hyperbolic parabaloid with a shell thickness measuring only 15 mm. It is a geometric structure that is self-supporting. The interpretation of this form is more clearly visible with the market hall in Royan, France, built by Simon et al. in 1955. The roof with its radial waveforms has a span distance of 52.4 m. Even broader is the wavy roof of the national circus in Bucharest, Romania, by Porumbescu et al. in 1960 (Fig. 5.27c). It spans a distance of 66.6 m.

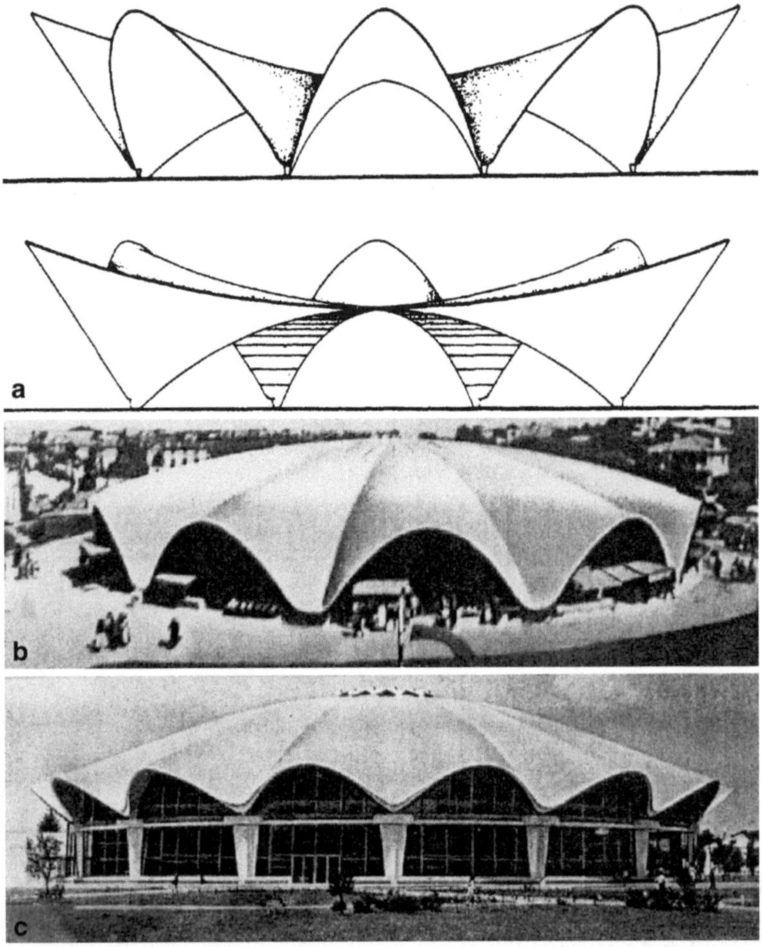

Figure 5.27 *Tridacna*-like shell structures, **a** restaurant, Xochimilco, Mexico, hyperbolic paraboloid, shell thickness merely 1.5 cm (!), **b** market hall Royan, France. Simon et al. 1955. Radially corrugated roof, span distance 52.4 m, **c** building for the national circus, Bucharest, Romania. Porumbescu et al. 1960. Radially corrugated roof, span distance 66.6 m. **a** from Blaser (Ed.) 1985 and **b**, **c** from Lebedev 1983)

With all of these shell structures a direct translation was not the ultimate goal, though it is known that the architects were inspired by the elegance of natural shell forms and played with these difficult to realize building forms.

Heinz Isler, as a modern representative, attempted long-spanning shell structures as well. He is major fan of gardens and had spent a lot of time in nature to become imprinted with the natural, botanical, zoological forms. His structures, which appear in Switzerland, for example as a roof cover for highway gas station, are exceptionally thin in comparison to their spanning length.

Shells consisting of concrete must be formed so that the supporting network of prestressed concrete does not deviate by more than a few millimeters from the plane of the self-supporting form. In the case of Heinz Isler, the form is even found by hanging chains; the coordinates of the perpendicularly cutting catenary curves were marked. Such a shell is—performed in the thought experiment by "welding" the hanging chains together at their meeting points and "flipping it over"—self-supporting. The structure can be entirely formed in this manner, so that it only develops compression forces at the supports. Of course, the function ultimately influences the form and therefore the design of the layout.

Figure 5.28 shows, extracted from the illustration by Patzelt 1974 and lightly supplemented, sections of shell structures with specifications of their diameters, shell thicknesses, and the ratio between the two measurements. Accordingly, a market hall in Algeria from 1955 with a shell thickness corresponds to about 1/1000 of its diameter and the 60 m spanning sport arena in Rome from 1956 (Nervi) with a relative shell thickness of 1/2400. The impressive building achievement from antiquity that is the Pantheon, built in the second century, must also appear here; its relative shell thickness is comparatively high with 1/44.

The proverbial chicken egg comes up short in comparison as well; its thickness ratio amounts to 1/112. The porous lime skeleton of the eggshell material is of course not an ideal building material. However, the form is so "refined" that it can withstand high compression forces. One cannot break an undamaged chicken egg

Fig. 5.28 Examples of early shell structures and their dimensions in comparison to a chicken egg: **a** market hall, Algeria, 1955, **b** palazzetto dello sport, Rome, 1956, **c** arena, Canada, 1958, **d** factory building, Jena 1923, and **e** shell of the chicken egg, F Pantheon, Rome, second century. (Adapted from Patzelt 1974)

Form	Diameter	Shell thickness	Shell thickness / Diameter
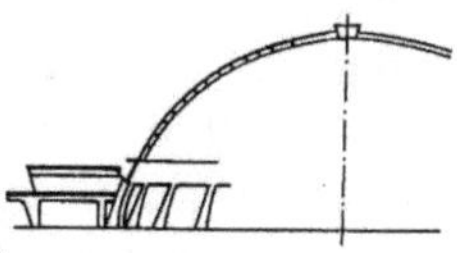 a	42 m	4 cm	1/1050
 b	60 m	2,5 cm	1/2400
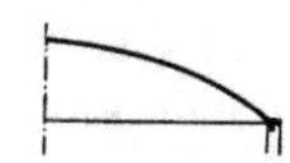 c	83,6 m	12,7 cm	1/685
 d	40 m	6 cm	1/665
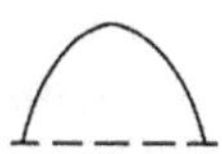 e	**0,045 m**	0,04 cm	1/112
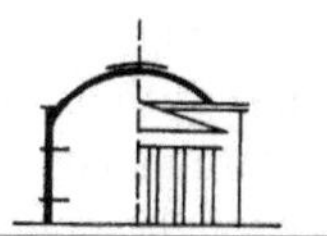 f	43,6 m	100 cm	1/44

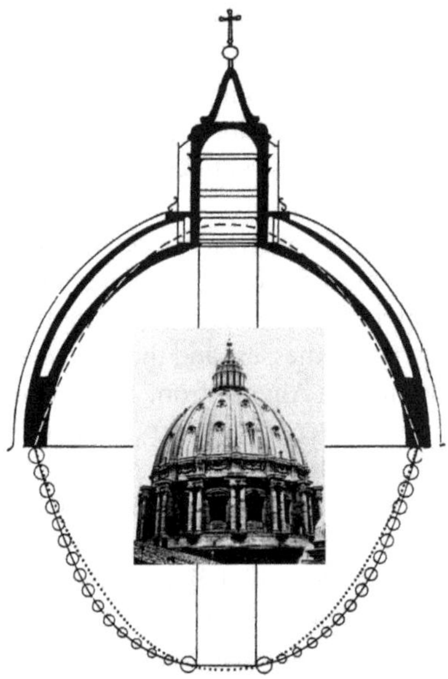

Fig. 5.29 Double shell of St. Peter's, Rome, Michelangelo et al., from 1561, in comparison with parabolic and catenary curves. (Adapted from Patzelt 1974, edited)

between the thumb and index finger. The "egg-shaped" envelope of the first nuclear reactor in Germany, the famous "atomic egg of Garching," was according to the architect not developed with the egg in mind. The form was the result of considerations of functionality, namely, how can the interior space with a given footprint be spanned with the lowest possible material expenditure. Due to this basic consideration an approximately "egg-shaped" shell form developed itself, which had been immediately classified colloquially as the "atomic egg." One can establish an analogy a posteriori, a similarity of form, as it often results in comparison of biological and technological structures.

The dome of St. Peter's in Rome originates from Michelangelo among others, built in 1561 (Fig. 5.29). "The eggshell is one of the few living structures that inspired major builders, as we know from historic dome structures" (F. Otto). The dome is formed as a double-layered shell, with the exterior layer and the interior ceiling formed differently. As the indication lines show, the dome is actually neither egg-shaped nor parabolic. Ultimately it was not formed as a catenary (inverted chain line). Despite extreme similarity, the egg is in this case not the exact inspiration and apparently not the catenary model as well.

5.5.3 Sea Urchin Shells → Inspiration for Structure

Sea urchins form very peculiar housings that must be structurally stable not only in their finished form but also during their "construction." Sea urchin shells and their qualities have already been illuminated in Sect. 2.1.5 under "Panel Structures." Their shape provided inspiration for the concept for the ice sport arena in Erfurt

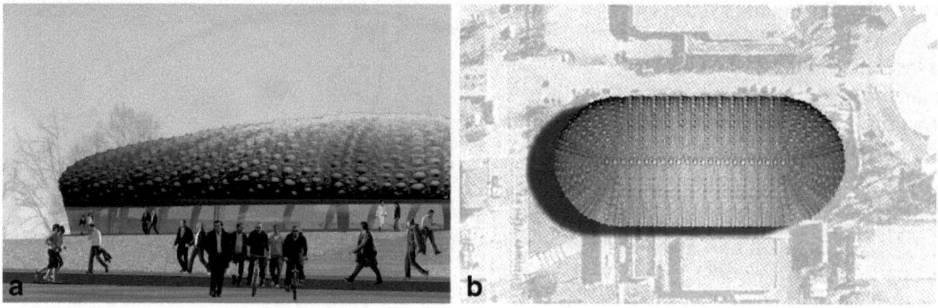

Fig. 5.30 Sea urchin shell and ice sport arena: **a**, **b** ice sport arena, Erfurt, Pohl Architects. (**a**, **b** Fig. J. Wilhelm, Pohl Architects)

(Pohl Architects; Fig. 5.30). Along with their composition as panels, sea urchin shells are spatially supported by rib forms. These kinds of spatial reinforcement structures provided the inspiration for the shell-like, reduced structural form of ribs that spans an area of over 80×200 m for the arena.

In Sect. 6.39 another project is described that had used the sea urchin shell as a precedent. The institutes ICD and ITKE located at the University of Stuttgart developed a pavilion according to the precedent of the "sand dollar" and thereby demonstrating the material-saving construction method in a field experiment.

5.6 Pneumatics: Buildings

The members of the collaborative research center 231 of the DFG, under the leadership of Frei Otto, have addressed natural structures as "pneus" provided they fulfill the definition of the three pneu elements:

1. an elastic membrane, which separates
2. an internal medium from
3. an external medium, so that the inner and outer pressures are different.

Most often the internal pressure is greater than the external pressure causing the membrane to expand. With a membrane of overall equal thickness it has equal tension at every location.

According to this definition the "water bubbles" of the ice plant *Mesembryanthemum crystallinum* are "pneus." The internal medium is in this case the intracellular fluid; the external medium is air (Fig. 5.31a). Correspondingly an air-supported structure is also obviously a pneu (Fig. 5.31b); in this case the internal and external media are the same. A potato sack stuffed with potatoes is likewise a pneu, even if this is not visible upon first glance.

Much has been discussed about Frei Otto's vocabulary term "pneu." In the living world the internal medium is almost always incomplete, the external medium is very often a fluid containing ions, namely water ("hydr"); the Greek word "pneuma"

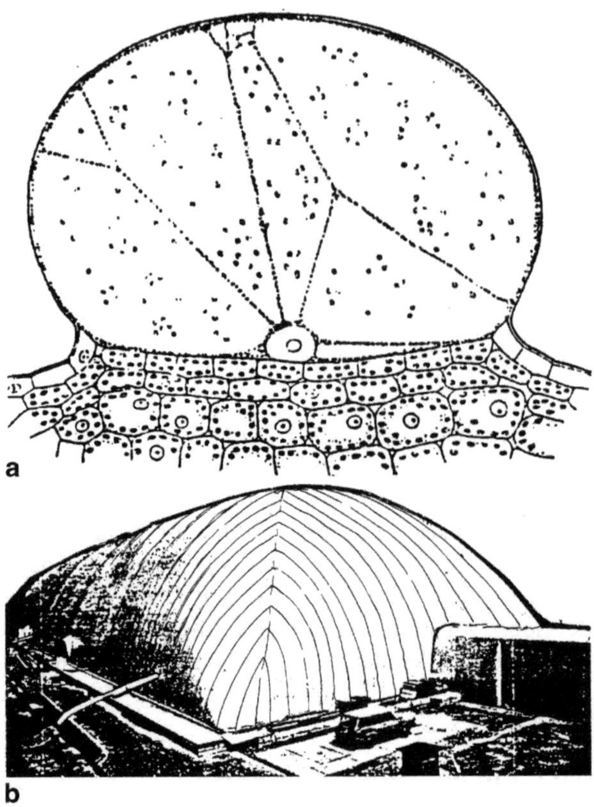

Fig. 5.31 Pneumatic systems in biology and technology: **a** "water bubble" of the ice plant *Mesembryanthemum crystallinum* and **b** air-supported structure. (Adapted from Nachtigall 1987 and IL-Report 12)

means however "air," as is generally known, but because the term "hydr," which W.N. had once jokingly suggested for this type of structure, does not sound terribly good when pronounced, it remained "pneu."

5.6.1 Biological Pneus → Technological Pneus

Wherever one looks in biology, one will find pneus. Typical forms are frog eggs, which conglomerate into spawn, but also their larvae contained within and ultimately the hatched larvae themselves (Fig. 5.32a, 5.32b, 5.32c, 5.32e). Each egg cell is also a pneu (Fig. 5.32d), and when they divide, both daughter cells are each in themselves a pneu (Fig. 5.32d). The eye during the development of mammals is a pneu, likewise the capsule of the skull, and additionally the amniotic sac (Fig. 5.32f). There are thus circumstances of a pneu within a pneu within a pneu within a pneu.

The laws of pneumatic formation processes, the forces in the elastic membrane dependent on the membrane form and eventual internal tension and other characteristics, are described in detail and can be checked, for example, in the IL report 9 of the former Institute for Lightweight Surface Structures ("IL") at the University of Stuttgart.

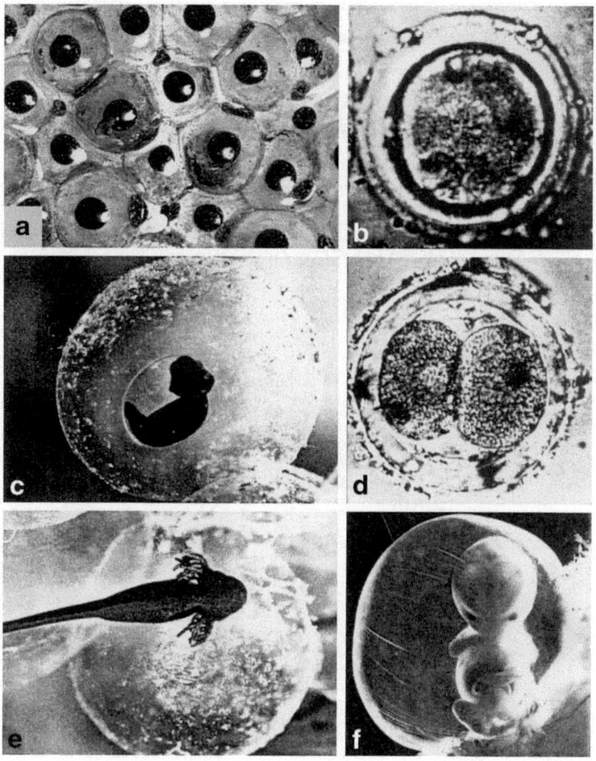

Fig. 5.32 "Pneus" in zoological ontogenesis: **a** frog spawn, **b** frog larva, **c**, **e** frog larva hatched, **d** egg cell after its first division, and (**f**) embryo of a human in the amniotic sac. (Adapted from various authors from Otto et al. 1982)

5.6.2 The Pneu as Key Element of Development

The drawings in Fig. 5.33 originate from F. Otto. Pneus are accordingly key mechanical elements, next to droplets, soap bubbles in air, and oil droplets in water. Microspheres in water belong in this category as well. Life could have developed from the latter—purely physical—aggregates. The important and surprising fact, even for subject biologists, is that, despite intensive investigation, one cannot find a structure in the biological world that does not at least in its development pass through a pneu stage, if it does not maintain the pneumatic structure throughout its life span, such as plant cells with their elastic cell membranes, vacuoles, and internal turgor pressure. The complete skull capsule of the human is obviously no longer a pneu, but during its embryonic development it forms as a bubble.

Bubbles of this kind also struggle against each other for space, reach an equilibrium of forces and thereby offer an essential mechanical basis for the development of an organism. The eye socket forms itself in this manner as the developing eyeball struggles for space in the skull. If the latter is missing due to an embryonic defect, the eye socket will remain much smaller, and the newborn will possess two unequally sized eye sockets.

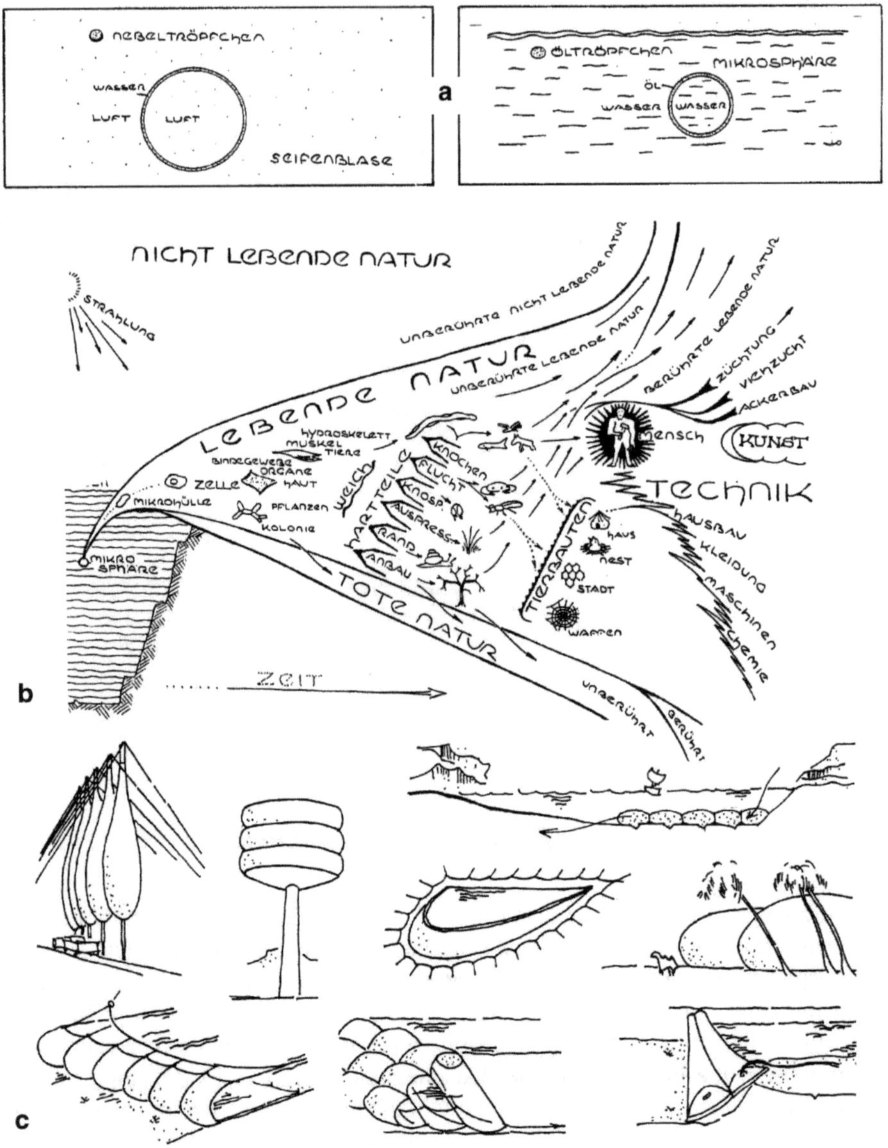

Fig. 5.33 The pneu as a building block for the development of life and pneumatic structures: **a** example of a pneu, **b** building blocks of development, and **c** example of a design for a pneumatic structures. (Adapted from Otto et al. 1982)

F. Otto is to be thanked for emphasizing the pneu as a universal building block of life. These structures and their lateral connections were already well-known and understood to the biologists of the nineteenth century; only the term “pneu” had not yet been used (Nachtigall 1986). Of couse, the biologists of the time were not aware that the structure represented—universally—the structural principle of life:

perhaps the most meaningful contribution of an architect to the field of "technical biology".

5.6.3 The Pneu as Technological Building Block

From the classical technological pneu, most represented in its simplest form by a balloon, technical-pneumatic constructions of the various sizes and purposes can be fabricated by using variable membranes (variations in thickness or stiffness), internal and external braces, and other relatively simple measures. They range from structures over artificial weirs for rivers to massive roof systems for entire cities in the arctic or deserts. Many of the proposals such as these were designed by the former Institute for Lightweight Surface Structures ("IL") under the leadership of Frei Otto (Fig. 5.33c).

Pneumatic structures admittedly still have an experimental and exotic character to this day. They are implemented more as eyecatchers than as functional lightweight structures, aside from air-supported sport arenas. Long-lasting structures require membranes that are equally lightweight as they are resistant to deterioration. As soon as such membranes are available, the architecture of pneus will doubtlessly make a great leap forward.

5.6.4 Tensairity: Connecting the Systems of Tensegrity and Pneu

The structural separation of tension and compression-withstanding building elements is often seen as the key for the development of effective structural systems. Cables and struts, provided they are placed exclusively under tension forces, can be designed slenderer than members under compression forces. Because compression members can buckle under stress, they should be thicker and have greater dimensions. With a tensegrity structure each element is ideally positioned so that every member assumes either the tension or compression load, but not both. Tension elements bear the tension forces; the compression elements are correspondingly thicker and more voluminous. A tensegrity system is best explained in model form (Fig. 5.34):

The stronger compression elements are linked with the tension elements, which also hold them in position. None of the compression elements meet at any point.

The idea of structural separation of tension and compression has been used for a development, which has trialed and tested in Switzerland. The so-called "Tensairity" structures are very similar to the tensegrity elements, and use however a pneumatic system that recalls the systems of natural pneus. In this technical, air-filled system a membrane of plastic is inflated with an internal pressure, corresponding to turgor pressure of plants, which forces the cell wall into form. The capability to absorb loads, comparable to a horizontal-lying beam or bridge, is achieved by this "balloon" by its elongated form and additional compression and tension elements (Fig. 5.35).

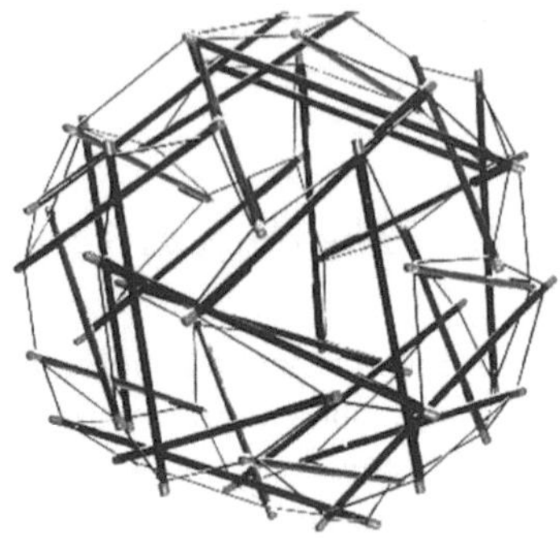

Fig. 5.34 Tensegrity model. (Fig. Luchsinger, lightweight structures with Tensairity)

With the Tensairity system, an air-filled pneu is subjected to an internal pressure that exceeds the external pressure. The material of membrane envelope experiences tension forces like a balloon. Ropes or cables support the membrane and the form against folding and are also under tension, the rods on the topside of the Tensairity system are under compression. In this sense the system can also be compared with an under-truss bridge (Fig. 5.36).

For the demonstration of its capabilities, a bridge was developed that was able to bear 3.5 tons with a span length of 8 m. The support structure consisted of two cylindrical Tensairity beams, each with a diameter of 50 cm. A standard PVC tissue material was used for the elastic membranes, as they have been used for example in the construction of stadium covers. The steel cables have a diameter of merely 6 mm. The compression rods mounted on top initially consisted of carbon fibers for test purposes; aluminum or steel would have also been acceptable. Wood boards were laid down for the "road surface"; however, they assumed none of the structural function. The Tensairity system of the two beams was inflated with an internal pressure of 400 mbar. The Tensairity bridge has a total weight of 2×98 kg = 198 kg, not including the wood boards (Fig. 5.37).

In comparison, natural systems are often provided with notably higher internal pressure. With *Equisetum giganteum* the turgor pressure amounts to several bars. Researchers at the Albert Ludwig University in Freiburg, Germany have been occupied for some time with the explanation of structural functions in plants. With the SFB 230 under leadership of the architect Frei Otto the research teams in Stuttgart

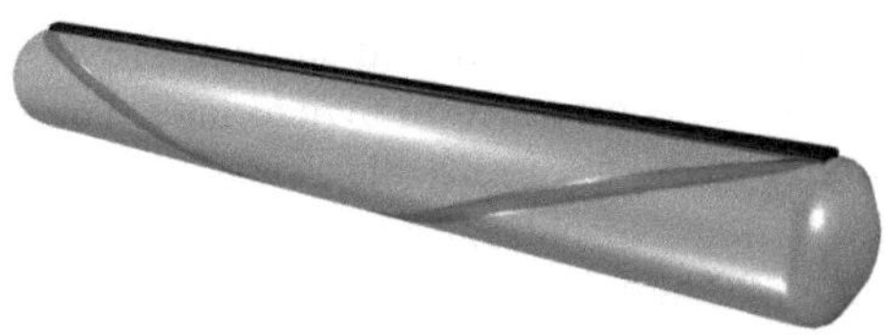

Fig. 5.35 Basic construction of a Tensairity system, consisting of a compression element lying on top, a cylindrical pneu body, and slanted peripheral tension braces. (Fig. Luchsinger, lightweight structures with Tensairity)

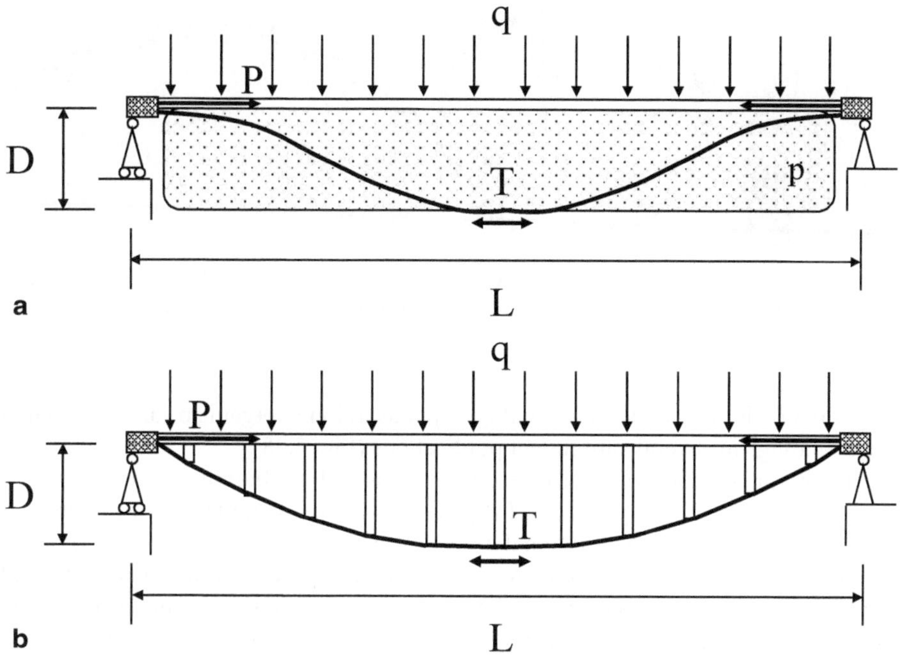

Fig. 5.36 Tensairity beams **a** in *comparison to an under-truss bridge* **b**. The vertical, rigid beams of the bridge are replaced by the internal pressure of the air-filled Tensairity system. (Fig. M. Pedretti, Tensairity, ECCOMAS 2004)

Fig. 5.37 Tensairity demonstrational bridge with 8 m span and a maximum load-bearing capacity of 3.5 t. (Fig. Luchsinger)

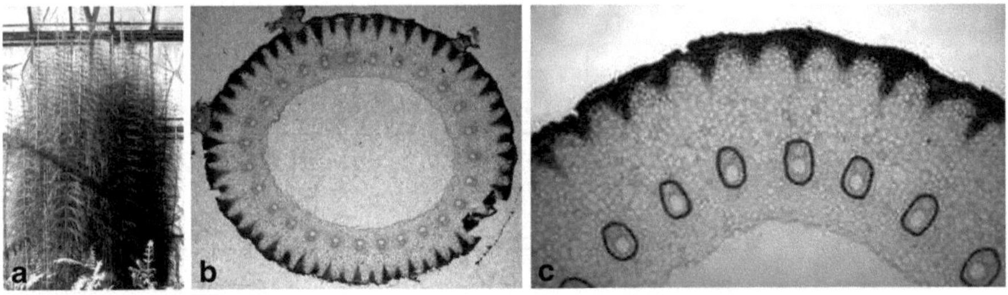

Fig. 5.38 a–c *Equisetum giganteum*, view, section, and peripheral detail. The turgor pressure amounts to several bars. (Fig. Plant Biomechanics Group, T. Speck, Albert Ludwigs University, Frieburg)

had already discussed the functionality of pneus in living organisms and distributed their findings in several subsequent publications (Fig. 5.38).

Today, pneumatic systems have been well tested for buildings: as a rule they consist of two or more layers of ETFE (ethylene tetafluoroethylene) that are pressed together and stabilized with internal air pressure. One of the most well-known built examples would be the façade of Allianz Arena in Munich (Architect: Herzog and de Meuron). These kinds of pneu systems for buildings are very similar to simple air balloons. Several balloons arranged in a row yield the façade structure. The possibility for the absorption of loads is limited; the construction of a bridge or free-spanning pneu structure is only successful with the pneu structures of nature or the pneumatic Tensairity system discussed here.

As one the first realized developments, a ski bridge in the French Alps was installed with a span length of 52 m. The wood planks and railing resting on the Tensairity cylinder once again do not perform any of structural functions. The Tensairity system is constructed so that the pneumatic internal pressure inside the membrane envelopes replaces compressive beams, which would otherwise be required for an under-truss bridge (Fig. 5.39).

The canopy of a parking garage in Montreux, Switzerland, completed in 2004 with a span length of 28 m, uses the Tensairity system as the structure for a membrane skin which is stretched like the webbing between the digits of some aquatic and flying animals. This system appears especially attractive thanks to its translucency and the potential to apply illumination (Fig. 5.40).

The current developments in Tensairity systems have attempted to translate the self-healing properties of living pneu systems, for example in plants, to the technical system. With light damage to the membrane of a Tensairity system the internal pressure drops slowly, which allows a certain amount of security. However, the exterior skin of the membrane cannot self-repair its defects as they are in natural systems. Research projects in Switzerland and Germany have attempted to understand the natural self-healing mechanisms (Sect. 6.57) and use them for Tensairity.

Self-repair in plants, particularly in the species *Aristolochia macrophylla*, has been investigated for the purposes of biomimetic applicability. In this plant wounds are generated during the growth process. The healing of the wounds according to

Fig. 5.39 Tensairity ski bridge with 52 m span in the French Alps, 2005. (Fig. R. Luchsinger/Charpente Concept SA, Barbeyer Architect and Airlight Ltd.)

Fig. 5.40 Tensairity roof structure for a parking garage in Montreux, Switzerland, 28 m span, 2004. (Fig. R. Luchsinger/Luscher Architectes SA)

Speck et al (2006) commences in four phases, of which the first phase appeared in the studies as the most interesting for applicability.

In this phase, parenchyma cells swell in the wound area and seal it. The actual healing occurs in the following phase, but is less of interest for a biomimetic application in a technical system. It is assumed that the sealing of the wound in the first phase of the process is mainly characterized by a viscoelastic/plastic deformation of the parenchyma cells, which are pneumatized by the internal turgor pressure. Similar phenomena were discovered in the rapid self-healing processes of *Phaseolus*,

Ricinus, and *Helianthus*, in which the healing occurs as a reaction to artificial induced damage (Speck et al 2006).

Tensairity® is a product of Airlight Ltd, Switzerland, developed with Prospective Concepts AG, Switzerland. The research activities take place at the EMPA, Switzerland; research for self-healing and translation to Tensairity systems at the Albert Ludwig University, Freiburg, Germany.

5.6.5 *Water Spider → Diving Bells*

The only spider that can survive underwater for a long period of time and appears adapted to this medium with its hair coat on its legs is the water spider *Argyroneta aquatica.* On aquatic plants under the water they form a silk-anchored web ball, which they fill with an air bubble. They achieve this by lifting their hind side above the water and pulling it under (Fig. 5.41a), thereby trapping air bubbles on the fine hairs on the rear of their bodies. The bubbles are then stripped from the hairs into the web. With repeated embedding of such bubbles the air-filled chamber can almost reach the size of a ping pong ball. This "diving bell" is stabilized by itself due its buoyancy against the anchoring fibers (Fig. 5.41b). In this bubble the spider lives out most its life, even for consumption of its prey and ultimately reproduction.

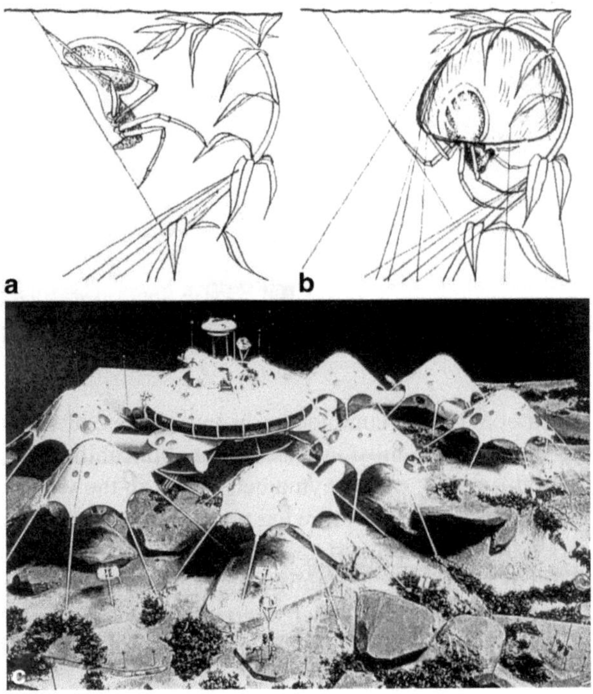

Fig. 5.41 Diving bells: **a** air collection, **b** formation of the bubble of the water spider *Argyroneta aquatica*, **c** concept model of an underwater housing project in the Caribbean, J. Rougerie 1974. (**a**, **b** from Freude 1982, **c** from Rougerie from Coineau, Kresling 1987)

A proposal to form diving bells with fibrous membranes originated from J. Rogerie (Fig. 5.41c). Similar to the web of the water spider, it would be self-stabilizing using the buoyancy of the imported air masses.

With this membrane entire underwater cities could theoretically be constructed. The principle is simple, requiring minimal material and mass, and the individual systems could be easily dismantled again. The only issue is its anchoring, as high tension forces would develop. Perhaps some day this vision will be realized, at the very least, as an attraction for tourists.

5.7 "Tree Columns" and Tent Structures

Frei Otto's Institute for Lightweight Structures also led exemplary studies on this subject, in which trees and spider webs were observed as parallels to "tree columns" and tent structures respectively.

5.7.1 Principles of Tree Structure → Tree Columns

The fundamental idea in this instance is as follows: Trees branch themselves more and more intensely towards their canopy as "bodies of constant tension".

If one considers tree-like structures, supporting a flat roof that must sustain environmental loads (wind, rain, snow), it may be that the dead load of this kind of structure is less than that of a traditional column system required for such a system. Additionally an area of the roof can then be supported at many points, thereby distributing the loads evenly across the roof instead of concentrating them at a few point columns.

Essentially, the branches of the columns with narrower diameters in the upper extents of the structures can weigh less in total than a normal, unbranched column. It depends therefore on where the branching moments are applied, not too early and not too late, so that a function of effectiveness must be formulated with the prerequisites "minimal mass with the given constraints of a load-bearing structure".

For the conception of tree-like supports self-formation processes were applied alongwith calculations and evolutionary strategies of optimization, as well as the study of forms in a soap bubble model. Portion **a** of Fig. 5.42 illustrates the principle, **b** demonstrates the soap bubble model, and **c** shows a support structure optimized for mass. Notably, the supports in the model for mass optimization branch at a point relatively close to the bases of columns.

5.7.2 Spider Webs → Tent Roofs

The most well-known work of the Institute of Lightweight Structures is related to the study of tent structures, and of these structures the Olympics Stadium in Munich

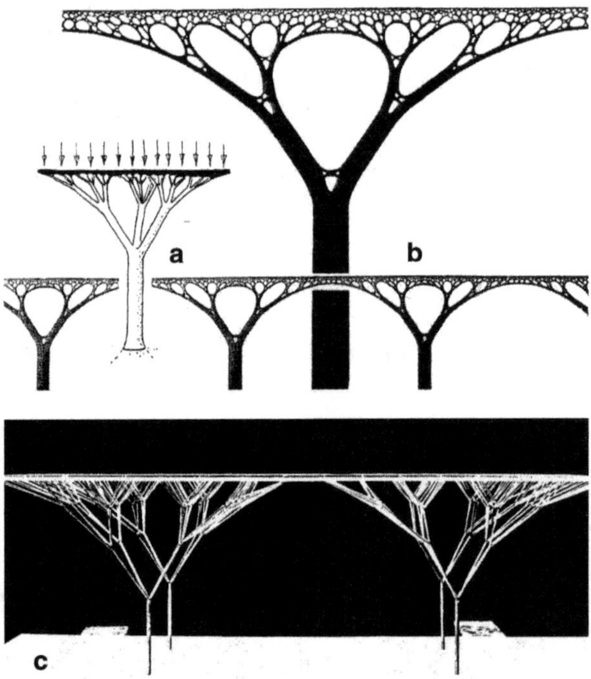

Fig. 5.42 Lightweight "tree columns" for spanning structures: **a** concept sketch with forces acting on the roof surface, **b** soap bubble foam model, and **c** stick model with support plate. (Adapted from Otto et al. 1982)

is probably the most famous, where the formfinding was significantly influenced by Frei Otto. A precursor can be recognized with the roof of the Expo Montreal, 1967 (Fig. 5.43b). Another predecessor had been dismantled and rebuilt in Stuttgart; it housed the Institute for Lightweight Structures for many years and still stands today. With each visit one is always surprised by how such an unexpectedly large space can be enclosed by a structure that appears smaller from the outside.

Issues of structures of this kind, structural as well as esthetic, are the simply enormous pylons required for their stability. The forces to be absorbed in the pylon are immense, and they must be centrally positioned and exactly parallel to the direction of compression forces, so that despite serious motion they do not buckle.

The webs of certain spiders (Fig. 5.43a) appear doubtlessly "tent-like" (compare to Fig. 5.43b), and in IL spider webs have been extensively studied as well in cooperation with the spider researcher E. Kullmann. Much was reported on the subject, more specifically in the IL Publication 8. However, F. Otto has repeatedly denied that spider webs and spider structures in general have inspired his technical web and tent structures.

In view of the intensive and interconnect study of both fields we believe this statement to be problematic. Inspirations cannot be avoided, and there is obviously nothing bad about that. In contrast, it was often emphasized that the comparison to technical tent structures would have caused the biologists to more closely observe spiders and other web constructions. That is without a doubt correct, and in

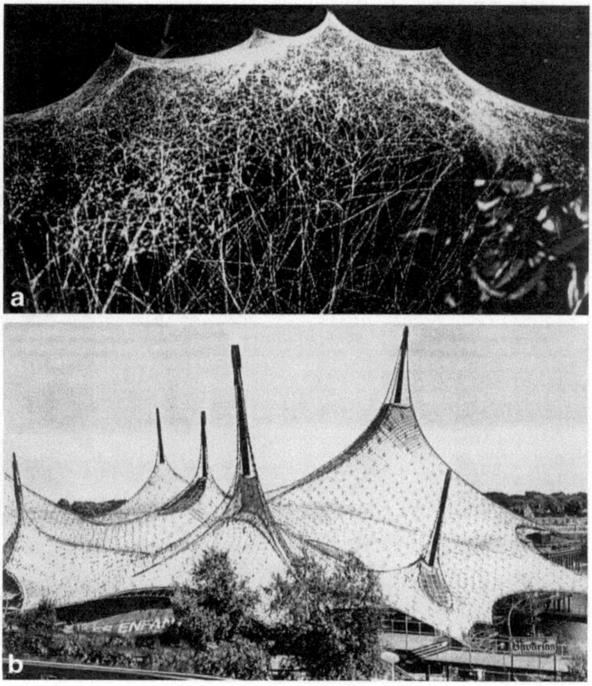

Fig. 5.43 Tent structures in biology and architecture: **a** "Tent roof" of a spider, **b** roof of the Expo Montreal, Frei Otto et al. 1967. (Adapted from Otto et al. 1984)

so doing, it was also discovered that the loops, introduced in web structures for the easing of the excessive point loads, between the actual web surface and the anchors at the ends of the pylons (Fig. 5.44a-c) can also be found in spider webs. This finding shows that the same or similar forms can be applied in construction or evolutions processes (which both contain elements of randomness) with the same or similar requirements.

The principle of mass increase on severely stressed nodal structures is another concept that holds for biology and technology. Spiders strengthen these nodes by either providing more silk threads or thickening the individual threads in the nodal regions (Fig. 5.44d).

5.7.3 The Variety of Tent Structures

The IL Stuttgart (Institute for Lightweight Structures) has developed an enormous variety of web constructions and tent designs, which can only be broadly described here. An extensive display can be found in the IL Report no. 8 (1975), "Webs in Nature and Technology." The IL Reports can still be acquired today from the Institute as reprints.

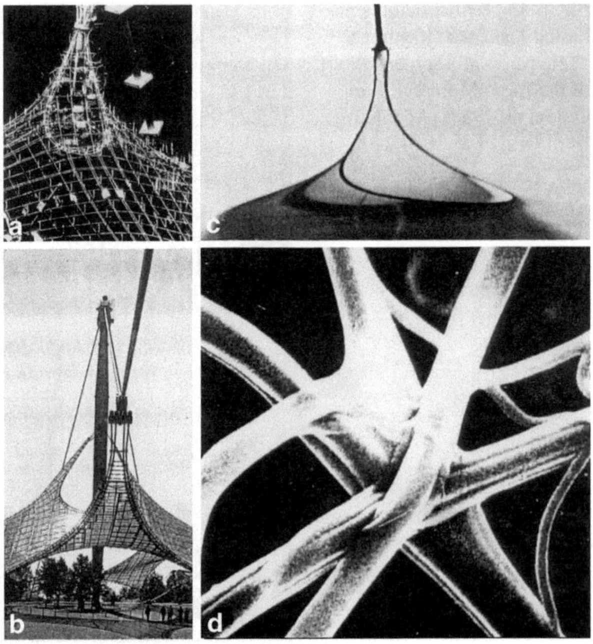

Fig. 5.44 "Eyes", **a**, **b** with tent roofs, **c** in soap bubble experiment, and **d** strengthening of the nodes in a spider web. (**a-c** from Oder 1982, (d from Kullmann 1975)

5.8 Moving Structures

Moving structures in nature have wielded major fascination among architects. They occur as autonomous or not autonomous movements. Particularly interesting are the material-technical insights, particularly those which suggest durability.

5.8.1 Non-Autonomous Movements

Non-autonomous organ movements of plants, that is, with leaves, occur in nature within a large breadth of movable forms and their underlying principles. These movements of plants are known as nastic movements. These movements have been categorized by their trigger: seismonastic or thigmonastic (plants' reaction to tremors or contact, i.e., *Strelitzia*), chemonastic (reaction to chemical stimuli, i.e., tentacles of the sundew plant), thermonastic (opening or closing in warmth, i.e., crocus or tulip flowers), and photonastic (opening or closing of flowers at different light intensities). There is of course an entire series of nastic movements, that is, traumonastic, as reaction to injury or damage, and hydronastic, as reaction to moisture. At the University of Stuttgart, specifically at the institute ITKE under leadership of Jan Knippers, multiple attempts were undertaken to translate these structures into promising architecture. One spectacular application was completed by Soma Architects for their design of "Thematic Pavilion 2012" at the Expo in Yeosu, South Korea (Fig. 5.45, Sects. 6.51, 6.52, among others).

Fig. 5.45 Soma. Analogous effects. Thematic pavilion 2012, Yeosu, South Korea (Fig. soma Architects)

5.8.2 Autonomous Movements

Movements that are driven by muscles, sinews, ligaments, or pneumatics have been likewise researched and developed by research teams according to analyses of examples in nature. Good examples have originated from E. Hertzsch and G. Pohl from biomimetics workshops at the University of Melbourne as well as from the Façade Research Group under Ulrich Knaack at the TU Delft (Sects. 6.49 and 6.50," thermoregulating envelope structures and ventilation systems for breathing building skins by Lydia Badarnah).

5.8.3 Responsive Movements

At the TU Berlin application areas are being researched for a natural phenomenon that would enable shovel or grabbing movements with low application of power or in another case would apply artificial muscles for lightweight bridges. Fish fins are driven on the one hand by musculature, but on the other hand they are formed by radial ligaments in the fins, so that they perform a shovel motion in the direction the fin is moving. This so-called "fin ray effect" was studied and published principally by R. Bannasch. The Institute for Civil Engineering of the TU Berlin, subject area design and construction, is developing applications for engineering under the leadership of M. Schlaich, for example responsive fastening systems for lightweight roofs. This institute is a pioneer with reactive movement technology, as well as the application of artificial muscles in buildings. Bridge structures have been developed whose vibrations are dampened by pneumatically driven "muscles." In Sects. 6.54 and 6.55 the systems are extensively illustrated.

Chapter 6
Products and Architecture: Examples of Biomimetics for Buildings

Nature has developed solutions for itself over time through complex networks. This strategy can be confirmed as successful in comparison to more technical, "linear" optimizations. On the contrary, natural optimization succeeds through reproduction, mutation, recombination, and selection, as well as the use of failures as a means of improvement.

The examples of biomimetics in this chapter cover realized examples as well as studies and idea sketches. The biomimetic method of abstracting a biological example for technological usage should be comprehensible and replicable. The examples in this book can never definitively show the possibilities of biomimetics and are considered more as stimulae and inspiration for one's own work.

It has become recognized that biomimetics is to be understood as a tool, as a resource, one that can stand alongside other classical design and development tools. To architects, planners, builders, and designers it will depend on obtaining an optimal result. Whether one uses only one method or combines multiple methods is for the end product the same. "Purely biomimetic" results will be shown in the following paragraphs, as well as results that merely represent analogous developments in technology and nature and do not follow the pure definition of a biomimetic process.

Above all we are convinced that the world of architecture is too complex and therefore no "biomimetic architecture" can theoretically exist. However, individual construction elements and materials or functions from nature can be interpreted in technology. A complex building consists of many elements, spaces, and functions that arose from a background of norms, traditions, and technological requirements. The German VDI guideline 6220 and the VDI guideline 6226 for the area of construction have already specified that a product can be only defined as biomimetic if its essential elements are developed biomimetically. Therefore a term like "biomimetic building" is clearly not consistent. We refuse the question as to whether or not a "biomimetic building" can exist, despite the debate of whether this or that building is biomimetic being present in varying publications. The term is often only used as a marketing strategy or arises from a general misunderstanding. Small structures, for

G. Pohl, W. Nachtigall, *Biomimetics for Architecture & Design*,
DOI 10.1007/978-3-319-19120-1_6

example pavilions, can be accepted as exceptions to the principle that no true biomimetic building exists. Such examples will also be introduced in the following pages. Therefore, the limits of biomimetics are not to be misunderstood: We are convinced that biomimetics has its place and is due to achieve a prominent meaning through the use of development potentials.

Biomimetics is justifiably depicted as a cross-section of disciplines, which will become clear in the following examples. Scientists often work beyond their subject area of knowledge, leading to expedient combinations of technological developments and biomimetic inspirations. *Biomimetics is in many cases an integrative working tool.*

In summary, the reader should be able to obtain with the help of the following examples an impression of the possibilities and depth of biomimetic approaches to design and be inspired to one's own ideas. The reader should also recognize where parallel developments on a technological foundation have already led to success without nature having had stood by as direct mentor.

Structure of the Sections in Chap. 6
The chapter begins in the first subsections with an extensive, though not all-encompassing, recount of the course of biomimetics research, showing possible development tracks through history and then discussing the results. The descriptions are kept relatively comprehensive, so that the research depth necessary for biomimetics is recognizable. However, despite the comprehensiveness of the descriptions they cannot entirely depict the scope of research, as they would overstep the possibilities within this book. In relation to the research examples, various exemplary developments and ideas in biomimetics will also be introduced, each limited to two pages of each subsection to simplify the comparison to one another. These sections begin with the biological precedent, clarifying the process of abstraction to technological realization and the possibility of utilization. The analogous developments of technology and their biological functioning counterparts will each be shown in the same manner. Further information about the authors, photography credits, and addresses for further research about the given examples are gathered in the end credits.

6.1 Biomimetics on the Basis of Algae, a Biological Example

Algae serve as the source of nutrition for many ocean dwellers and represent among other forms of life the lowest level of the food chain. Discolorations found in ocean water known as algal blooms are a well-known effect of this organism. Lesser known is that the single-celled algae are co-responsible for atmospheric carbon dioxide production on Earth, playing a larger role than the rain forests, for instance. The number of single-celled organisms in our oceans is compared with the number of celestial bodies in the universe and is estimated at 10^{22}. The pigment composition responsible for the coloration of the oceans is less interesting for biomimetics as are the microscopic and extremely manifold construction of the algae themselves. The fine structure of the exoskeleton and its abutting plasma

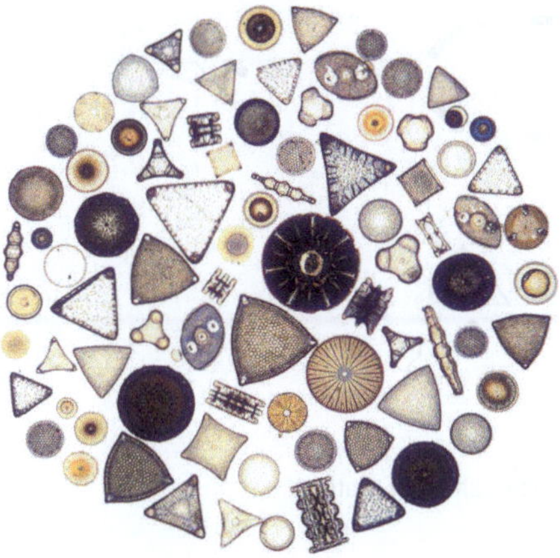

Fig. 6.1 Fossil marine diatoms from the Oamaru-Deposit in New Zealand (late Eocene), arranged by Alfred Elger, circle diameter ca. 500 µm

layer separates the cell contents from the ocean water.

These skeleton-like structures are built up through the use of different structural principles and various materials: coccolithophores use calcium carbonate ($CaCO_3$); diatoms, radiolaria, silicoflagellates use silicates (SiO_2*n H_2O). Complex geometric structures are built by Foraminifera through the use of calcium carbonate, by Acantharia through the use of celestine ($SrSO_4$), and by Radiolaria, also with the use of silicate. The diatoms form the largest group with approximately 25% of the total. The shells of the diatoms have developed a highly geometrical complexity (Fig. 6.2), upon which the individual types differentiate themselves, the number of which is projected at 100,000. Their rib, honeycomb, and pore structures have sizes of about 1 µm, 150 nm, and 20 nm and shape the body mass into a triangular, cylindrical, needle-like, or into other further geometries, in often fractal, self-repeating structures. Many prominent scientific investigations are currently underway, which is important also for the further research in biomimetics.

The discovery of the attractive forms of the microscopic diatoms and radiolaria (Fig. 6.1) led to their use in

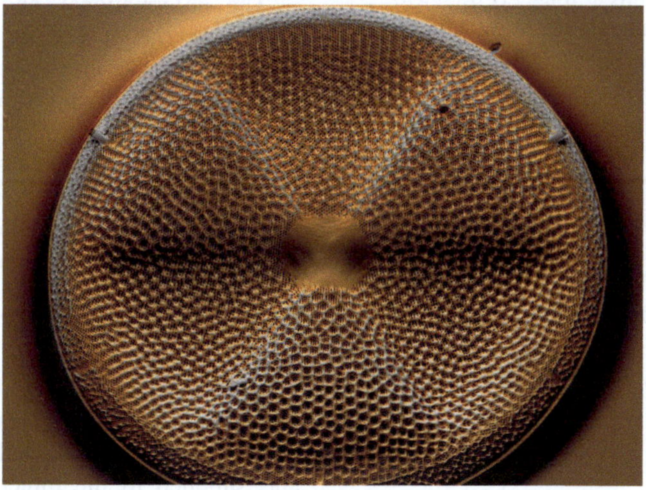

Fig. 6.2 Typical representative of the diatoms with clearly visible petri dish form: *Actinoptychus*

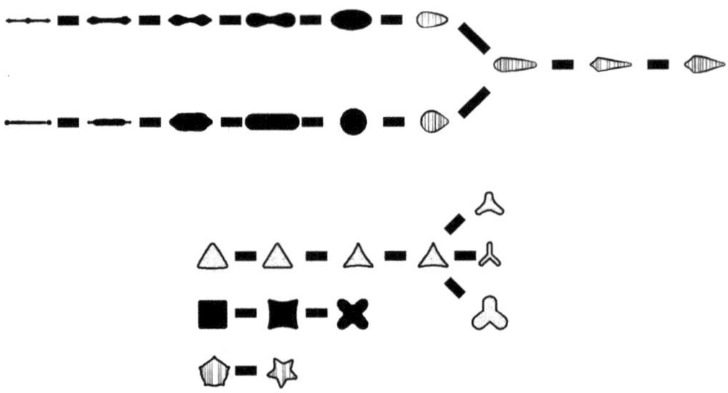

Fig. 6.3 Different exterior forms and symmetry relationships of diatoms

academic salons in the nineteenth century, where educated society was able to view the preserved specimens in microscopes and philosophize about the beauties of nature (Fig. 6.1). Around this time the widely read, although today antiquated work of Jena Biologist Ernst Haeckel appeared: "Art Forms in Nature." In the 1970s and 1980s academics such as architect Frei Otto, known for his light, tent-like structures, botanist Johann-Gerhard Helmcke, and plant physiologist Anne-Marie Schmid further analyzed the origins of the forms of diatom husks (Fig. 6.3).

In the meantime, knowledge of this subject was able to be refined and exacted. Present-day scientific insights have shown that the shell structures of diatoms fulfill the high demands of static stability and mechanical load-bearing capacity. Furthermore, the shells are optimized against attacks from Copepods (*Copepoda*) and their silicate-coated oral apparatuses. For protection, the diatoms use a hard, though delicate, shell of bio-silicate that is so finely structured that the smallest pores in the silicate hull occupy the same semipermeable characteristics as a membrane: Certain particles of matter are allowed into the cell body, whereas others are excluded.

6.2 Pool Research as Biomimetic Method in Application

The shell formations of diatoms are ideally suited as subjects of investigation for lightweight constructions, a subject that G.P. has concerned himself with for years.

The informational and investigational material of G.P. at the Alfred-Wegener-Institute in Bremerhaven from the research activity by PlanktonTech is available here. In the frame of the international research project PlanktonTech, a virtual institute of the German scientific Helmholtz Society, biologists occupy themselves with the basis research on plankton as well as architects and engineers with the question of technological feasibility of products in the areas of architecture and design. With the biomimetic method "Pool Research" scientific insights were collected, evaluated, and supplied to the direct prototypes

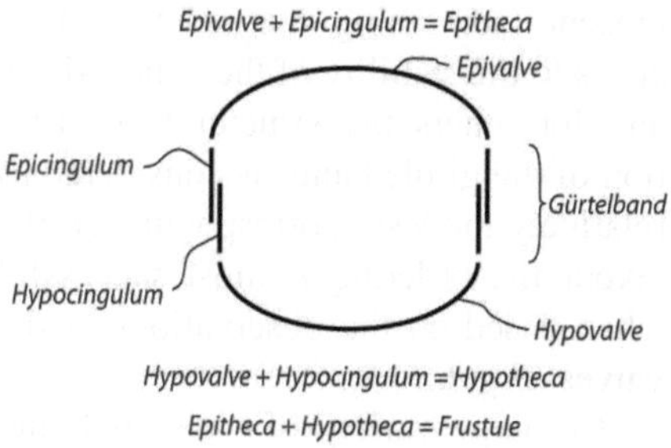

Fig. 6.4 Construction schema of a diatom shell

from the research series PlanktonTech (compare: COCOON_FS, introduced in another chapter of this book) and made available to the industrial wood construction development within the framework of BOWOOSS (BOWOOSS is a biomimetics research project on the use of shells in wood construction).

6.3 Pool Research: Abstraction Through the Classification of Biological Precedents

6.3.1 Classification of Diatom Species

Diatoms consist of two interlocking shells, the hypotheca and the larger epitheca, that surround the smaller hypotheca. The shells link together in the connective region, known as the girdle band, to form a larger mass known as the valve (Fig. 6.4).

The focus of the classification of diatom types led by research project BOWOOSS rested on the investigation of particularly outstanding examples (Fig. 6.5), which were considered to have particular application for the con-

SPECIES	IMAGE	ANNULUS/ CHANGE IN STRUCTURE AT THE CENTER	STRENGTHENING RIBS	HIERARCHICAL STRUCTURING	ORGANIZATION	UNIQUE QUALITIES	POTENTIAL FOR TRANSLATION
Thalassiosira	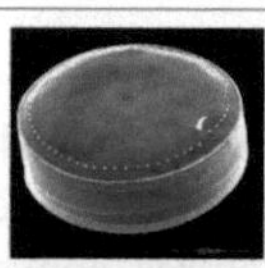	+, very large, frontal mapping	-	+	central-planar	regular openings	weight savings
Actinoptychus	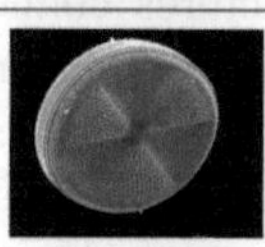	+, closed	+	+	centrales	radially wavy	increased stability by concentric curvature change
Arachnoidiscus	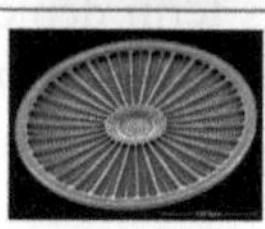	+	+	+	centrales	Interior side: clear, radial strengthening ribs, hierarchical, concentric ribs, while the edge appears very massive. A change in the structure to a completely different form occurs in the center.	shape, dimension and orientation / relationship of the stiffening ribs to each other. Pressure ring / pressure zone in the center.
Trinacria	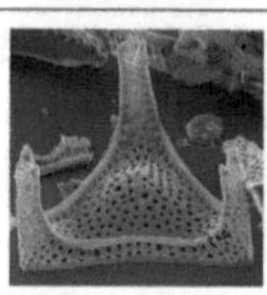	-		+	planar	regular round / oval openings	weight savings

Fig. 6.5 Extract of classification of diatoms

struction industry. With the help of this classification, the researchers were able to successfully isolate and compare different solutions in nature with structural problems.

The foundational organizational shapes were divided into categories based on radial or ctenoid appearance, according to Round et al. This division appears insufficient in light of the present knowledge; as a supplement to those categories diatoms with perforation are also included (Fig. 6.6).

The classification relates overwhelmingly to morphological and topological characteristics of the valve, because for some the girdle bands are often only fragmentarily or not at all present (something that admits inferences to the stability of the connection), and for others the structural constitution of the girdle bands is considered as relatively modest. Correspondingly, the taxonomic ordering is most successful when based on the observations of the valves (Fig. 6.5).

The large variety of types of living and fossilized diatoms (estimates range from 10,000 to 150,000, compare: IL 28 S. 42) and the consequent variety of shapes and structures could be suitable for wide-ranging approaches for their interpretation in architecture.

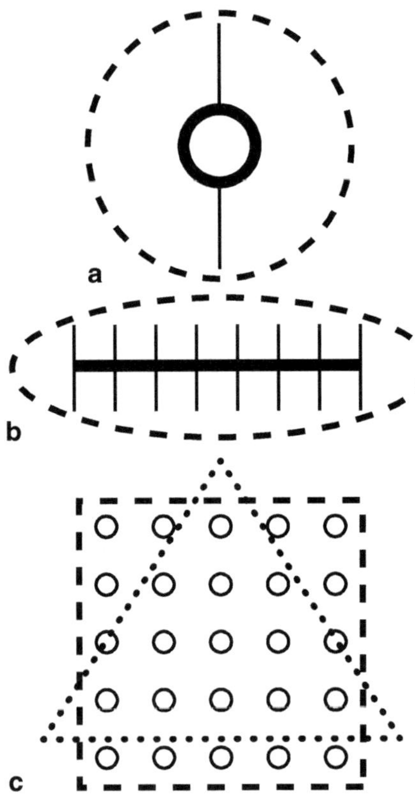

Fig. 6.6 Fundamental forms of diatoms

6.4 Pool Research: Analysis and Evaluation

The ability to analyze the morphological construction of diatoms lies in the observation of their shell structures (Fig. 6.6). In related investigations, distinctive features have been recognized for their similarities to structural members of architecture.

Hierarchical Ordering of Members
Diatom shells consist of several differently scaled structures connected with one another. The hierarchical ordering (Fig. 6.7) of these structures is an essential noteworthy aspect in the shell structural system. This term depicts the dissolution of a load-bearing structure to a system of individual elements that are in turn always further subdivided into a substructure. This subdivision often follows a diminishing, self-repeating pattern. The corresponding increase of surface moment of inertia and the reduction of weight and material usage can be considered efficient when compared with "monolithic" structures. Often

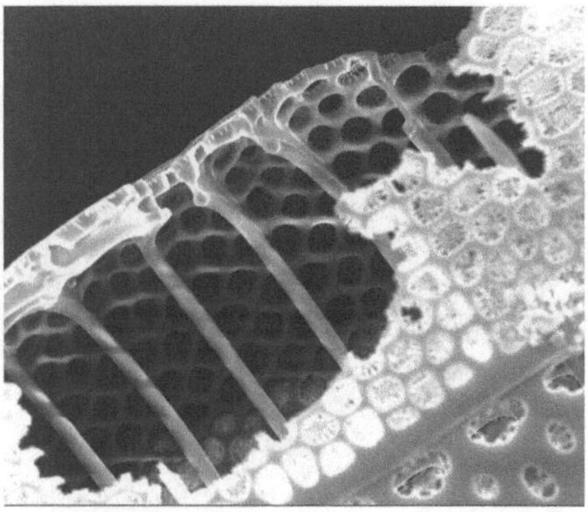

Fig. 6.7 *Isthmia*

Fig. 6.8 *Arachnoidiscus*

closed, honeycomb-structured cavities called "bullulae" (i.e., in Aulacodiscus) are encountered, which produce a foam-like substance (compare: II 28 DIATOMEEN I, p. 56). Also to be found are spherical pockets and two-dimensional mesh networks (IL 28, p. 80).

Strengthening Ribs

Many diatoms, in particular those belonging to the type *Pennales*, exhibit on the underside of the valve a pronounced rib structure, often consisting of a parallel or radial system, to which the lesser members are connected. Reinforcement on the outer edge is present in virtually all types in order to absorb tension forces on the shell (i.e., Figure 6.8 *Arachnoidiscus*).

Symmetry

Equally striking is the observed symmetry in the specimens. The *Centrales* consists of a radially symmetric structural system of mostly round or polygonal shells. As opposed to the pattern of *Pennales* the substructure of *Araphidineae* has a pronounced dual-axis symmetry.

Separated Shells

In many types (i.e., *Actinoptychus* Figs. 6.9, 6.10, 6.11) it is to be noted

Fig. 6.9 *Actinoptychus*

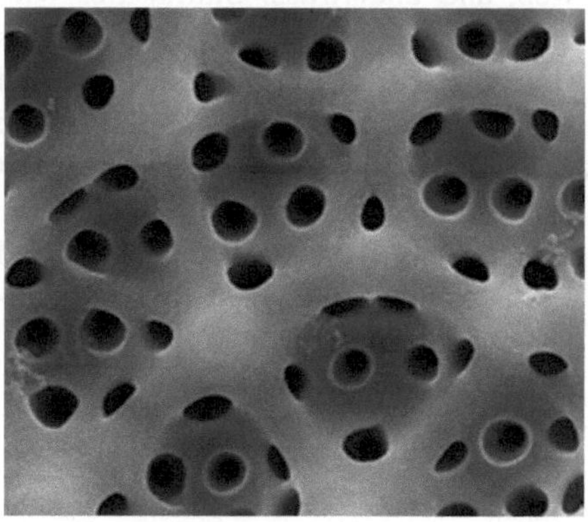

Fig. 6.10 *Actinoptychus*

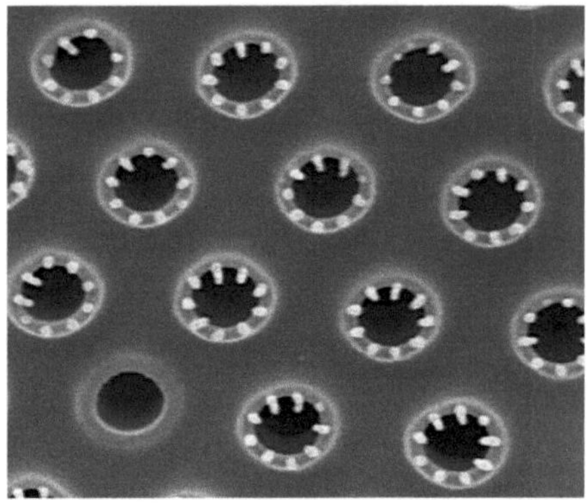

Fig. 6.11 *Actinoptychus*

that their shells are constructed from two morphological and completely differentiated layers, whose structures seemingly bear no relationship to one another (compare: IL 38. DIATOMEEN II, p. 90 ff.).

Parallels to Architecture
The similarity to architectural construction is particularly noticeable in centrally chambered diatoms. The individual elements of the shell, namely the outer boundary layer, the lateral boundary layer, and their interstitial connecting members, can be easily compared with the architectural terms overtruss, web, and undertruss. Similarities with double-layered frameworks are obvious (IL 28, p. 288 ff.) in more decomposed shells. The rib structures can also find a counterpart in architecture. The comparison is most clear in *Arachnoidiscus*, whose shell recalls a cement-ribbed vault, thanks to meridial ribs and sublayered concentric ribs.

6.5 Pool Research: Abstraction of Geometric Principles

The significance of exploring more complex geometries of morphological building methods lies upon the notion that avoidance of geometric complexity in technological developments has up to now been the rule. However, complex structures of biological systems have already proven a superior performance.

These types of structures require developments in support systems, connections, and overall complex production and assembly chains for the supply of building parts to be used for architecture. The further investigations of diatoms within the framework of BOWOOSS and the translation of the developed forms into computer-aided design (CAD) models require an abstraction of the discovered principles. In light of their symmetrical characteristics, the simple constructions of the *Centrales* and the *Pennales* appear to be especially advantageous for the task of abstraction and translation into architectural forms (Fig. 6.12).

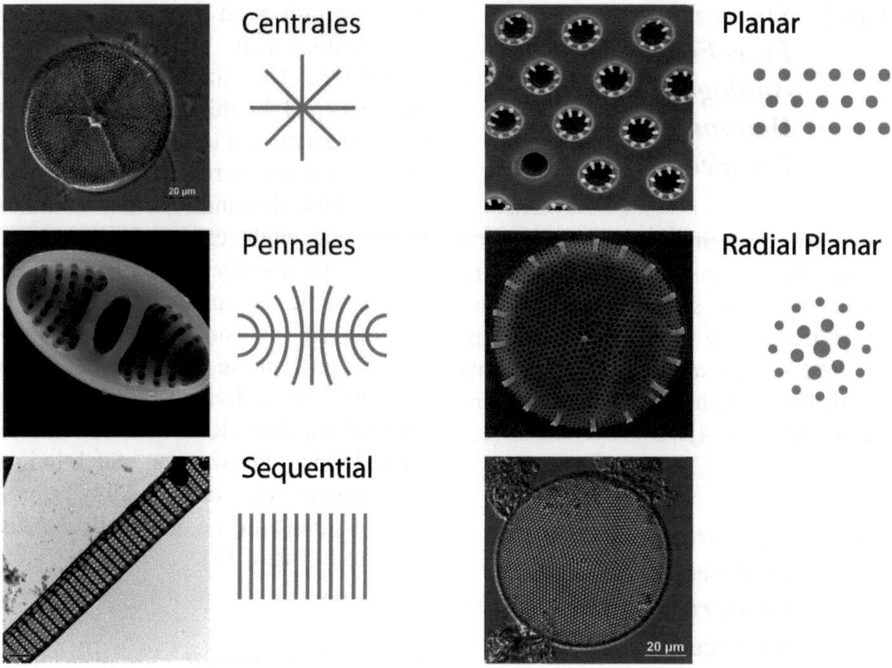

Fig. 6.12 Organism and abstraction

6.6 Pool Research: Translation into CAD Models

In the later stages of the research project BOWOOSS the first translated patterns were investigated to verify the suitability for later implementation. In the first instance, potentials were sought that could be realized in wood construction. The interpretation of the three-dimensional (3D) models, though within the limits of CAD's "drawability," yielded recognizable forms that could already be discerned for their use for later production.

6.6.1 Structuring of a Free-Form Surface Analogous to the Centrales

For this CAD model a regular, repeating hexagonal structure following the precedent of the diatom was applied to a free-form surface. The orientation of members follows a polar coordinate system, which lead however to a heavy distortion in the polar region, affecting the breadth of the struts (Fig. 6.13).

6.6.2 *Structuring of Free-Form Surface Analogous to the Diatom Species Craspedodiscus*

The basic form of this CAD model is a completely asymmetrical, double-contorted free-form surface. The structure is inspired by the spiral-patterned openings of *Craspedodiscus*; similar patterns can be found on the flower heads of sunflowers (Fig. 6.14).

6.6.3 *Segmented, Radially Symmetric, Double-Contorted Free-Form Surface*

By structuring this surface with a concentric pattern many meridial members meet together in a center, resulting in an unrealistic “fusion” of members at the intersecting point. This problem could be evaded by a tapering of these members, but high costs would be expected for a constructable implementation (Fig. 6.15).

6.6.4 *Structuring of a Free-Form Surface Analogous to the Pennales (Araphidineae)*

A right angle-derived, dual-axis symmetrical free-form surface, whose direction of curvature differs along the length of the surface, was used as the basis for the following variants. The CAD models were processed for calculation and optimization with the FEM Program SOFiSTiK. The size of the building members and the thickness of the structure were assumed to be for a comparatively small structure (Figs. 6.16, 6.17).

The high demands on the manufacturing and engineering of these components are already obvious with these first, simple computer models, even without the inclusion of assembly details. Even if these models were to be divided into individual elements for production, each element would still be curved along the two axes. Such building elements could only possibly be prepared with the means of a costly 5-Axis CNC mill (Fig. 6.16).

6.6.5 *Evaluation*

Nature as structural precedent can be modeled with the means of CAD programs, but in reality this method ensures that the developed geometries can be constructed only through the use of complicated methods. For example, the radially symmetric models, whose meridial members meet at a central point, could only be realized as load-bearing elements if they behaved similar to nature, that is, tapering or fusing together. The common technological practice of producing prefabricated parts and assembling them in construction reaches its limit. More advantageous would be freely formable, literally growing structures. The synthesized abstraction efforts for free-form surfaces (Fig. 6.17) based on the analysis of diatom structures only illustrate the "directly" translated ideas.

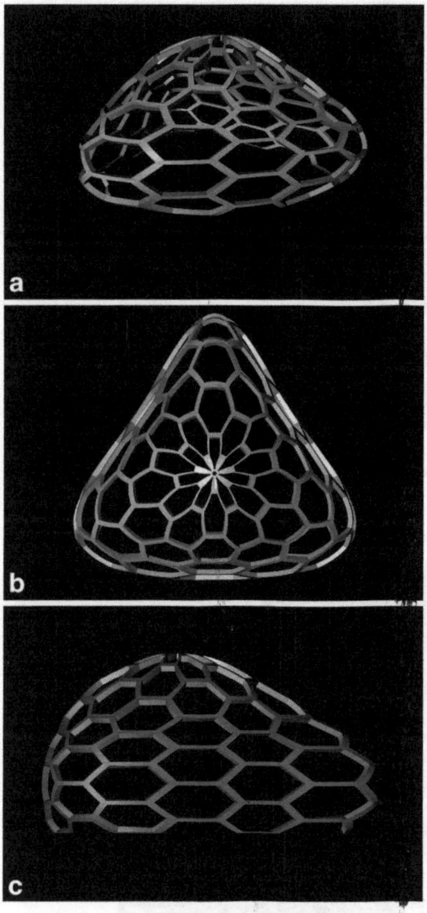

Fig. 6.13 Models of geometric abstractions

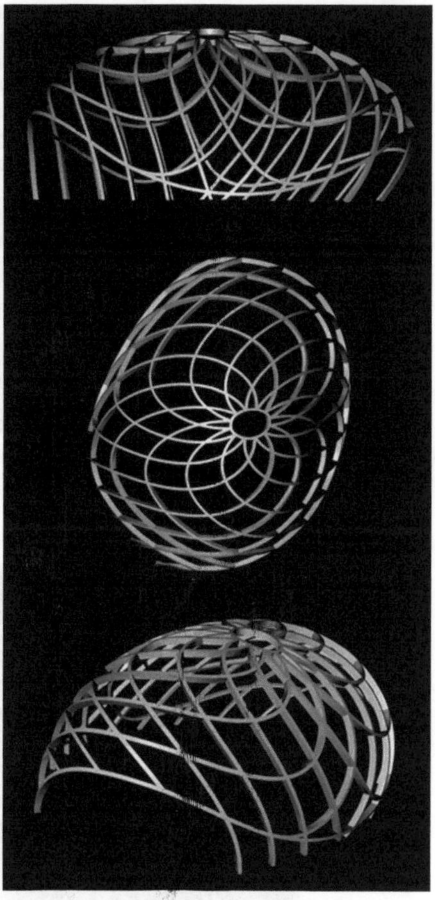

Fig. 6.14 Models of spiral abstractions

Fig. 6.15 Models of radially symmetric, segmented abstraction

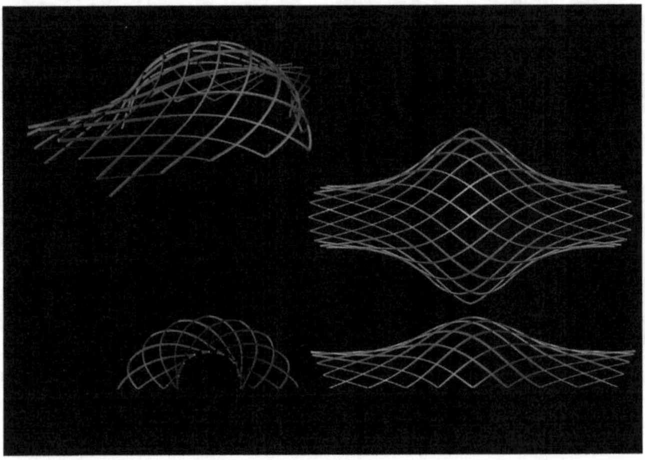

Fig. 6.16 Abstracted free-form surface analogous to the *Pennales Araphidinae*

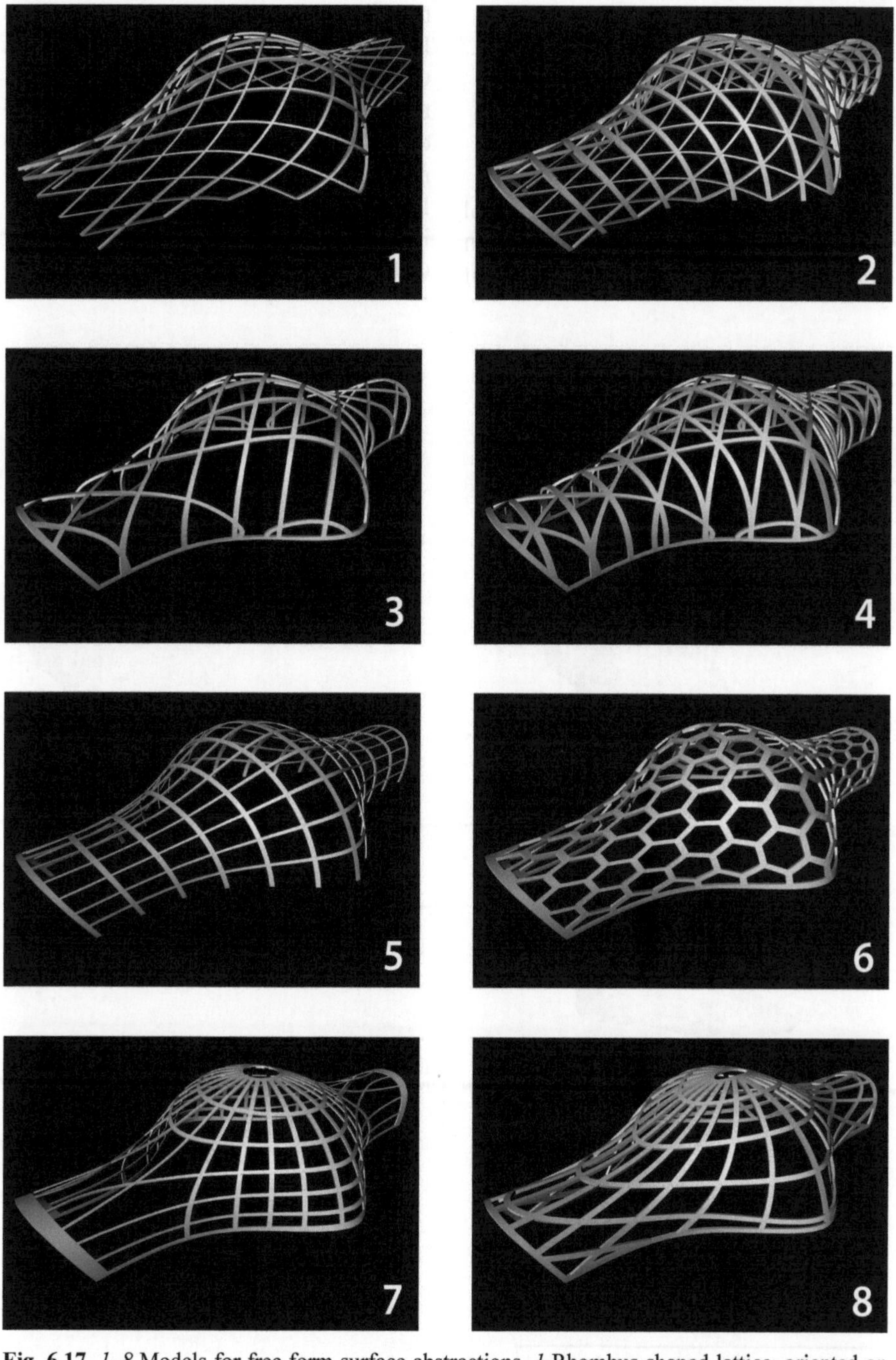

Fig. 6.17 *1–8* Models for free form surface abstractions. *1* Rhombus-shaped lattice, oriented on internal surface coordinates. *2* Arch construction with diagonal bracing. *3* Projection of a regular rhombus-shaped lattice in plan view. *4* Like *3* with additional arches along the sectional axis. *5* Regular orthogonal lattice along the uv-coordinates. *6* Hexagonal pattern along the uv-coordinates. *7* Projection of a concentric pattern in plan view. *8* Orientation of a concentric pattern along the uv-coordinates

6.7 From Pool Research to Applied Research

Data Processing for Analysis and Construction

The insights supplied from the Pool Research at this stage could have been further developed on various different tracks; however, in this case the focus lies on the utilization of the data for a CAD–CAM (computer-aided manufacturing) process. The processing of the data by the FEM program (SOFiSTiK) runs parallel to its construction in the DXF or IGES format. The translating process was prepared for eventual CNC-driven fabrication (Fig. 6.18).

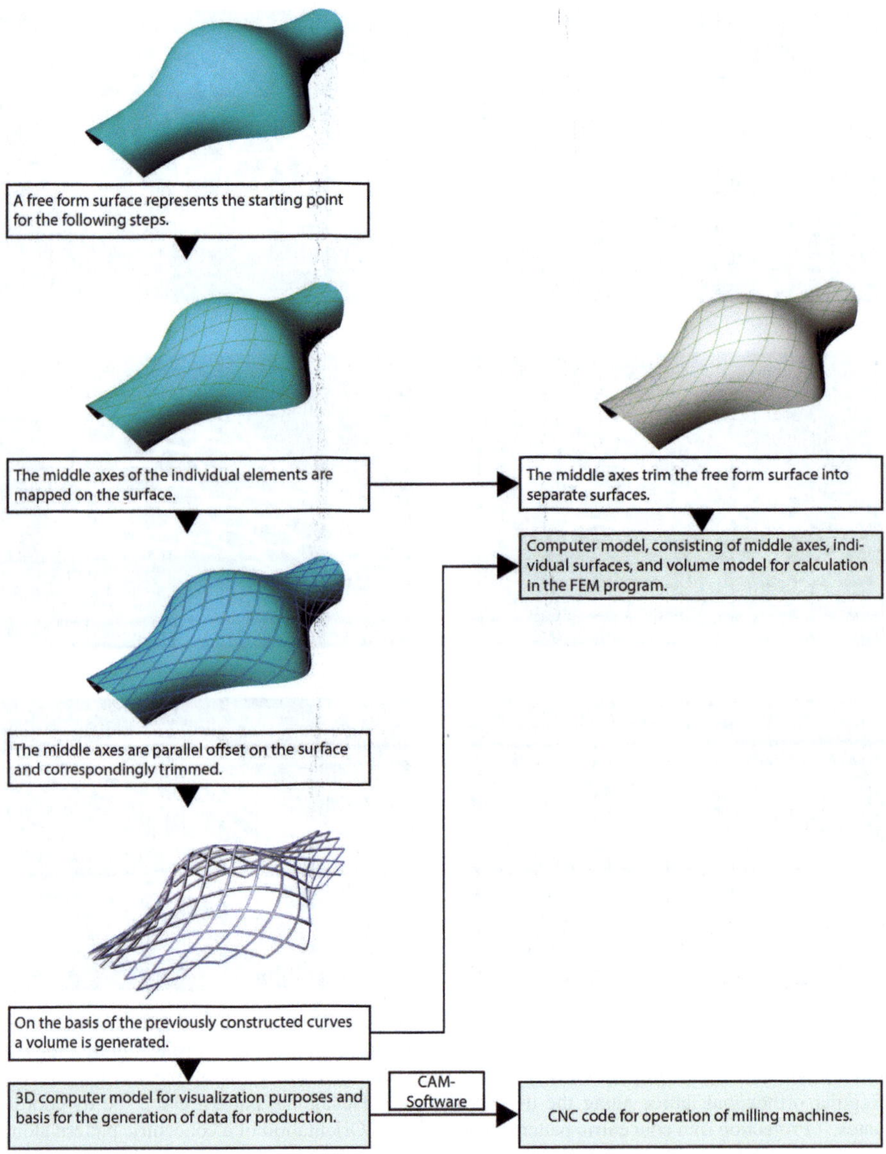

Fig. 6.18 Process scheme of data preparation for analysis and construction

6.8 Generative Design

The technical possibilities offered by computers and software increasingly influence architectural design. With the means of generative design, a CAD process can use the capabilities of modern data manipulation. Parameters describe functional or geometric contexts or requirements and connect the specifications together in such a way to aid the user with design decisions; simply stated, a linking of design, mathematics, construction, and also function. Therein lies the ability to quickly consider and experiment with many iterative variations of development; the number of which depends only on what the user considers to be a sufficiently executed study. As a rule, parametric design tools are used to find optimal solutions or to weigh different design variations. However, these instruments of generative design are not only implemented in the design and concept process, but have also found usefulness in later planning phases, for the visualization of different conditions, and successions of different variations. The parametric descriptions allow a high variability within the dependencies of the given conditions, in which nearly limitless possibilities are generated. For example, path routes can be simulated, spatial confinements described (i.e., building gaps or building space), or static, use-conditioned (necessary free openings, passage ways, etc.), as well as zoning requirements integrated with each other, serving as a "specification guide" for the project model. Further capabilities could be considering various load-bearing systems for a structure, observing different effects of daylight and building transparency, or comparing facade designs and skins (Figs. 6.19, 6.20, 6.21).

Fig. 6.19 Inzell speed skating hall

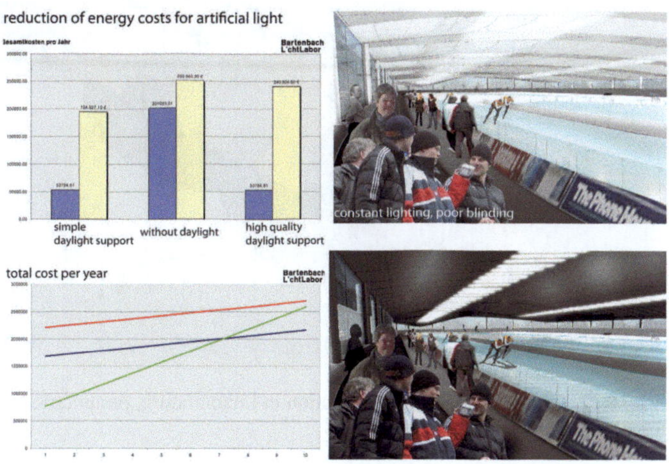

Fig. 6.20 Inzell speed skating hall: typical study of daylight intensities

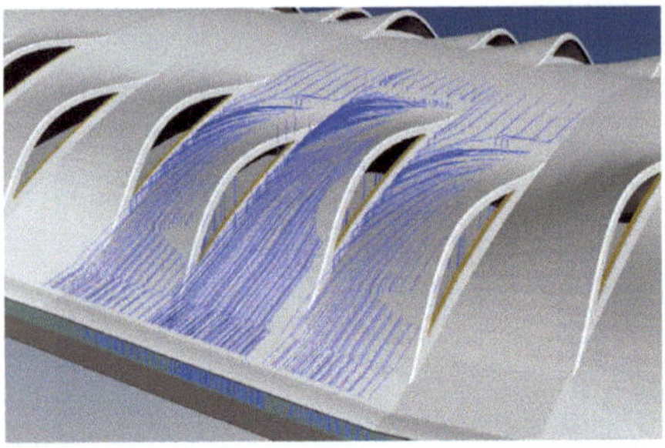

Fig. 6.21 Inzell speed skating hall: typical study of rainwater flow

For the recently constructed speed skating venue in Inzell (Figs. 6.19, 6.20, 6.21) the project team of Behnisch–Pohl Architects developed a large, cantilevered roof form with the use of generative programming and, in cooperation with lighting specialists from Bartenbach Lighting Lab and climate engineers from Transsolar, were able to optimize daylight penetration and indoor climate.

A further example of the use of generative design tools is the competition project for a new convention center and neighboring train station in Luxembourg (Figs. 6.22, 6.23). Pohl Architects parametrically programmed 3D iterations to weigh different functional optimizations and design appearances, whose process is described next.

Parametric 3D Design for the Development of Constructional Principles The roof support system for the exhibition hall of the convention center in Luxembourg is based on a 6 m raster, which could be adapted to the grid pattern of the entire complex as well as to grid of the roof envelope itself. This envelope and its geometric formation had to support the overall design concept of "Solar Plus" pertaining to solar energy production for the complex. Its folded form is a result of the exploration of specific parameters (Fig. 6.24, 6.25), so that the roof surfaces with embedded solar panels are placed at the optimal angle with respect to the Sun; at the same time en-

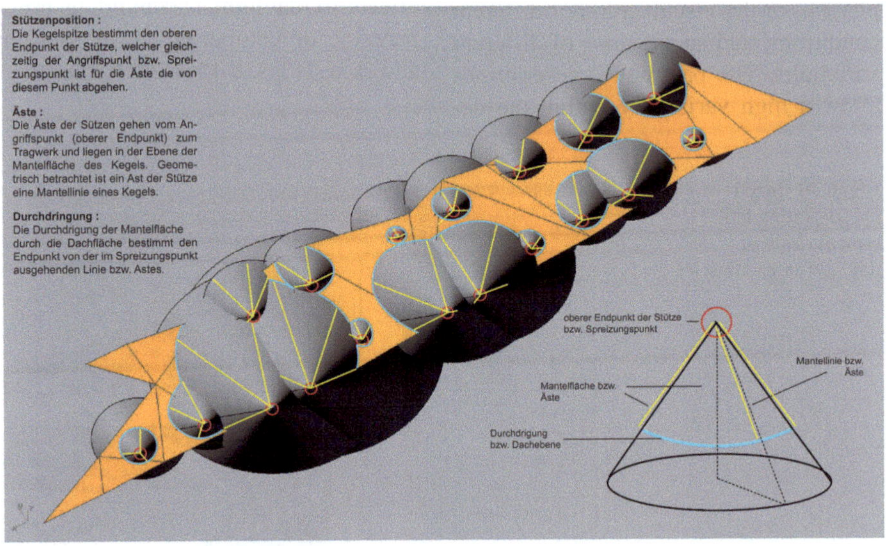

Fig. 6.22 Convention center train station in Luxembourg, parametric model of column placement relating to transportation surfaces

Fig. 6.23 3D model of Luxembourg convention center and convention center station

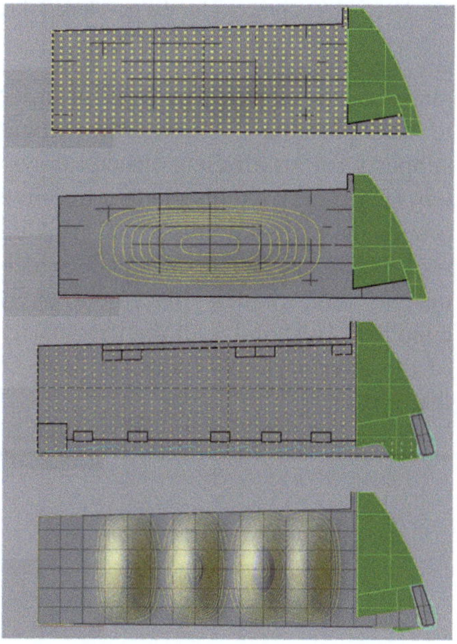

Fig. 6.24 Luxembourg convention center, parametric programming of the envelope

abling the opposite, skyward surfaces to allow the most glare-free, indirect light into the interior. The multifunctional folding shape functions equally as part of the structural system while retaining geometric developability and regularity, so that the outer sealing could be economically implemented and with a high degree of prefabrication. The parametric optimizations defined the final dimensions and slopes of the folds and their basic configuration, owing in part to the fold principles of Japanese engineer Miura. The transition from the horizontal roof to the vertical facades was likewise parametrically modeled and developed in 3D (Fig. 6.25).

An undulating facade for the envelope of a planned high-rise structure as part of the convention complex was also determined using parametric constraints. With the capabilities of the parametric tools, the designers could achieve a unique, shifting facade despite

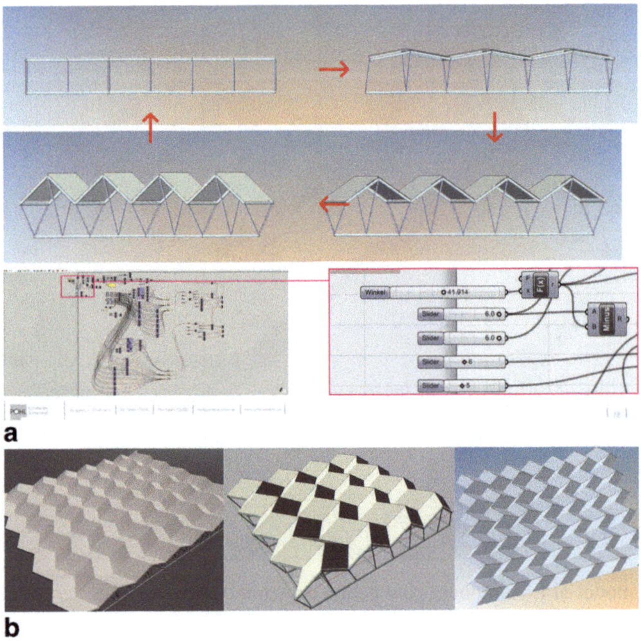

Fig. 6.25 Luxembourg convention center, parametric development process for the envelope folds

demands of structural loads and the economic feasibility of reproducibility and fabrication. The entire building volume was reshaped by these rigorous constraints.

Parallel to the convention center building construction, the user requirements of the neighboring train station, that is, track beds, train platforms, escalators, and footbridges, needed to be mapped and modeled as well, in order to find possible positions for vertical supports for an efficient structural system (Fig. 6.22). In this manner tree-like branching columns under a crinkled roof landscape were designed, which despite their regular placement managed to correspond with the irregularity of the interior activity and the form of the roof itself.

6.9 Physical Models

Following the technical experiments, simple methods were imagined for the realization of biomimetic constructions through the means of physical modeling and without the aid of computer technology, that is, drawing materials, cardboard, scissors, and glue. Contrary to computer models, whose graphic visualization only simulates structure, physical models give immediate feedback to this property. The materials to be implemented for a future building often demand further constraints (i.e., material-immanent characteristics, accessibility, production and assembly costs, limitations and realities of transporting large elements, behavioral tendencies, durability, etc.); however, abstracted physical models are well suited for first approximations of these requirements (Fig. 6.26).

The implementation of biomimetic discoveries into physical models was executed within the framework of a student workshop at the B2E3 Institute for Efficient Building at the School for Architecture in Saarbrücken, shown in these images. The students were tasked with the development of a skin structure for a small pavilion following a precedent in nature. They were to test biomimetic work methods and develop

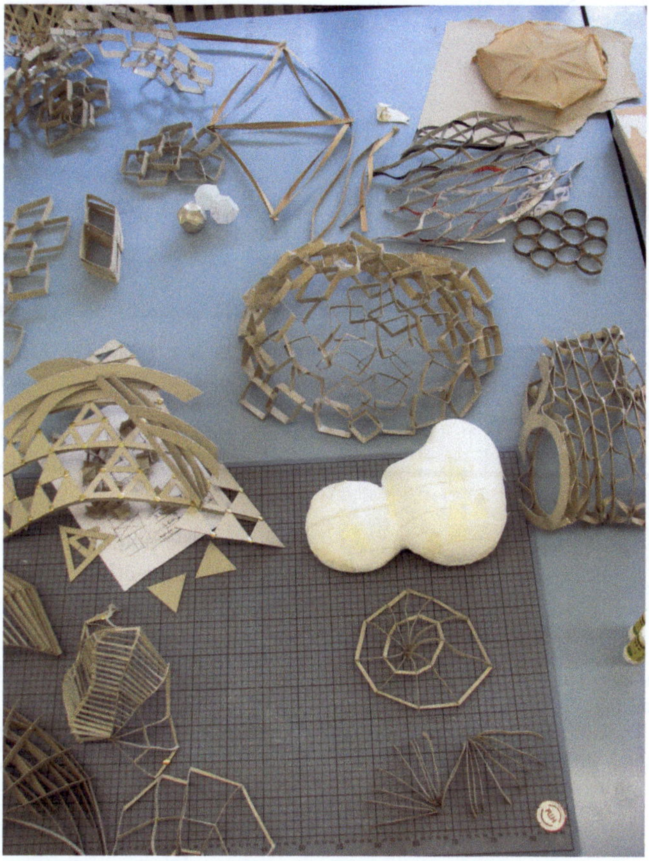

Fig. 6.26 Simple modeling attempts of different forms and assemblies

Fig. 6.27 a–d (*above to below*) a and b (*above*): Working on simple physical models. c and d (*below*): Model by Frei Otto for an experimental structure at the Institute for Lightweight Structures at the University of Stuttgart

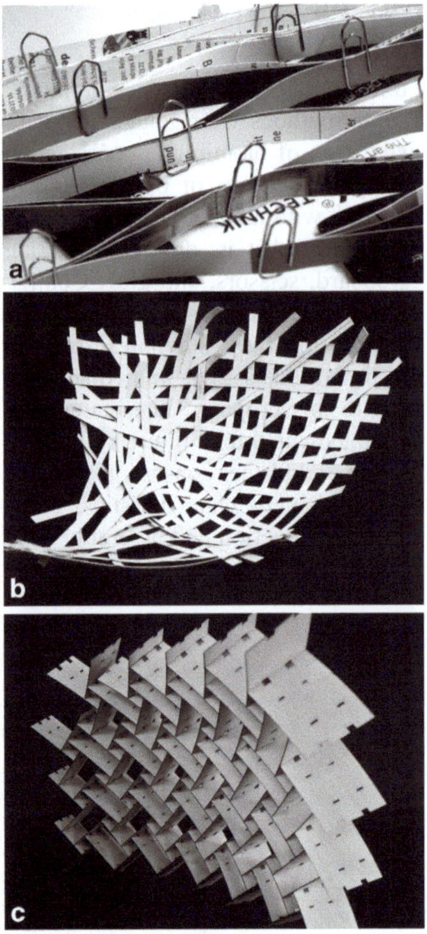

Fig. 6.28 a–c (*above* to *below*) a (*above*): Model of bent paper strips for a free-form structure. b (*below*): Paper strip model of an irregularly woven, dome-shaped skin. c (*below*): Model of offset wood panels

new creative potentials, while researching and abstracting inspirations from biology. Simple designs of physical, analogous structures were then drawn by hand and fleshed out in models of cardboard, wood, or fabric.

The accumulation of insights that underlies this process also suited for students is found in the discovery of "natural" solutions, without the back-

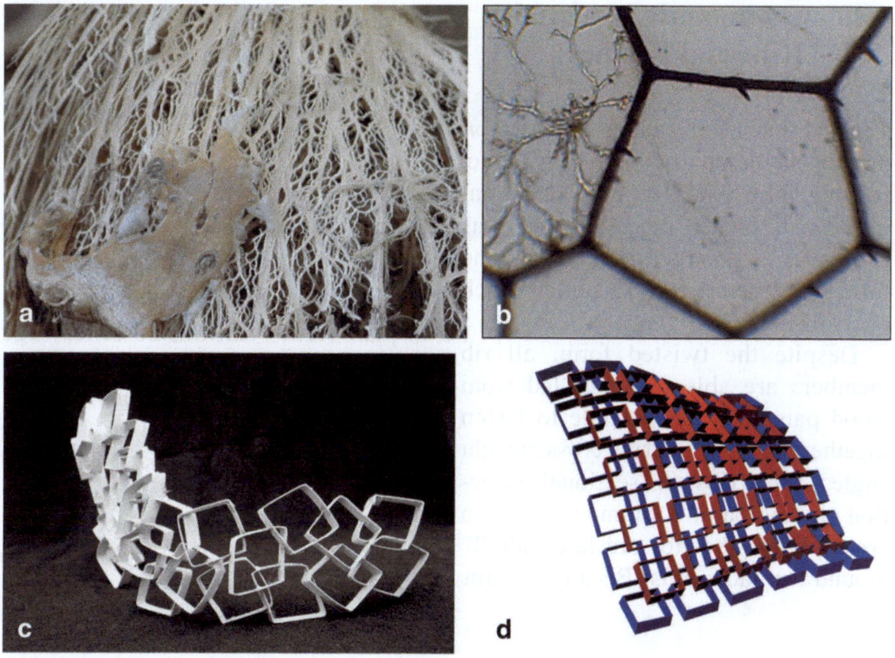

Fig. 6.29 **a–d** (*above* to *below*) **a** (*above*): Sclerenchyma of an *Opuntia*. **b** (*above*): Vein structure of a dragonfly wing. **c** and **d** (*below*): Implementation of a skin structure from the precedent of the dragonfly wing: paper model and computer model

ground knowledge in technology. The biological examples offer recurring design stimulae, give clues to new methods, and are partners in the struggle for creative stimulation.

In the next step, the hand-drawn and modeled discoveries were translated into CAD models. The computer modeling simplified the process of building of complex, 3D physical models. The advantage of this step lies on being able to identify realistic and feasible proportions on the computer and then comparing and considering them with actual, physical models.

This potential of modern CAD systems for developing and testing prototypes was not available for Frei Otto's early fabric and textile models of experimental tent constructions at the Institute for Lightweight Structures in Stuttgart: The modeled form was photogrammetrically displayed and replicated onto further modeling attempts up to the final 1:1 construction. The geometry revealed itself retroactively (6.27 c,d).

As shown in the student work at B2E3 Institute in Saarbrücken in Figs. 6.26 and 6.27a, b and 6.28 and 6.29, biomimetic-inspired building elements were developed for a wood skin. Their complex structures were worked out from simple cardboard models and hand drawings, as well as from intricate 3D CAD models, whose ability for complexity further intensified the designs (Figs. 6.27, 6.28, 6.29).

6.10 Biomimetic Potentials: Ribs and Frames

The additional skin studies are based on the considerations of how a dynamic, flowing form could be produced from individual, planar pieces. Wood was chosen as a construction material. The form and structure recall a curved conch or snail shell.

Despite the twisted form, all rib members are able to be milled from wood panels and are simple to fasten together, as all of the ribs cross at right angles. The ribs in sectional direction are laid parallel to one another; in cross-section the ribs are laid radially around a middle axis. Because of this pattern, the rigidity against twisting is increased.

In architecture there is a series of examples for this type of construction process, though, as a rule, they are often built of ribs perpendicular to each other, as for example in the project "Metropol Parasol" in Sevilla by architect Jürgen Mayer H. In this example, a modeled volume was sliced with an even grid in plan view. With this so-called "egg slicer" method the fiber direction of the wood is often not taken into account, resulting in less correlation between structure and form (compare: Kraft and Schindler 2009. Digital carpentry in: Sabine Kraft et al. (pub.): ARCH+193: Holz, September 2009, pp. 94–95.) (Figs. 6.30, 6.31, 6.32).

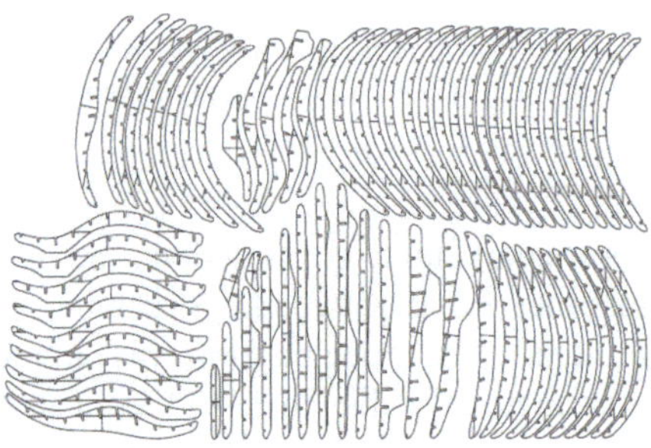

Fig. 6.30 Cutout patterns

Fig. 6.31 Biological precedent: Ribs and framework of *Arachnoidiscus*

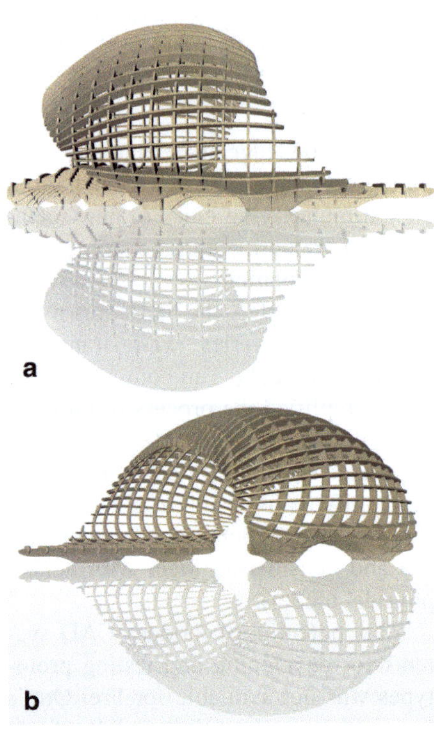

Fig. 6.32 Ribs and framework for a roof structure

6.11 Biomimetic Potentials: Rectangular Frames

The basic form for this construction is patterned according to tortoise shells, which are built up from two layers: the outer layer of scales and the underlying

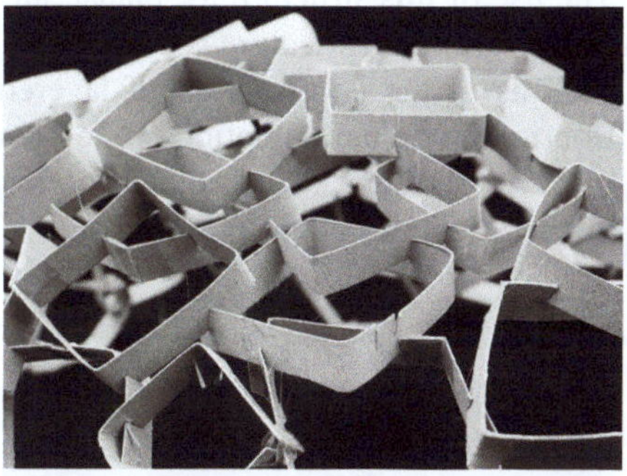

Fig. 6.33 Fragment built of similar square modules

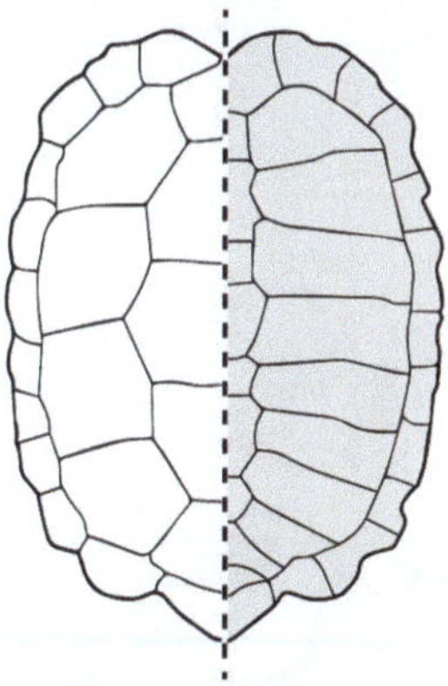

Fig. 6.34 Biological precedent: shell of a marsh turtle, left scales, and right underlying bone plate

bone plate. The stability of the shell is reliant upon this layering (see: Westheide, W./Rieger, G. (pub.), 2010. Spezielle Zoologie. Teil 2: Wirbel-order Schädeltiere. Heidelberg: Spektrum, p. 365). For a constructable interpretation two or more layers of slotted, square frames were linked together to form a stable structure pattern, of which many variations could theoretically exist. Models of some different variations were tested in CAD and cardboard models. The patterns were able to range from nearly regular to chaotic systems. The structure consists of the same elements throughout; the variation lies in the changing positioning of slots that connect neighboring elements. The CAD models were generated partially with the help of "Paneling Tools": Because the panel connection slots intersect at complex angles, a 5-Axis CNC mill is necessary for production, as otherwise the production effort and cost would rise because of increased manual labor. The square frames are then mounted in complete assembly and connected by means of biscuit joints and screws (Figs. 6.33, 6.34, 6.35, 6.36).

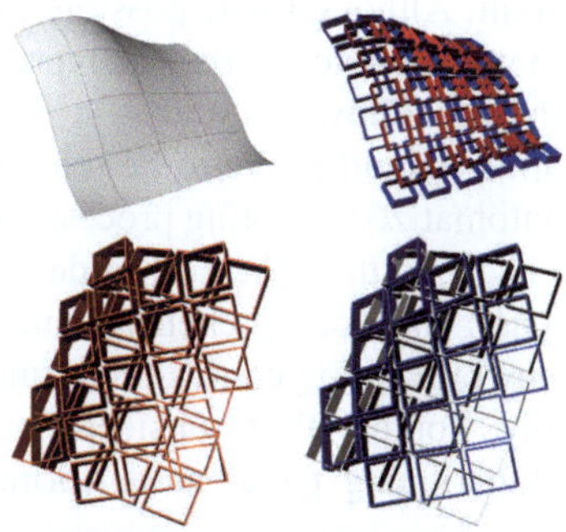

Fig. 6.35 a–d Application of the structures to a free-form surface

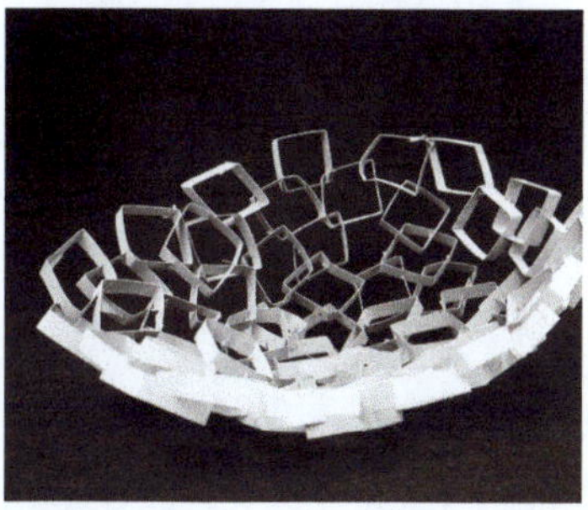

Fig. 6.36 *Dome-shaped shell* of similar square modules

6.12 Biomimetic Potentials: Layered structures

This structure consists of a further development of the earlier described systems. The essential difference here lies in the pieces being bent perpendicular to the surface. The arrangement of the elements in two layers on top of one another allows the surface moment of inertia to be increased. Similarly, the pretensioning of the elements and the layering of two hexagonal grids heightens the stability of the system. For materiality, curvable strips of material, that is, plywood, were considered. The individual parts could be produced with minor expenditure with a 3-Axis CNC mill; however, the issue in this experiment is the application of the system to free-form surface, because each part is slightly different. Although it is possible to develop multi-curved parts—as represented in the following study—a manual assembly would prove to be too costly. An automatized scripting process would lend itself, in this instance, to designate the parts in a comprehensive manner, thereby preventing confusion during assembly. For the first considerations to this functioning paper model principle the connections are kept as simple as possible for implementation in small-scale models. The elements are fixed at the nodes with screw joints; the entire form then automatically conforms itself to the most efficient structural position, a particularly important step for curved surfaces. Only when the structure finds this position can it be glued together. The use of this structure for planar surfaces beyond architecture is imaginable, such as in the construction of layered flooring systems, vibration protection of HVAC systems, and soundproofing in sport complex (Figs. 6.37, 6.38).

Fig. 6.37 Abstraction and model

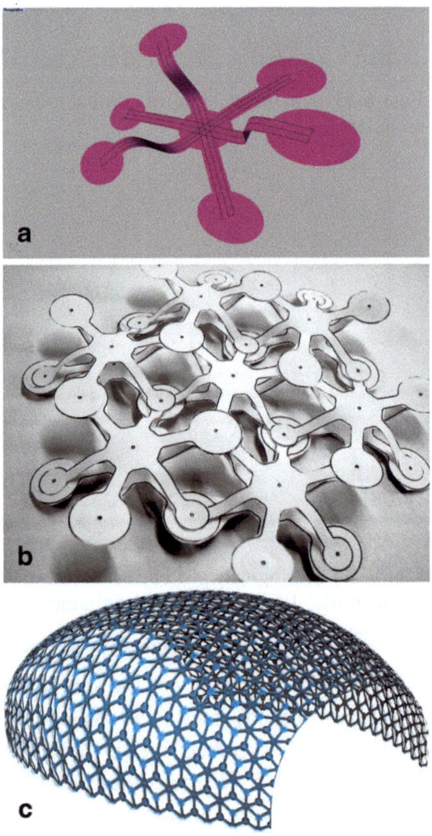

Fig. 6.38 Biological precedent: layered structures with regular geometries in *Isthmia*

6.13 Biomimetic Potential: Offset Beams

The study of offset beam constructions follows the precedent of *Staurosirella* diatoms. The first generated forms were made possible with the help of the Paneling Tool in Rhino. A 3D basis module is generated on a previously given free-form surface. With the input of a duplication with a certain factor (here four) and the establishment of the *X*- and *Y*-axes, an offset of the elements along the surface is possible. An earlier constructed parallel surface determines the depth (*Z*-axis) of the basis elements and, in turn, the structure. A previously defined raster dimension (here points on the surface) determines the dimension of the grid. After several studies a particular variant was chosen to further investigate its structural capability and rigidity in model (Figs. 6.39, 6.40).

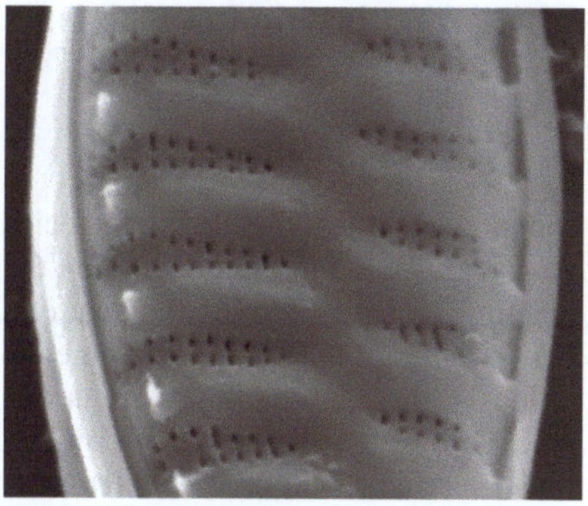

Fig. 6.39 Offset beam structures in the diatom *Staurosirella*

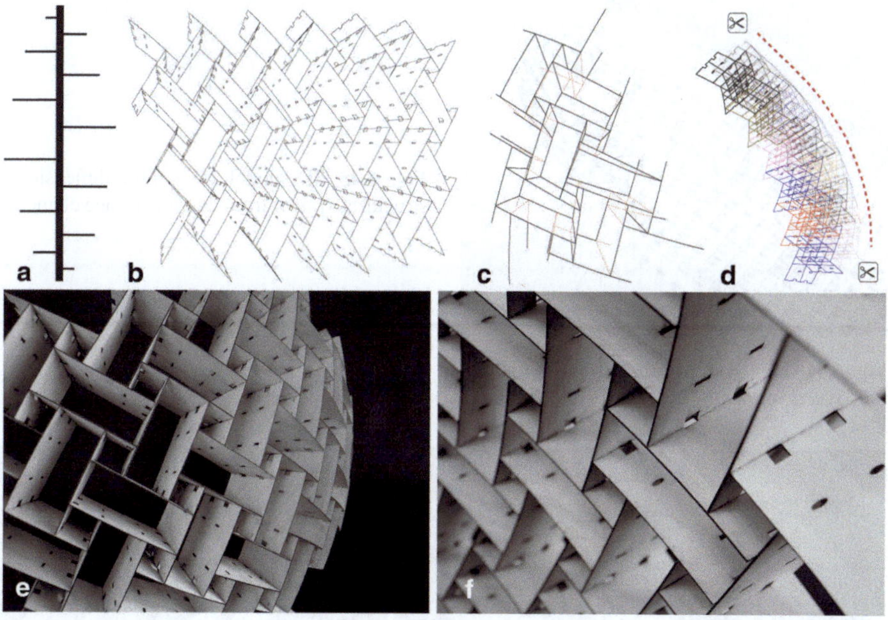

Fig. 6.40 **a–f** Offset surface structures are applied in a truss-like system—abstracted and **e–f** implemented in the model

6.14 Biomimetic Potentials: Incisions and Curvature

The precedent of the diatom species *Synedrosphenia* form the basis for the experiment sequence of bent planar elements. Similar to the other experiments, variously shaped modules were developed and generated on a determined model surface. Using the bent elements, a space defining structural system was created, whose load-bearing members mutually support one another. The surfaces can be constructed from plywood strips and acquire a structural stability with appropriate dimensioning and connections with the neighboring surfaces, as well as through the geometric curvature (Figs. 6.41, 6.42, 6.43, 6.44).

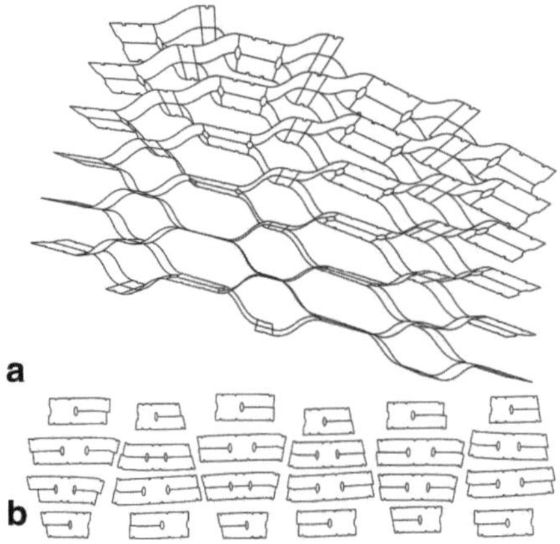

Fig. 6.42 **a** and **b** (*above, below*) Model of abstraction of forked and bent surface-forming structural elements

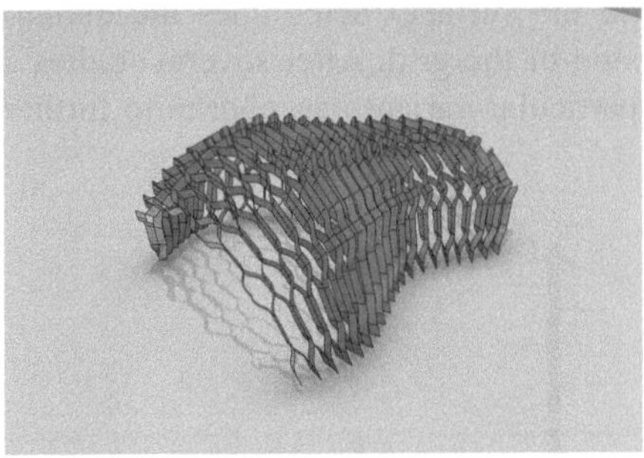

Fig. 6.43 3D model of a three-dimensional structural system from curved surface elements

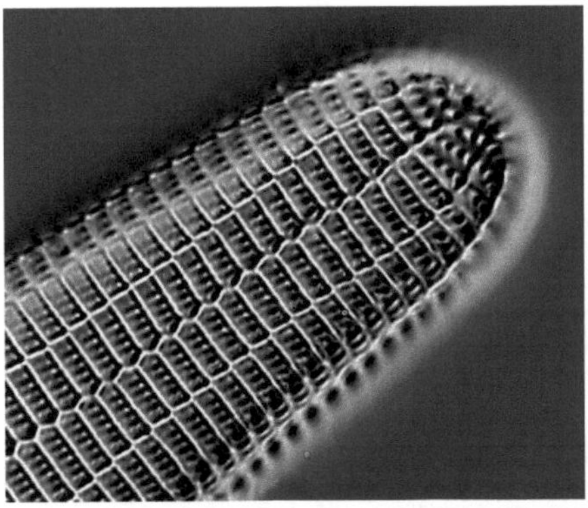

Fig. 6.41 Forked structural system in the diatom *Synedrosphenia*

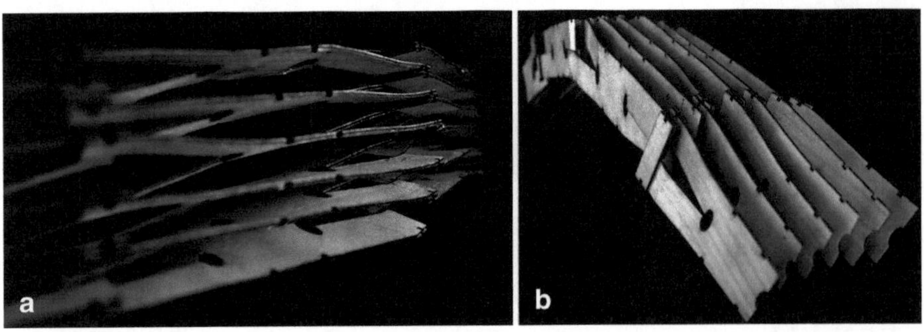

Fig. 6.44 **a** and **b** (*below*) Model in plywood

6.15 Biomimetic Potentials: Curvature

The subsequently described structure is based on the sclerenchyma skeleton of the *Opuntia*, a cactus species. This form is built principally from parallel longitudinal members connected to sinusoidally curved braces, formed from elastically deforming parts that can be held in position by the parallel elements. Particularly well-suited for the modeling of this form is birch plywood, which exhibits a strong rigidity in curvature. This building method is recognizable in many natural structures, notably in fibrous structures, which contain pressure-resistant cells between their tension-resistant fibers and form supporting "pillows" of fluid (Figs. 6.45, 6.46).

Fig. 6.45 **a** and **b** *Opuntia* as precedent for curved elements: *right,* an intact stem; *left*, sclerenchyma skeleton

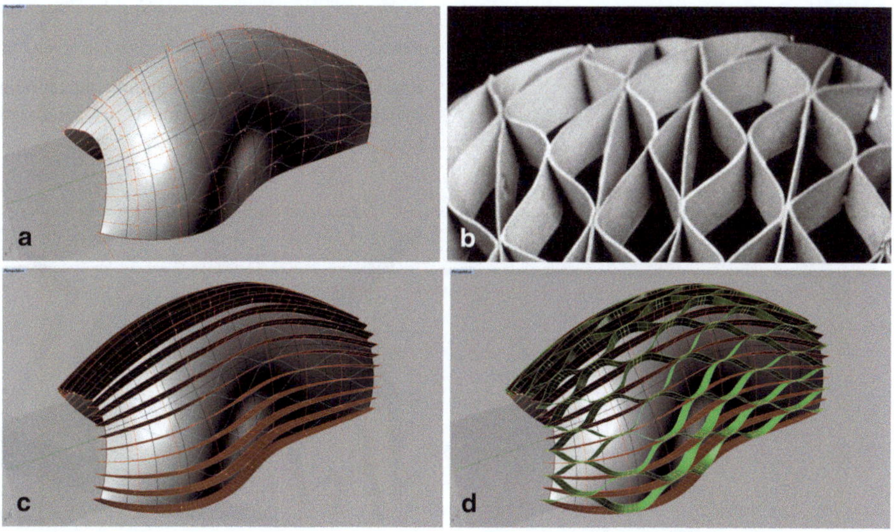

Fig. 6.46 **a–d** Curved structural members. **b** Paper strip model. **a**, **c**, and **d** Modeled in computer on a free-form mass

6.16 Biomimetic Potentials: Hierarchical Structures

The diatom species *Actinoptychus* is well suited as a precedent for hierarchical structures. In this example, ever finer substructures (secondary and tertiary structures) diminish from a larger primary structure. In the biological precedent these structures, on the one hand, take on the role of separating the cell body from an outside medium (sea water), on the other hand the role of mechanical (protection against natural enemies) and static functions (general body structure). The bio-silicate used by the diatom is efficiently implemented and represents an astoundingly stabile framework with a complex spatial 3D structuring (Figs. 6.47, 6.48, 6.49, 6.50).

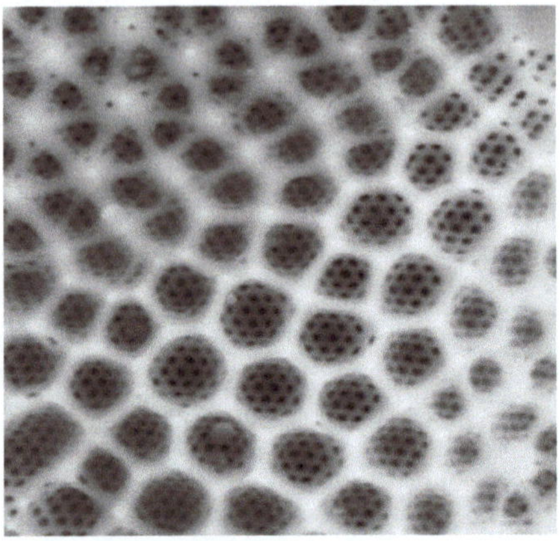

Fig. 6.47 Hierarchical structuring of the diatom *Actinoptychus*

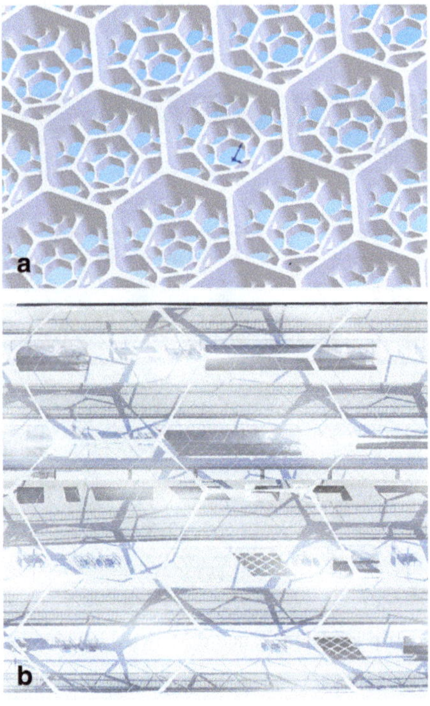

Fig. 6.48 **a** and **b** Biological precedent. Abstracted above. Below 3D model of a roof support system with hierarchical construction, technologically interpreted from the abstracted precedent

Fig. 6.49 Model of abstraction: hexagonal structures form support elements in a sandwich plate

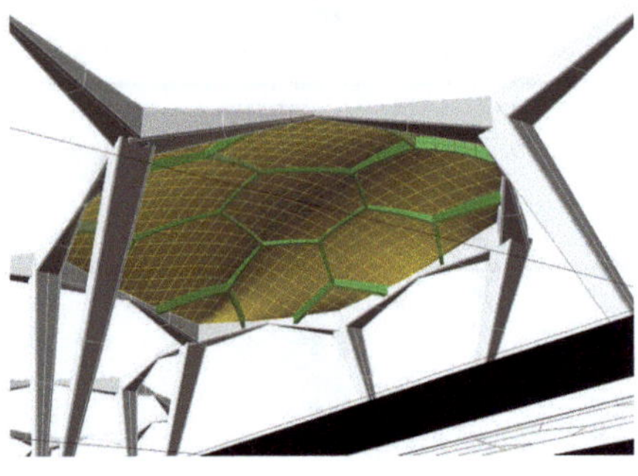

Fig. 6.50 Roof construction modeled in 3D with the use of hierarchical structures

6.17 Biomimetic Potentials: Fold Systems

Folding is a widespread principle found in nature for increasing the rigidity of surfaces. For example, folding structures are to be found in insect wings, tree leaves, and sea shells. Cactuses often have folded forms for, among other reasons, the increasing of surface area.

With the construction of free-form surfaces in a folding system, that is, the technological interpretation of biological precedents, a geometric problem presents itself, which naturally growing structures do not to need negotiate: the surface must be fragmented into planar shapes. The fragmentation of a surface with a regular pattern (tessellation) occurs without problem with the use of triangles, because a plane can always be constructed from three points in space. Various 3D formats are based on this principle. However with four-sided shapes the points do not automatically lie on a plane. With help from the plug-in "Paneling Tools" for the CAD software "Rhinoceros" it is possible to approximately describe a multi-curved surface with quadrangular planes.

The forming of the individual panels represents the next problem. The edges here would exhibit irregular angles to one other, shapes only technically producible with the aid of a 5-Axis CNC mill. This aspect and the building joints in general were disregarded in the subsequent models. The model in this instance merely describes the outer surface of the skin (Figs. 6.51, 6.52).

Fig. 6.51 Foldings in leaves

Fig. 6.52 a–e Paper strip models are able to be developed with the aid of triangle structures. The folded masses are more or less irregularly formed and consist of planar individual surfaces, thus easing a technical realization

6.18 Translation and Technological Implementation in the Example of the BOWOOSS Research Pavilion

Before the potential for biomimetic-inspired solutions can be tested in larger, complex systems, it is necessary to experiment with differently suited building elements in a smaller scale. Simple, single functioning spaces, similar to the first small-scale physical models, are qualified for this task, such as pavilions. These spaces serve the investigation of appropriate design tools and interfaces through the production, selection, and testing of materialities, as well as feedback to their preparation, transportation, and construction, and their overall capability. The test structure is an enterable and experiential ambassador for experimental construction.

The research project and the BOWOOSS Pavilion is a joint project under the funding guidelines of the German Ministry for Education and Research BMBF (Bundesministerium für Bildung und Forschung der Bundesrepublik Deutschland). The research emphasis of biomimetics was promoted by the national government as a high-tech strategy for sectoral drivers of innovation in environmental technologies under the title "BIONA—bionische Innovationen für nachhaltige Produkte and Technologien" ("Biomimetic Innovations for Sustainable Products and Technologies"). In this frame of research, within which the overlapping disciplines of biology, architecture, civil engineering, industrial design, various scientific disciplines of technology, including process engineering and optimization strategies, economic sciences, sociology and other research subjects all converged, renewable raw materials and lightweight constructions played a special role in the area of architecture. The focus on resource-saving building methods was highlighted in all submitted research projects as a special emphasis. Within this context, the subsequently described research project BOWOOSS investigates the subject matter of sustainable building systems of biomimetically inspired wood shell constructions.

6.18.1 The Research Project BOWOOSS as Example for Research and Development

The acronym BOWOOSS stands for "Bionic Optimized Wood Shells with Sustainability."

The research project occupied itself with the implementing of insights from biomimetics for sustainable wood shell structures. Project partners are B2E3 Institute for Efficient Constructions at the HTW Saar, Germany, University of Applied Sciences, chair of building construction Göran Pohl, chair of structural planning at the Bauhaus University Weimar, Germany, Jürgen Ruth and the firm Stephan-Holzbau, as well as Alfred Wegener Institute Bremerhaven and the Lightweight Construction Institute Jena, all based in Germany. For the implementation of biomimetic discoveries the following preliminary considerations were set forth (excerpt from the description of the research project BOWOOSS):

> In view of the growing demands on the CO_2 and energy balance and on the recy-

cling capabilities of construction, building materials and parts from renewable raw materials will become more important. Keeping pace with the climbing number of requests for renewable building materials, the price of these materials will also climb; the availability will be limited by the renewable potential. Material efficiency is becoming one of the prominent themes in the research of lasting building systems.

Fig. 6.53 *Isthmia* with rib and pore structures

Currently, the construction industry uses essentially heavy and bulky building parts. This can also be observed in construction with renewable raw materials, such as wood construction. Materially economical building parts are supplied a lesser role, demand a complex developability, and, in the end, are defeated by conventional products as long as the material savings are so cost intensive.

Fig. 6.54 *Isthmia nervosa* with clearly visible hierarchical structuring

In contrast, contemporary architecture increasingly orients itself on shell-like and biomorphic structures. Along with difficulties of replicating their forms, high production, assembly, and construction costs associated with current shell construction methods are seen as too extravagant and exceed the value of cost savings of conventional building methods.

Material efficiency is in nature the effective intercourse with "expensive" to obtain metabolic products. Nature has developed particularly effective lightweight shell and fold constructions and elements that can grow and are stable nonetheless. Their potential is to be fathomed for technological use. Examples are shell constructions of muscles, urchins, etc., but also fold constructions of surface structures in leaves: hornbeam, various types of palms, etc.

Fig. 6.55 *Arachnoidiscus* with hierarchical structuring of the primary and secondary ribs

The research capabilities with help of biomimetic approaches have the aspiration to attain translatable technological

Fig. 6.56 Form–structure–space–stability: model studies of volumes with the use of folding methods

solutions in construction of wood shells and to durably and economically realize them in the marketplace. Modern form generation and optimization tools are to be applied within the research approach. The numerical translation of these results for fabrication will be likewise computer based (CIM) directly on the basis of the optimized result. An optimizing and complex approach can be recognized with shell constructions in nature and will be modeled and investigated as biomimetic potential for technological derivation.

The building material of wood carries an interesting potential within these considerations both in view of its possibilities in curved volumes and in its material characteristics.

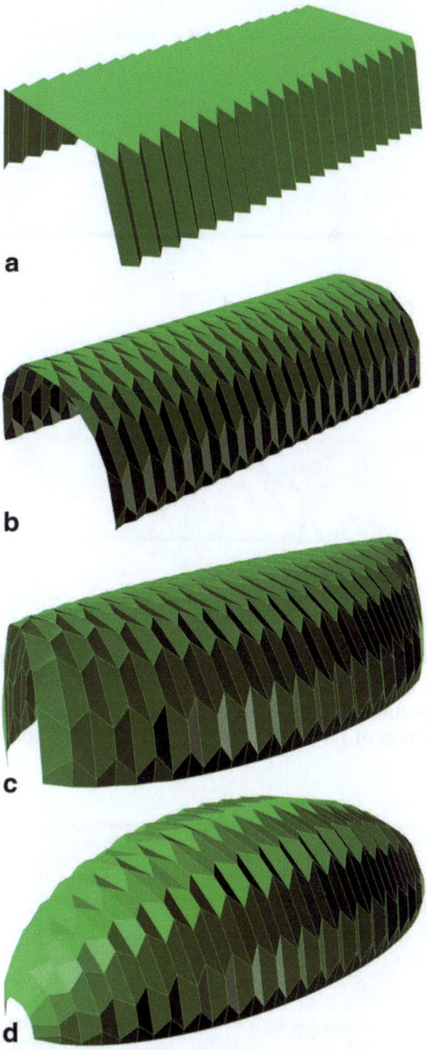

Fig. 6.57 **a–d** Form studies of the BOWOOSS Fold Pavilion: development series from **a** to **d**

Fig. 6.58 Form studies of the BOWOOSS Fold Pavilion: development series

6.18.2 Process Method of the Biomimetics Research Project BOWOOSS

The preliminary considerations were implemented in research project BOWOOSS in the following modules:

Biomimetic Inspiration
The basic form is inspired from many comparative studies and the basis of research insights into the variants of shells in nature and their technological implementation of the diatom species *Isthmia nervosa*, *Actinoptychus*, and *Arachnoidiscus*. The previously gained insights into the biological examples of folding, rib structure, and hierarchical structure appear to be better suited among the other insights for a translation and implementation (Figs. 6.53, 6.54, 6.55).

Envelope—Functions
The envelope provides protection against environmental influences. It is a filter that regulates internal illumination, ventilation, and visibility. An extensive weather protection is, however, not provided in this experiment; the research project is to be developed as a summer pavilion.

Form
The form emerges from the basis of the parameters: number of inhabitants–usage–area–height of space. After various form studies a mirror-symmetrical basic volume was developed for the BOWOOSS Pavilion. BOWOOSS is symmetrical, outward sloping, and tapered in plan.

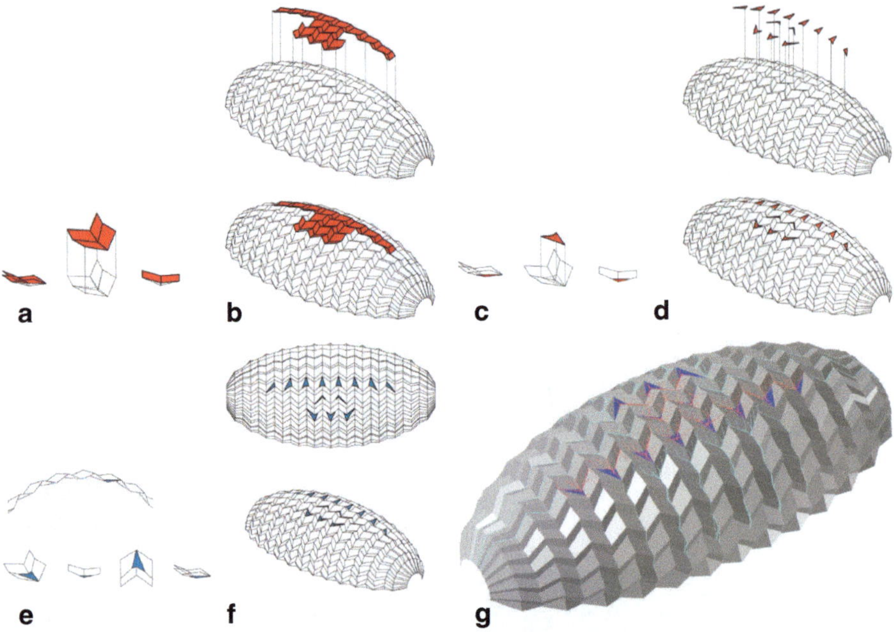

Fig. 6.59 **a–g** (from *top* to *bottom* and *bottom left*) Parametric studies of folding systems for water flow and removal. Thanks to these studies, the problematic instances were localized and improvement strategies discussed (*red*: partial elevation). In result the folds were overall better optimized for water removal. *Bottom left*, the direction of flow of water determined with a computer simulation

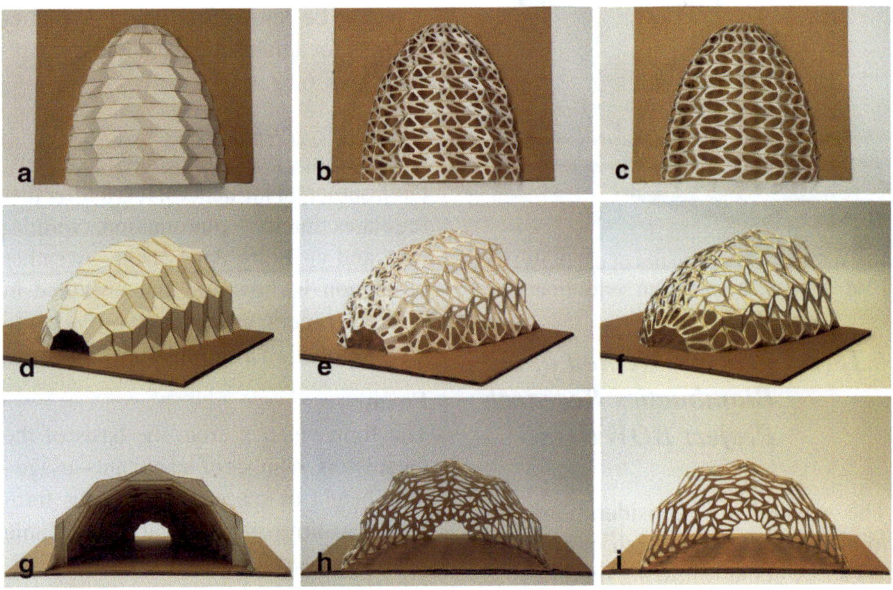

Fig. 6.60 **a–i** Model studies for various opening patterns

System
The shell retains an entrance at the widest point. The halves were pushed apart from one another and completed with barrel vault shell modules. BOWOOSS is flexible in its dimensioning (length) and can be adaptably used and built.

Cost Effectiveness
The highly varying members of the end sections require complex fabrication. A component of the research effort is to gain insight into the interfacing of CAD with CAM, often referred to as "design to production."

Translation to Computer Model
The volume/function studies of plan variations in the computer led to further envelope variations. The resulting, geometrically complex form followed from the background research goal of gaining insights into the realization of geometrically free-formed volumes (Fig. 6.61).

Investigation of the "Ideal" Fold Structure
The simplification of the curving volume into planar surface pieces should bring about repetitions in pieces, thereby easing the fabrication process. Nevertheless, it is expected that a wanted complexity for the computer data generation will remain thanks to the myriad of geometrically different pieces (every fourth piece is identical) and the myriad of differently angled chamfer joints. The generation of various, generic, fold typologies occurs in light of later investigations with respect to structure and functionality (Figs. 6.57, 6.58).

Functional Comparisons
Water drainage and water stagnation are subject matters to be investigated, even though the pavilion is not to be weather protected in the actual sense and only used as a shade structure and accessible space (Fig. 6.59).

Physical Models
In the next steps the computer model developed basis volume was compared with test models. For this purpose computer-aided section models were generated. Different folding patterns on the basis volume were subjected to static calculations and comparison studies for vibration behavior (natural frequency) (Figs. 6.56, 6.60).

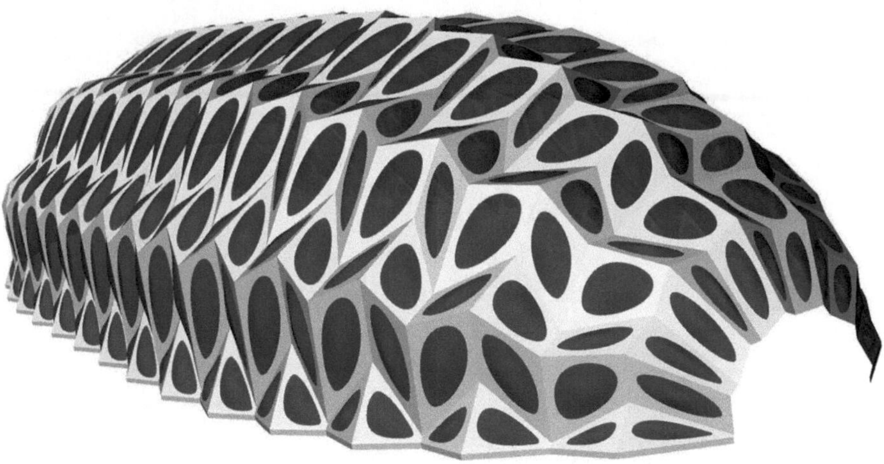

Fig. 6.61 Computer model

6.19 BOWOOSS Research Pavilion: Methods and Results of Building Biomimetics

With much iteration an integrated structure and envelope system was developed within the frame of the comparative analysis and according to the previously discussed biological precedents of *Synedrosphenia*, *Actinoptychus*, and *Arachnoidiscus*. Of the calculational models, one particular combination and spatial arrangement of major and minor ribs was proven to be the most effective. Originally, a "traditional" process of pure structural planning favorited a parallel rib construction (conventional frame construction), but was discarded after screening and investigation of the biological precedents and testing of their verifiable system improvements. The

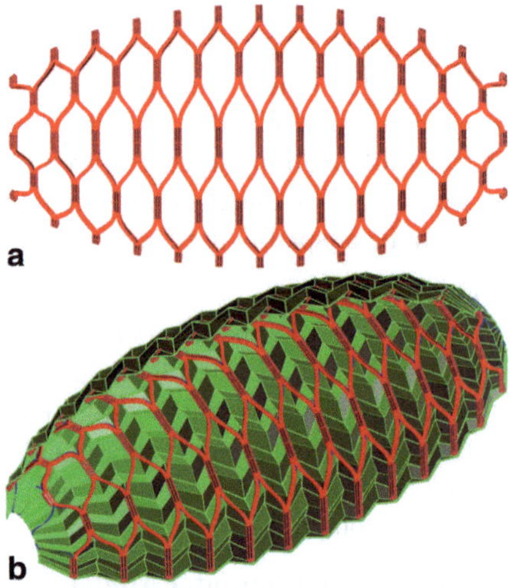

Fig. 6.62 **a** and **b** Structure and envelope system form an integrated shell

rib supports consist of shaped, laminated wood elements, which form the main and subsupport beams yet also work in spatial harmony. Each beam is tapered

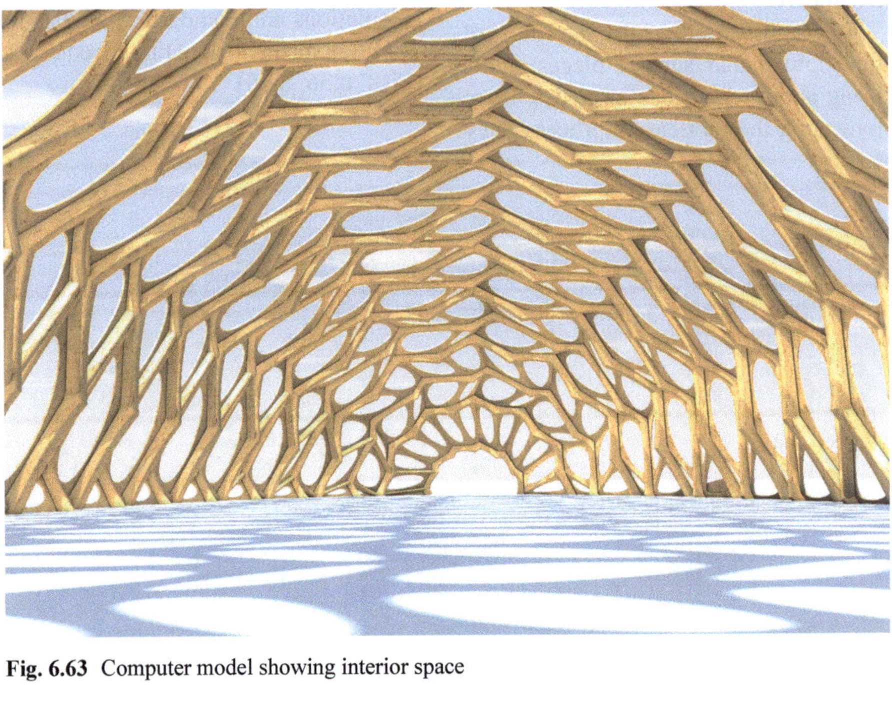

Fig. 6.63 Computer model showing interior space

in the middle; those in the area of most strain, at the crossing points, are more heavily formed. This principle follows from natural flexion-optimized growth forms. The "organic" ribs are coupled to the folded shell with 30-mm laminated veneer lumber. Thus a rib-supported, folded shell emerges in a hierarchical network (Figs. 6.62, 6.63).

The hierarchical system is, like *Actinoptychus* and *Arachnoidiscus,* multilayered: pores, which can boast several levels in living precedents, were formed after the arrangement of the rib system. The complex system of BOWOOSS follows the principle of structure and envelope united, as in the biological precedents. The openings in the wood folds of the pavilion are generically determined and optimized: material can be removed in the nonstructural surface areas. The openings allow air circulation and reduce structural load and material, which reveals itself clearly in the lessened weight of transportation and assembly. The sizes of the openings between the ribs were established after static-structural tests. The openings are rounded out for avoidance of stress points, in which maximally sized openings with minimal rounding was submitted to a "mock-up" to test visibility through the structure (Fig. 6.64). An oval opening was found to be the most aesthetically pleasing and structurally sensitive form.

The cutout oval pieces were further reused for interior seating in the pavilion, thereby keeping material waste to a minimum (Fig. 6.65).

Fig. 6.64 1:3 scale mock-up

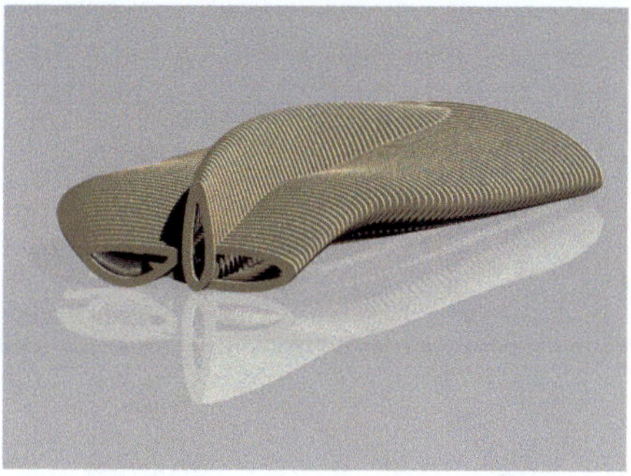

Fig. 6.65 Seating structure of leftovers from cuts

Fig. 6.66 **a** and **b** (*below*) Transporting

Fig. 6.67 Biomimetic-inspired foldings and hierarchical structuring lead to a perforated envelope. Folding, structure, and opening all guarantee stability. Construction and lighting are integrated members of a materially justified building form

The development of the biomimetically inspired support structure and hierarchical system of the BOWOOSS Pavilion led to improvements in methodology. Important were the experiences gained from a computer-generated production of complex 3D data and their further use in CAM fabrication. A frictionless data delivery from the 3D digital basis to the material world of fabrication had needed to be developed and tested; afterward was able to submit to iterative improvements. Production processes for fabrication engineering and fabrication technology experienced a valuable impetus for the future development of software and collaborative education as well as methodology and fabrication technology themselves, which, next to the technological advancements, is seen as high profit. The ability to produce and use demandingly complex geometries was proven time and time again against the backdrop of this experiment. Changes to the final construction were

Fig. 6.68 Looking into the structure

Fig. 6.69 Volume of the BOWOOSS Pavilion amounts to $l = 16$ m, $b = 8$ m, and $h_{max} = 4$ m

Fig. 6.70 View at night

able to enter into the design process on the basis of fabrication, which, instead of compromising and complicating the process, led to an efficient result.

Through biological inspiration the planning process discovered new sources and potentials, which flowed directly into the development of the BOWOOSS Pavilion and will positively influence other working methods. With the biomimetic method "Pool Research," an immeasurable wealth of ideas was gained, whose worth can only be properly appreciated in the implementation of future projects. This wealth certainly affected not only design inspirations, but

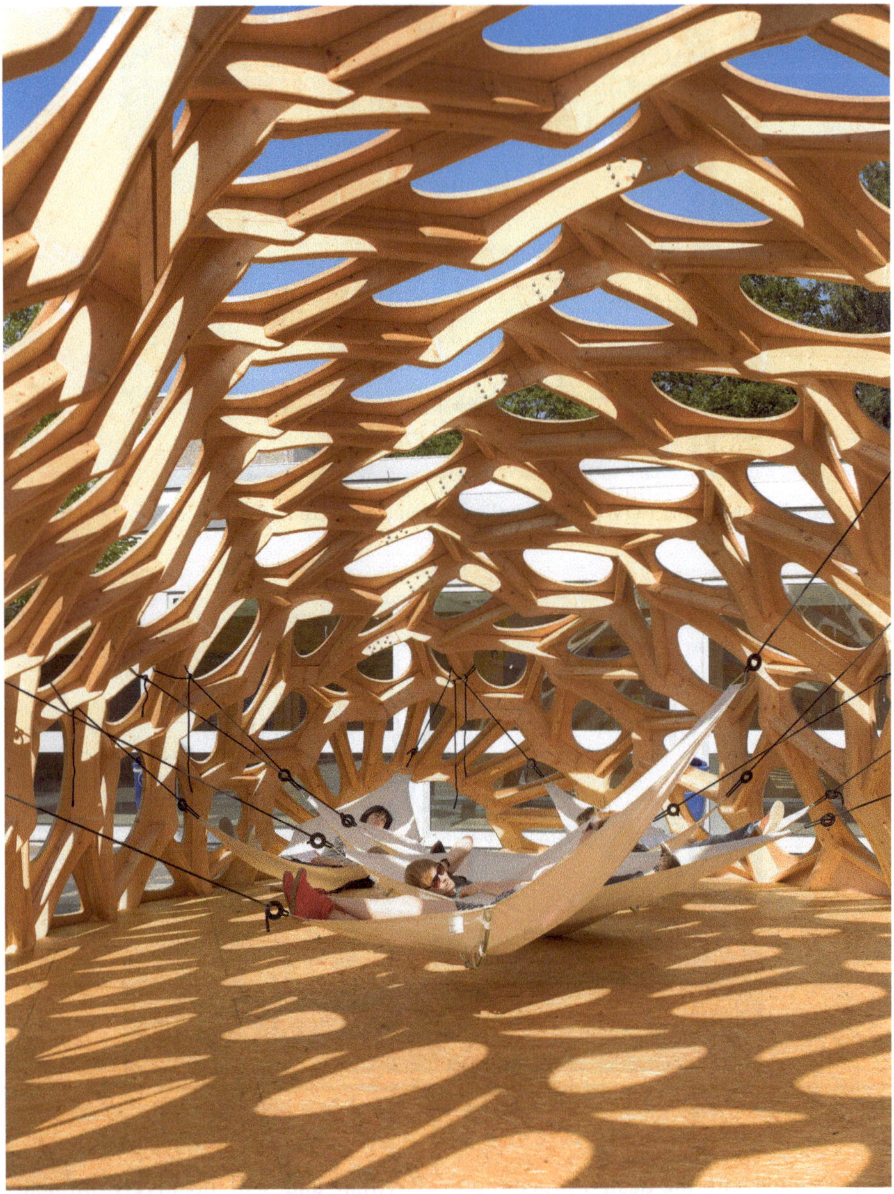

Fig. 6.71 Interior

also in greater measure the knowledge of nearly infinite approaches to solutions for structural and constructional problems in building envelopes. Application lends itself primarily not only to small and large spanning structures, but also to facade structures and in the process of design development. With the goal of a material-efficient lightweight structure and consideration of biological precedents, the research method led to many construction approaches, which

Fig. 6.72 Folding

bred, combined, mutated, recombined, and selected prior developed "species" according to a generative design process. The result is in no way indebted to a linear development process, instead emerges from the basis of the biomimetic discoveries. The material and weight optimized lightweight construction tested by BOWOOSS managed without the aid of steel members, representing an enormous knowledge gain for future shell projects (Figs. 6.66, 6.67, 6.68, 6.69, 6.70, 6.71, 6.72).

6.20 Building Biomimetics in Examples: Biomimetic and Analogous Developments

Now that the preceding sections have addressed what biomimetics can be and demonstrated with a concrete example the process and the ramifications of biomimetic working methods, the following subsections will detail the references stated prior. The contents of these sections are structured in such a manner so that the optimization methods are elucidated first, followed by the subsequent results of research on the subject with examples illustrated by contextual, large-scale projects, and also individually standing, small-scale systems. These subsections unfortunately cannot grasp the entire breadth of biomimetic research. Each quintessential example of a biomimetic development and idea is limited to one double-sided page on account of better summary and comparability. The upper half of each topic is devoted to a brief characterization of the biological precedent, followed by its subsequent abstraction and description of its technological interpretation. In most cases suggestions will be included as to their further development for potential products and tools. The analogous development in technology is likewise illustrated. Importance was laid on succinct visualizations for the retention of clarity. All collected information for further study of the topics as well as authors and photo credits can be found in the appendix (Figs. 6.73).

Fig. 6.73 Giant water lily *Victoria regia* has inspired English architect Paxton to biomimetic developments, in the construction of a greenhouse especially for this species of water lily (1837) and subsequently in the construction of the Crystal Palace (1851) for the World Exhibition in London

6.21 Structural Optimization

Strategies of Structurally-Adaptive Growth

Growth in nature is not only defined in terms of mass increase; growth can be differentially affected by the external effects of structural loads. For example, bones show an increased formation of spongy bone, or trabeculae, in strained regions. In trees, trunk growth is mechanically stimulated; more growth is directed in regions that are particularly burdened. Even the smallest plankton organisms form, in a similar manner, strengthening ribs in their shell cladding for protection against natural enemies

Fig. 6.74 Structurally adaptive tree growth

Structurally Adaptive Growth of Human Femoral Bone

Bones form themselves through adaptive mineralization. It is a materially optimized process: They can strengthen and build themselves up, or likewise reduce mass in particular regions to favorably reduce the weight of the overall bone without compromising the structure (Figs. 6.76, 6.77, 6.78).

The "Soft Kill Option" method (SKO method) from Claus Mattheck was developed at the Karlsruhe Institute for Technology (KIT), Germany,

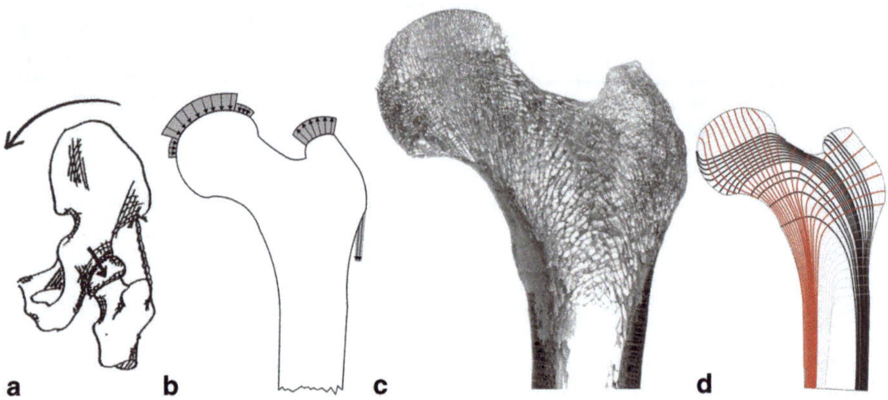

Fig. 6.75 Human femoral bone: **a** edge conditions, **b** structural load, and **c** visualized structure functions (trajectories of major tensions, red pressure, and black tensile)

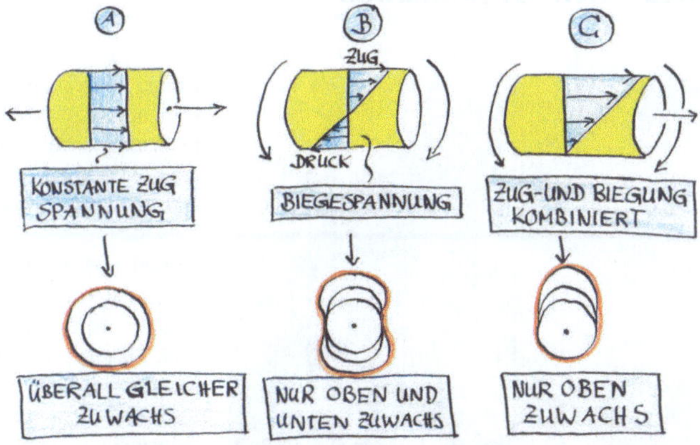

Fig. 6.76 Effect of structural load on a root section of a tree. Equal tension (**a**) leads to radial growth, unequal tensions bring irregular growth. High tensions produce a thicker root in section (*right*). (From KIT, C. Mattheck)

Fig. 6.77 By SKO optimized hooks, student work at the University of Magdeburg-Stendal, Germany

and simulates this principle of adaptive bone mineralization: Heavily burdened regions have increased rigidity; less burdened regions are reduced in mass. By now this method has been acknowledged in science and technology and is used in engineering to develop structurally optimized, lightweight tools and structures with less mass.

At the University of Magdeburg-Stendal under A. Mühlenbehrend, and in cooperation with Sachs Engineering, industrial design students developed designs for consumer goods, optimized using the SKO method. The results, as here illustrated with designed hooks by S. Biller (Fig. 6.77–6.78), distinguish themselves from conventional design approaches through the considerable savings of material, weight, and cost.

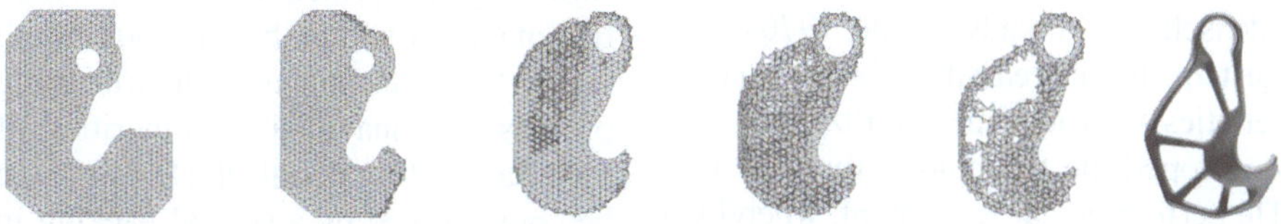

Fig. 6.78 Process of optimization on a heavy-duty hook, student work at the University of Magdeburg-Stendal, Germany

6.22 Self-Organization

Living beings are not simply developed in arbitrary accordance to their genes, but also respond to mathematical and physical principles, which are used to obtain the highest structural stabilities and lightweight structures.

Growth and structure formation in natural development start from genetic determination through assumingly numerous processes of self-organization. Through the millions of years of evolution of a tortoise shell, nature accomplishes the transformation of a rectangular pattern to multi-dimensional, more rigid, hexagonal and pentagonal patterns.

Fig. 6.79 Tortoise shell

The researchers P. Green, Stanford University, USA, and A. Newell, P. Shipman, University of Arizona, USA, showed how macrostructures in the tissues of plants can emerge through self-organizational processes and how they can be simulated with mechanical calculations (based on the Karmansche equation). Comparable approaches can be discovered in nonliving nature, which can be described through mathematical and physical laws. Self-organizing processes arise through the emergence and overcoming of instabilities, so-called bifurcations; they develop then macrostructure formations in the thin-walled shells. The German researcher Frank Mirtsch had already in the 1970s suggested the increased stability characteristics of vault structuralizations and developed the technology for their use. He performed an experiment, whereby a pipe section was supported on the inside by a rigid spring and applied pressure

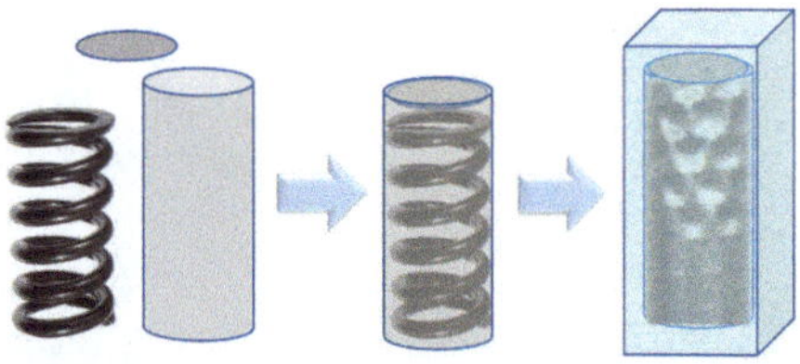

Fig. 6.80 Basis principle of self-organizing, quadratic bulge structuring

from the outside. This action resulted in regular, offset, quadratic structures in the pipe wall (Fig. 6.80). If instead of the rigid spring an elastic support element was used, hexagonal bulges (Fig. 6.81), called "vault structures," would emerge following the principle of minimalization of energy within the thin and smooth walls. In the technical vault-structuring process the unavoidable impurities of material and material thickness must be compensated by a special backing in order to achieve a regular pattern and structure (Figs. 6.80, 6.81, 6.82, 6.83).

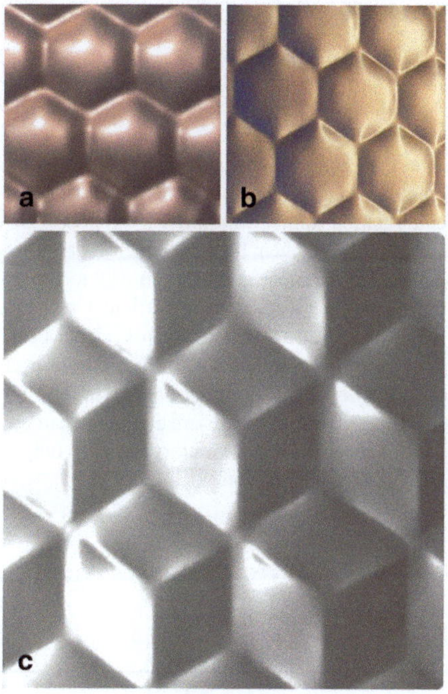

Fig. 6.81 Macrostructure formation in fine sheet metal—vault and cube structures

Fig. 6.82 Utilization potential for vaulted sheets: Dr. Mirtsch GmbH

Fig. 6.83 Utilization potential for vaulted sheets: roofing sheet metal for an athletic complex in Odessa, Ukraine

The calculation of the biological macrostructures in comparison to the technological rests on the same nonlinear differential equations. The essential characteristic shared by natural and technological structural vault formations consists in the occurrence of only flexion and pressure membrane forces on the basis of energy minimizing and self-organization in relation to stiffening fold structures. The necessary membrane pressure for self-organizing vaults in a shell is generated in biology by an enzyme (a harder shell grows faster than interior tissue) and in technological vaulting techniques by a prestressing of smooth material by excess pressure on the outside. In the result the material is, however, not thinned or weakened by the manipulation process, but actually highly strengthened even while retaining its surface area properties. Therefore, long-fiber-reinforced materials can potentially be three-dimensionally strengthened without danger of thread tears in this process that is found both in nature and in technical applications. With such arising technological macrostructures in forms of thin, vault-structured, level or warped walls, applications emerge for surface-refined sheet metal (diffuse, low-glare, light reflecting sheet metal) as well as for sheet metal with stabilizing or tension-equalizing vault-structuring, all without damaging the surface area properties.

6.23 Evolutionary Design

Principles of Evolution

The principles of natural evolution, such as breeding and inheritance, mutation, genetic recombination and selection, lead to species that are adapted to their habitat: either broadly capable, and therefore adaptable in different environments, or specified for a particular niche. A broad disposition in nature often leads to a sum of capabilities beneficial for survival fitness, rather than a development of one particular ability or speediness. Human beings are one such example. Niche specialists on the contrary are often only fit in their particular region.

Fig. 6.84 Indian rhinoceros

Evolutionary Computer Tools
A high creative potential develops itself for design with the use of specific computer tools with generative and generic algorithms. The dynamics of reproduction, mutation, competition, and selection, utilized as strategies of design, find solutions like the natural precedents, that is, broadly capable or niche-adapted. "Morphogenetic Design Experiments" at the Institute for Computer-based Design (ICD, A. Menges) of the University of Stuttgart, "research for the furthering of evolutionary computer tools for the development of performative material and construction systems. Similarly lies the emphasis on the investigation of certain efficiency

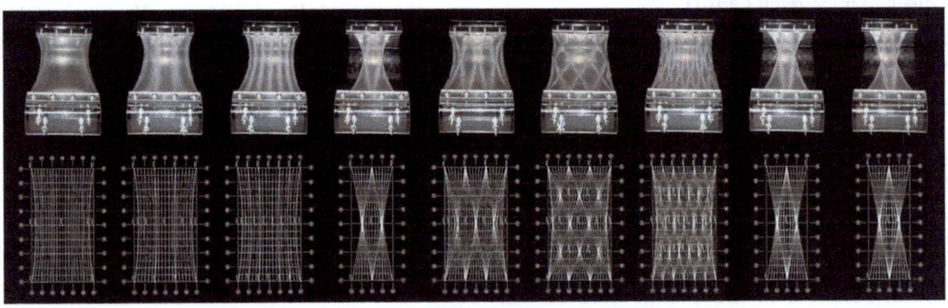

Fig. 6.85 Complexity through versatile morphology with a constant basis

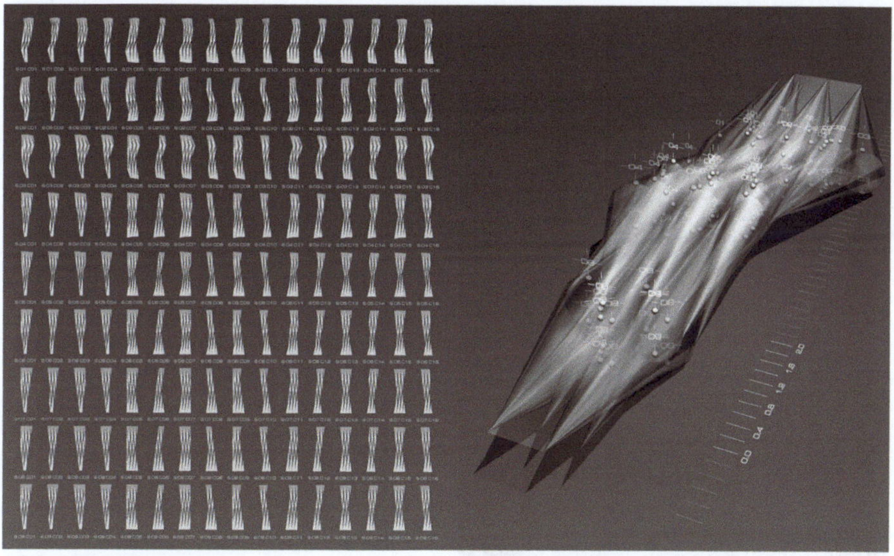

Fig. 6.86 Variants through evolving computer methods

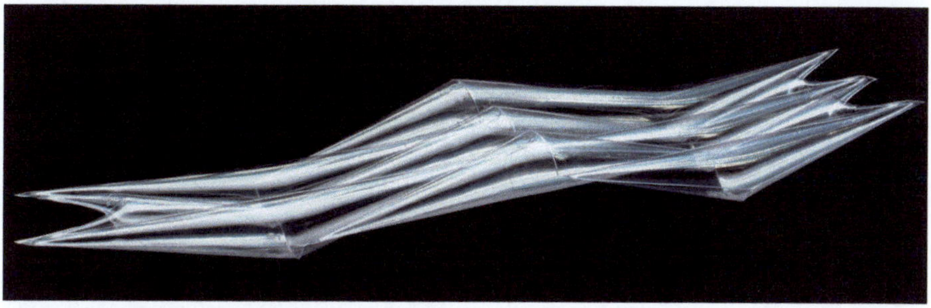

Fig. 6.87 Pneumatic structure

and behavior patterns, which develop themselves automatically in population systems, potentially over several generations." (Achim Menges, Morphogenetic Design Experiments). The studies dealt with the development of a pneumatic module system. Starting from a pneumatic module on a trapezoidal base and constant conditions for pneumatic forms, different ability-criteria were defined. After 600 generations adhering to all of the pneumatic conditions, the evolution process resulted in a number of different systems, thus confirming the creative potential (Figs. 6.85, 6.86, and 6.87).

6.24 Morphogenetic Design

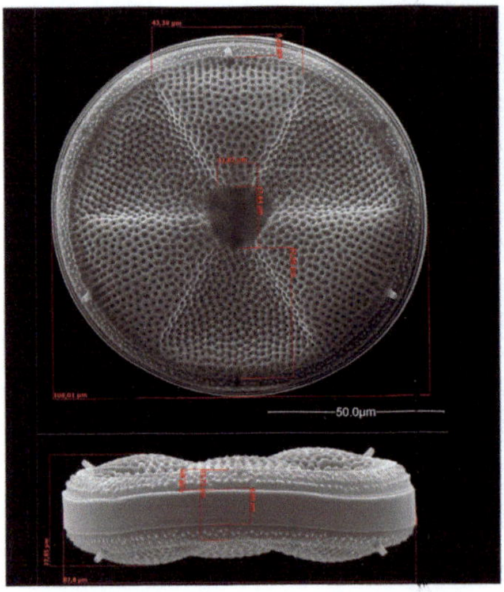

The Greek term Morphe means "shape", "form" and Genesis "beginning", "emergence".

Morphogenesis is understood in biology as the study of formation of structure and form. Morphology describes the macro- and microscopically visible characteristics, from structural systems and organs to organelles and other smaller structure differentiations. The taxonomy of diatoms and plankton life-forms is oriented on the characteristic rib-structures (valves). Constructional criteria consist of simple radially-symmetric structures (Centrales, generally of fossil origin) and more complex multi-polar symmetric structures (Pennales).

Fig. 6.88 *Actinoptychus senarius*

Morphogenetic design is the development of structure and form with consideration to differentiations such as shape, subdivision, and fine detailization. Diatoms, such as the species *Arctinoptychus senarius* studied at the Alfred Wegener Institute in Bremerhaven, show morphological peculiarities in their bio-silicate structures. These peculiarities result in astoundingly stable, yet lightweight structures that exhibit most significantly a material-saving construction even in the details. The accumulated silicate in the skeletons of diatoms must be produced at the cost of food ingestion; therefore, better material efficiency is an evolutionary advantage for the life-forms. The crystalline hulls protect against hunters and are therefore designed especially stable: The more stable the hull, the more protection it offers. The best performative characteristics occur in the combination of the highest protection with the least material consumption. Building elements in architecture, provided that they

Fig. 6.89 *Isthmia nervosa*, detailed capture of the structural membering. (Courtesy of Christian Hamm, Alfred Wegener Institute, Helmholtz Centre for Polar and Marine Research)

are predominantly load bearing and must fulfill protective functions, can profit from knowledge about morphological features of natural structures. In frame of the international research network "Planktontech" of the German Helmholtz Research Association, researchers

Fig. 6.90 COCOON_FS

of the Lightweight Structure Institute Jena (Leichtbau Institut Jena) and practitioners from Pohl Architects abstracted the natural precedents of diatoms and translated them to building elements. The goal was to develop lightweight envelope structures that yield maximal stability for building envelopes with minimal material expenditure. For the realization, fiber-reinforced plastic offered itself as particularly useful, as it is ideally suited for anisotropic construction. The research teams processed the geometric constraints and the materially immanent specifications by computer and iteratively refined them, out of which Julia and Göran Pohl developed the prototype COCOON_FS, an accessible exhibition space, as well as landmark conceptualized for application in various outdoor spaces. COCOON_FS (FS stands for "floating system") has been offered ever since in low volume production for art and exhibition purposes (Figs. 6.89, 6.90).

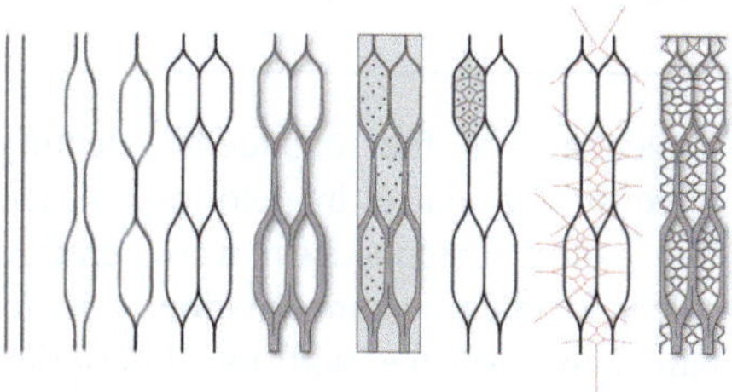

Fig. 6.91 Structural morphogenetic design

Fig. 6.92 Constructional morphogenetic design

The steps of morphogenetic design break down into the following echelons: (a) structural morphogenetic design (Fig. 6.91: generic development steps of the hierarchical facade and envelope structure) and (b) constructional morphogenetic design (Fig. 6.92: Material and fabricational optimization)

6.25 Geometric Optimizations: Sectional Optimization

Column cactus

Cactuses (Cactaceae) are known for their frugality and bizarre growth patterns through their optimal adaptation to various niches. Alongside the thermophilic types there exist as well cold-resistant types, commonly known as "old man cactuses" (Cephalocereus senilis) named after their coat of gray hair, which store certain substances in their water circulation system to deter freezing. Among cactuses there exist also ground-covering, broad-branching types, or some with leaflike and bushy branches (i.e. B. Opuntien), as well as some with considerable growth, like some Cereus.

Fig. 6.93 Column cactus

Collectors in the eighteenth century were so fascinated by exotic cactuses, that some greenhouses were erected solely for their accommodation. A large, ball-shaped species with the description "Mother-in-Law's Cushion" (*Echinocactus grusoni*) was named after the cactus collector Hermann August Jacques Gruson of Magdeburg, who allegedly possessed the largest collection of cactuses in Europe. In South America, cactuses were used for everyday applications (fishing hooks made from thorns) and still today for medicinal purposes and consumption; dead cactuses found application as building material. The Aztecs performed sacrificial rituals on large cactuses.

Today cactuses, with their tall growth and corresponding wind exposure, are of particular interest for scientists in aerodynamics. The species known as "Column Cactuses" can reach above 20 m in height. Sections through the plant show a middle ring of vascular tissue (xylem). From the inside out, the centrally located water-retaining tissue (in barrel cactus up to 1 m) is followed by the chlorenchyma (responsible for photosynthesis); the under-skin (hypodermis) lends the skin a high sturdiness and the over-skin (epidermis) excretes a wax layer (cuticula). Their morphological construction and the form lend the cactus stability (Figs. 6.94, 6.95).

In the investigation of the column-shaped cactuses scientists under Mike Schlaich at the University Berlin have found that these cactuses behave particularly well in wind on account of their

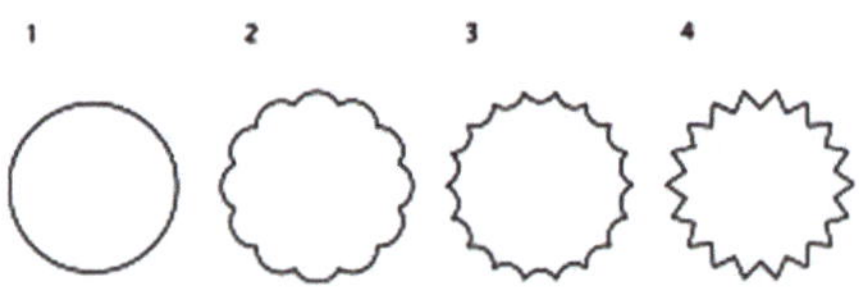

Fig. 6.94 Geometric abstraction of the sectional shape of column cactus

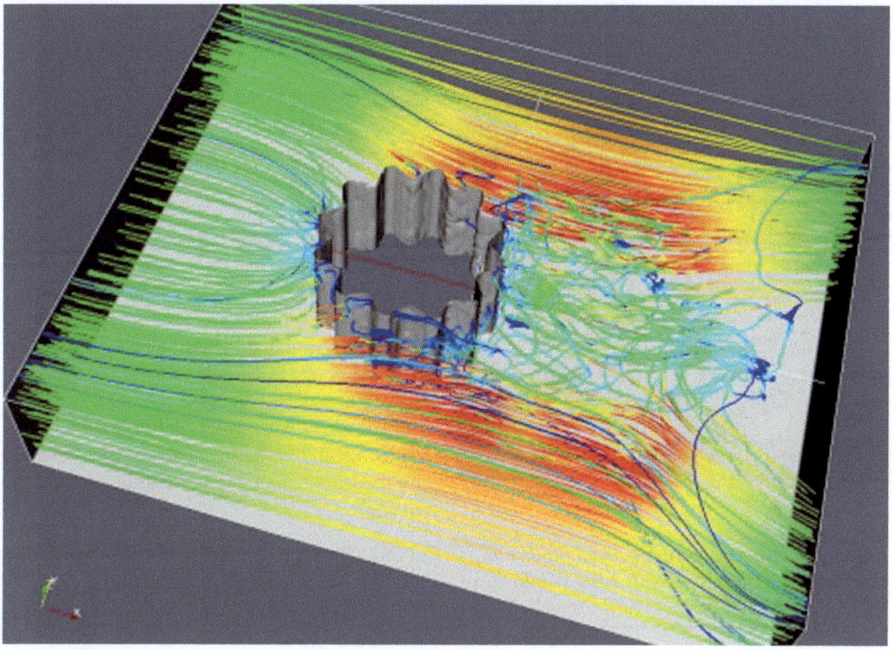

Fig. 6.95 Section of a column cactus in the wind channel model

rib-like formation. The ridges cause boundary layer turbulence in the wind and beneficially influence the formation of eddies and thus the vibration behavior without increasing wind resistance. The sustained wind forces, which the plant structure must endure, are thus minimalized through the geometric form of the plants. These characteristics can be translated for instance to the cladding tubes of cables for cable-stayed bridges to reduce the susceptibility to vibrations. Scientists are also researching, alongside the application to steel cables, the possibilities of application for high-rises that are minimally affected by wind (Fig. 6.96).

Fig. 6.96 Burji al Khalifa in Dubai (SOM). The geometric structuring of the 828-m tall building was defined following the results of wind tunnel studies and shows similarities to the geometric disposition of the column cactus: The arrangement of the building results in a branching of the tower that in turn forms wind eddies to minimalize the occurring wind forces

6.26 Hierarchical Structures

Segmented Silicate Shells

Diatoms build up finely segmented silicate shells for an outer skeleton (exoskeleton) that must be particularly stable: they protect the plankton life form from their natural enemies. For this purpose the shells exhibit rigid ribs and further branching structures (Fig. 6.97). The constructive build of the exoskeleton is structured so that the large ribs are supported by ever finer substructures, resulting in exceptionally stable system. This formation is called "hierarchical structuring". A further task of hierarchical substructuring is the semipermeability for the exchange of substances and foodstuffs, another essential function for the plankton.

Fig. 6.97 Rib structure in *Arachnoidiscus* is compared in literature with rose windows of gothic churches. This "artistic similarity" is a product of lightweight constructions

In many diatom species (Fig. 6.98), the hierarchical structuring of the silicate shell shows hexagonal ribs or round openings in a very geometrically patterned construction in hierarchical gradations.

The investigation of this type of functional construction discovered a strong integration of all substructures for the benefit of a reduced number of main ribs. The abstracted translation to a technical building part is exemplified in the following structure (Fig. 6.99).

The sketched technological interpretation shows the development of a support and envelope structure for a large-spanning canopy following the example of the hierarchical structuring of diatom shells. This roof, developed by Pohl Architects and SteinmetzdeMeyer

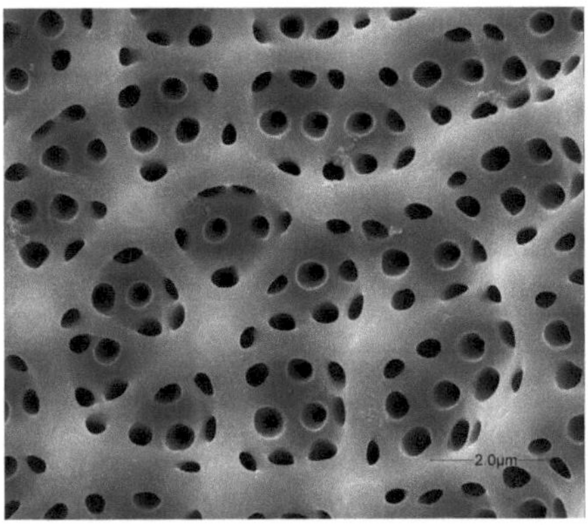

Fig. 6.98 Rib structures of *Actinoptychus*

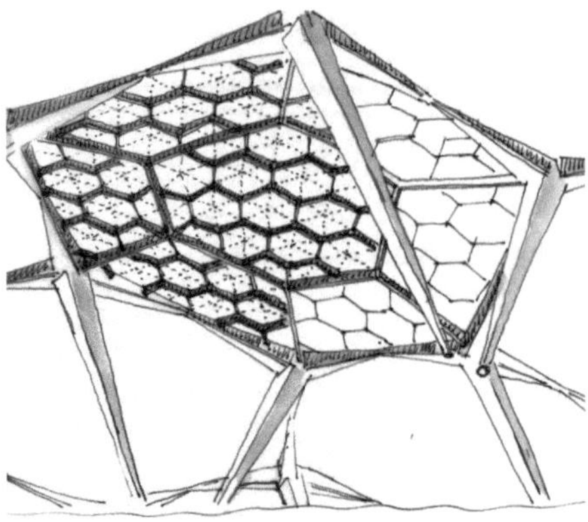

Fig. 6.99 Abstraction, geometric transformation

Fig. 6.100 Train station roof, Luxembourg, Cessange

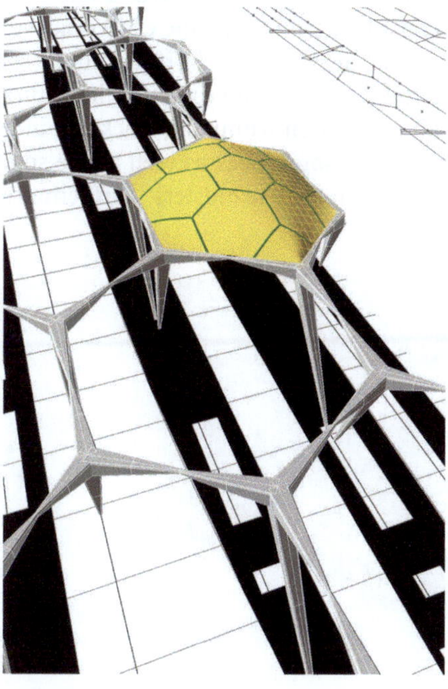

Fig. 6.101 Development of a structure and envelope

Architects with Knippers-Helbig Advanced Engineering in a competition for the design of a roof system, is modularly constructed, so that the size of the roof can be flexibly extended for program demands. A hexagon module provided the basis for the roof system. The spanlengths are negotiated by efficient dimensioning of crossbeams; over-dimensioned, heavy building parts are avoided. Simultaneously the hexagon module enables the flexible geometric adaption to the alternating track and platform distances. Primary and secondary structures shaped spatially to a dome form, so that a pressure-resistant shell structure emerges (Figs. 6.100, 6.101).

6.27 Evolutionary Urban Planning

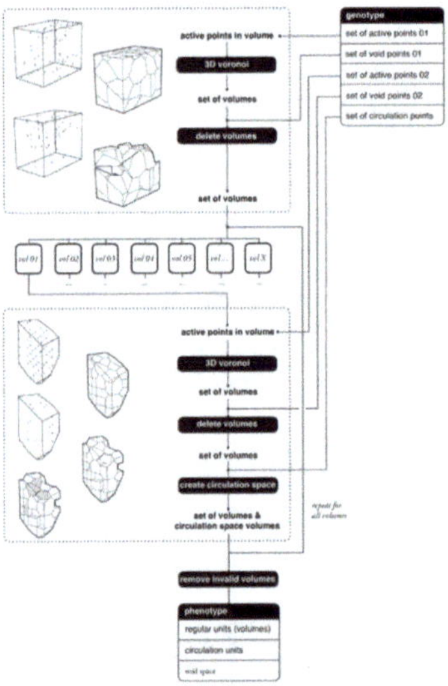

Natural evolution distinguishes itself by the optimizing of genetic individuals for a niche through mutation, recombination and selection. Planners can capitalize on this type of process to produce various solutions to a problem and to be able to compare them with one another under defined parameters. To achieve this process, planners use computer-aided methods (Fig. 6.102) similar to natural processes: they breed, combine, mutate, recombine and select.

Fig. 6.102 Design process of evolutionary urban planning

Achim Menges of the Institute for Computer-based Design (ICD) at the University of Stuttgart, Germany, describes the development of an evolutionary and climate-oriented design process at the scale of the city block: "At initialization approximately 40 random 'genetic individuals' are generated and studied in consideration of climatic criteria as well as the provision of infrastructure. The climate analysis investigates the natural air circulation within the block and individual living spaces as well as the solar entry into the use clusters. Furthermore,

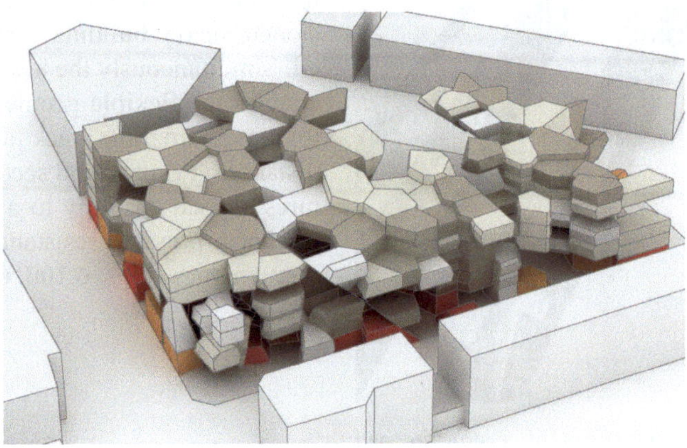

Fig. 6.103 Result of a block with different use-cells, following the climate-oriented conditions

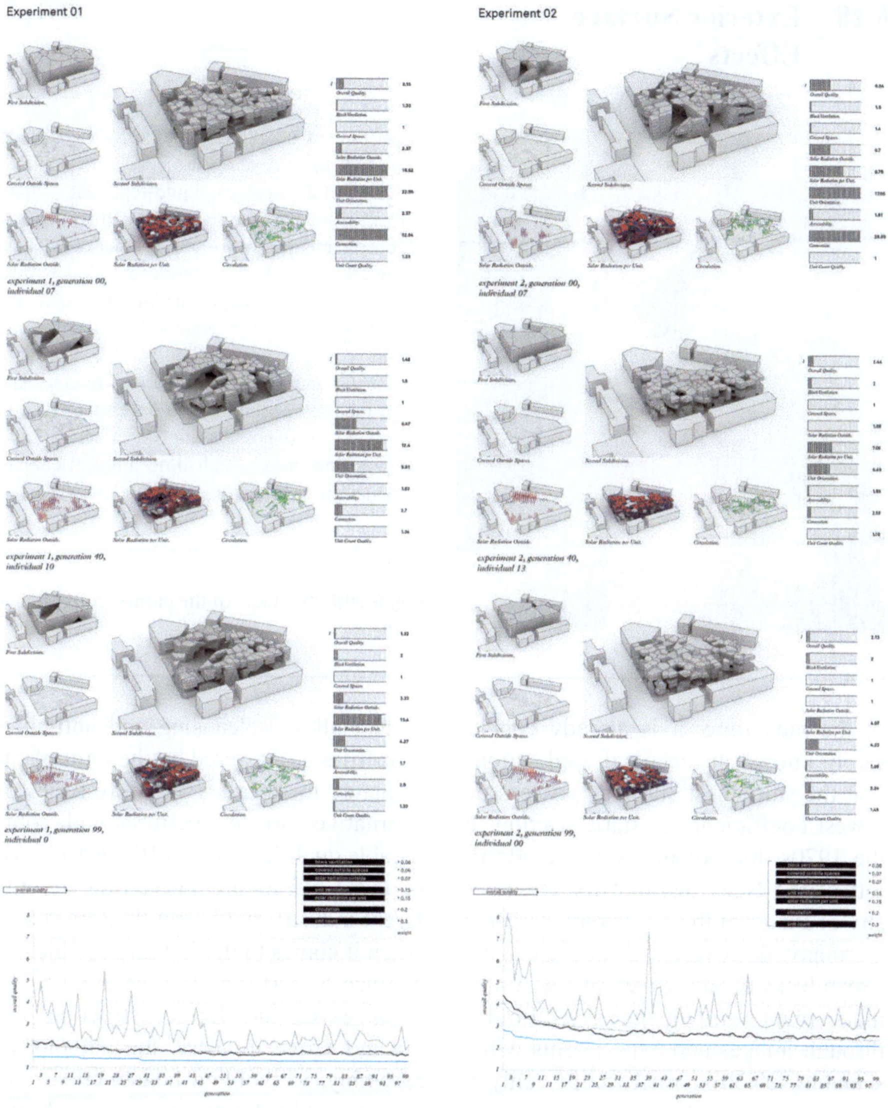

Fig. 6.104 Evolutionary generations of different "individuals"

the quality of public space is evaluated in consideration of sunlight and protection against precipitation. Concerning the infrastructure, the accessibility of the individual units is tested over infrastructure cells. As a result, the structure of the infrastructure here evolves, instead of resorting to common typologies. Additionally, the number of usable units are evaluated and compared with an initial freely definable goal value. On the basis of this evaluation the provided variants are assessed and correspondingly sorted to their fitness in consideration of the described criteria." (Figs. 6.103, 6.104)

6.28 Exterior Surface Effects

Lotus flowers (Nelumbo nucifera) are marsh plants that have developed a special capability: the leaves of the lotus are liquid-repellant, so that water drips off and removes filth. On account of this characteristic mushrooms or other organisms cannot adhere to the plant surfaces. Similar surface effects for the capturing of foodstuffs are formed by the pitcher plant (Nepenthes): they possess a permanently wetted surface on which insects slip and fall into a digestive funnel. The desert-inhabiting ìsandfishî, a species of lizard, and sharks also possess highly effective friction-reducing skins, which allow a low-frictional movement in the mediums of sand and water respectively.

Fig. 6.105 "Funnel" of the carnivorous pitcher plant

In the meantime, it is already conclusively known in technical applications, that the smoothest surfaces possess the lowest coefficients of static friction. In the 1970s, the botanist Wilhelm Barthlott of the Nees Institute for Biodiversity of Plants at the University of Bonn, Germany, discovered the self-cleansing capabilities of the leaves of the Asiatic marsh plant *Nelumbo nucifera* ("Lotus") through images and experiments with a scanning electron microscope, only to be confronted with incomprehension to his assertion of finding a plant surface that is smoother than a Teflon-coated steel panel. Consequently, the now highly endowed scientist had to accomplish long years of work convincing the others of his discovery, so that it could succeed in being technologically translated. Since then the surface properties under the brand name "Lotus Effect" have been an economic success.

Not all self-cleansing and anti-cling properties are inspired by the lotus plant. In nature there exists an entire series of alternative surface structures with comparable qualities (Figs. 6.106, 6.107). All of these effects are interesting for different industry fields and their products, when it comes to the lowest possible adherence to surfaces. Use of these properties exists for ship construction, the air and space industry, the automotive assembly, in building construction, and also generally for the pigment industry (for pigments and coatings). Similarly material scientists have attempted to design surfaces with microstructures for the least amount of friction.

The trap of the carnivorous pitcher plant is equipped with tiny bumps upon which a liquid film clings. Insects then slip on this surface from the brim of pitcher plant mouth into the interior, where it is digested in a nutrient solu-

Fig. 6.106 "Lotus Effect" on a blade of grass

tion. This characteristic inspired Joanna Aizenberg and her group of material scientists at Harvard University in Cambridge, Massachusetts to develop self-cleansing surfaces. According to the precedent of the lotus flower these surfaces should be theoretically superior: The researchers moistened finely dimpled surfaces with fluorinated fluids that can mix with neither water nor oil.

Ingo Rechenberg and Abdullah Regabi El Khyari in the subject area of Biomimetics & Evolutionary Technology at the Technical University Berlin demonstrated with experiments on the "sandfish" (*Scincus albifasciatus*) that its skin exhibits a lower friction than glass or Teflon-coated surfaces: The sand slid off the technical surfaces at a slope angle of 28°–30°; off the preserved skins of the sandfish at 21°. In investigations of shark skins, paleontologist and zoologist Wolf-Ernst Reif noticed under the microscope that the scales possess fine longitudinal grooves that run in the direction of flow. These so-called "riblets" have the resulting effect: The finer and distinct they are, the faster the shark can swim. In the 2010 America's Cup the BMW-Oracle Team competed against the Swiss Alinghi Team. The winner was the American sailboat whose hull was coated with a riblet film. In 1996, the 700-m² riblet film was adhered onto an Airbus A320. A test flight resulted in about 1.5% reduction in fuel consumption, but the films were however not (yet) sufficiently durable.

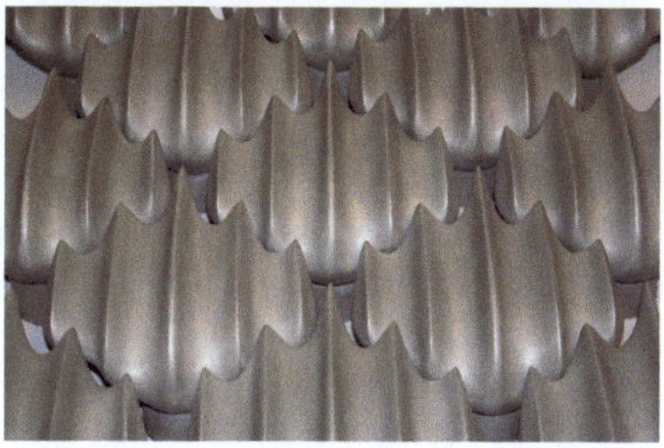

Fig. 6.107 Modeled scales of shark skin

6.29 Fundamentals of Resource-Efficient Facade Technologies

Efficient Resource Usage

Biological organisms have developed strategies in diverse, sometimes extreme climate regions to enable maximal functionality with minimal resource application. The result is not always lightweight or energy efficient construction. Nature is much subtler: natural organisms frugally use those resources that are painstaking to access or transport but achieve at the same time an "abundance" of resources by using locally and easily accessible materials and energy sources. (Fig. 6.108)

Fig. 6.108 Growth rings of a tree stump, with which moist (nutrient-rich) and dry (nutrient-poor) seasons are recognizable

With the research project "BioSkin" at the AIT, Austrian Institute of Technology, research potentials for biomimetic-inspired, energy efficient facade technologies were developed as the basis study within the frame of the promotional program "House of the Future." Susanne Gosztonyi, AIT, stated the difference between technological and natural systems: "Sensory and actuator technology, adapting and filtering characteristics, etc. are the inherent qualities of biological organisms [...]. With adaptive growth and the capability of self-organization [...] a highly complex function system of an organism is developed, which remains in permanent communication with the environment in order to reach an optimal functionality.

Technological systems are, on the contrary, composites of monofunctional, singular components, which form themselves in closed systems [...]. Formation and function cannot react self-adaptingly to changes in conditions."

For the BioSkin study, abstracted interrogations were developed on the basis of conditioned function characteristics for energy efficient and adaptive facades. Analogies in nature were sought to their abstraction for development of technological concepts. The partial results of all stages of development were assembled in catalogs as the basis work for further research and development.

The results of "Bioskin" demonstrated that the methodology of "Pool Research" occupies an important position

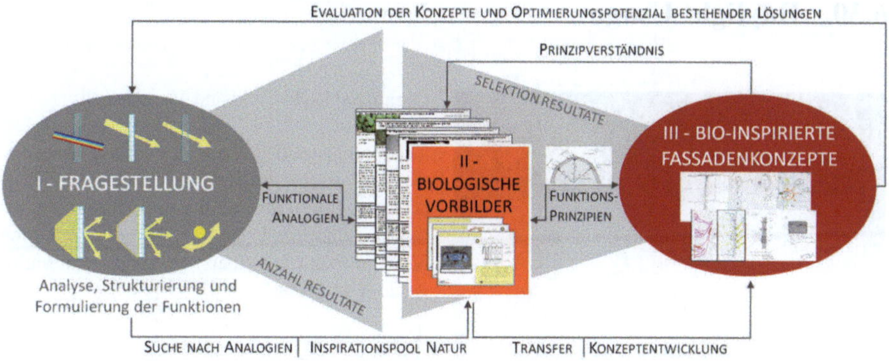

Fig. 6.109 Pool research with BioSkin project

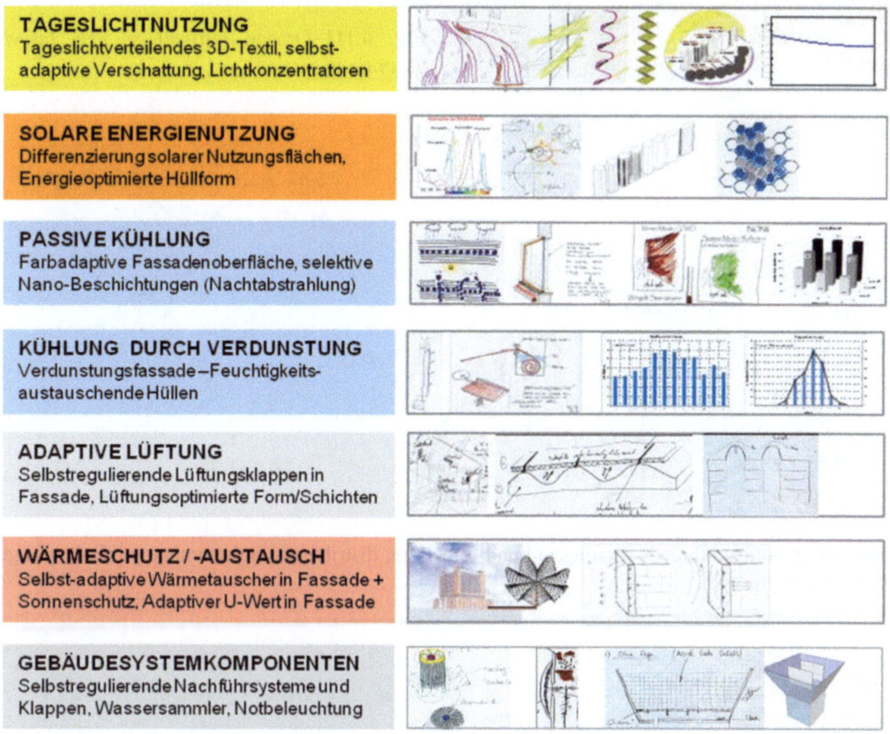

Fig. 6.110 Bio-inspired concepts

among biomimetic work methods. The foundation gained by the research project does not need to immediately lead to application; the principal purpose is to deliver a solid starting point for later developments. This type of "groundwork research" delivers the pool as such for other future product developments to use (Figs. 6.29.2 and 6.29.3) (Figs. 6.109, 6.110).

6.30 Daylight Usage

Optimized Light Yield and Management

The orange puffball sponge (Tethya aurantia) dwells in deeper waters. Their noteworthy characteristic lies in their ability to perform complex light distrubition: Light transfer and distribution occur through bio-fibers, crystals, and multi-mode-light fibers.

Fig. 6.111 Orange puffball sponge *Tethya aurantia* in section

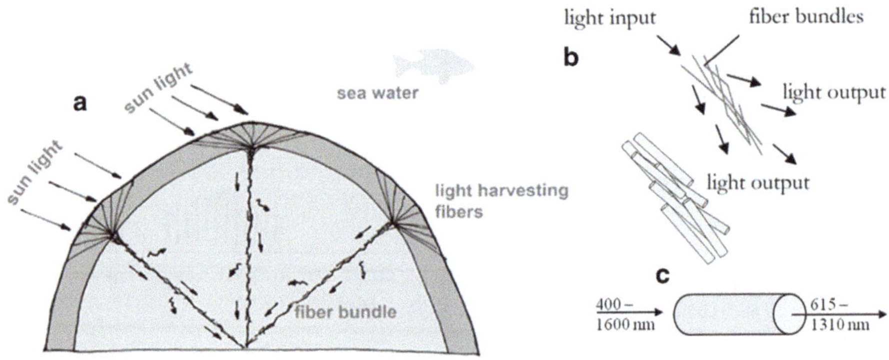

Fig. 6.112 Silicate threads running in bundles as light distributors in the orange puffball sponge *Tethya aurantia*

USE OF SUNLIGHT	Selection of biological principles						
technical functional goal	Colors in structure	Anti-reflective layering	Light transmission through lenses and facets	Light transmission through fibers and crystals	Selective light control using pigments	Structural shading devices	reversible actuator systems
maximum light transmission	x	x	x	x	x		
selective light transmission	x	x			x	x	x
light direction	x	x		x			
Examples of potential biological precedents	Cyphochilus beetle, Hercules beetle	moth eyes	Venus's Flower Basket	Orange Puffball sponge	Compound eyes of insects	Cactus growths	Leaf movements, orientation of the heliconia

Fig. 6.113 Selection of biological principles and precedents for daylight usage

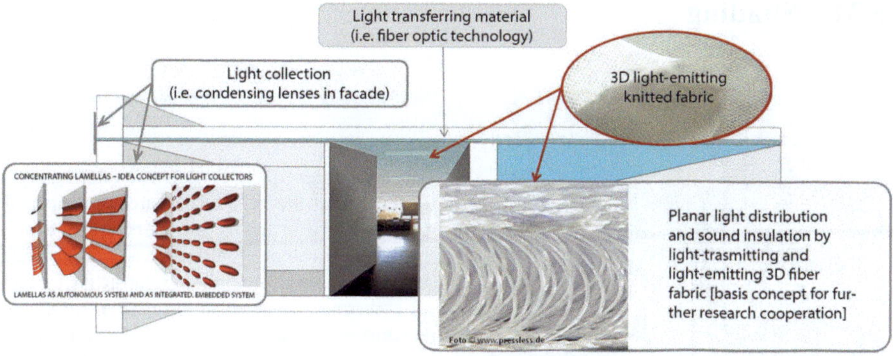

Fig. 6.114 Conceptual idea, light distributing tissue

In the frame of BioSkin at AIT, the sponge *Tethya aurantia* was identified as a potential precedent for day light usage on building facades and tested for possible application areas for building design. The sea sponge uses funnel-arranged, bundled silicate fibers for the collection of light on its outer surface (Fig. 6.111). Silicate fibers in clusters lead and emit light in the interior of its body (Fig. 6.112). The fibers appear to function as a high-pass filter or, respectively, a low-pass filter (Fig. 6.112c).

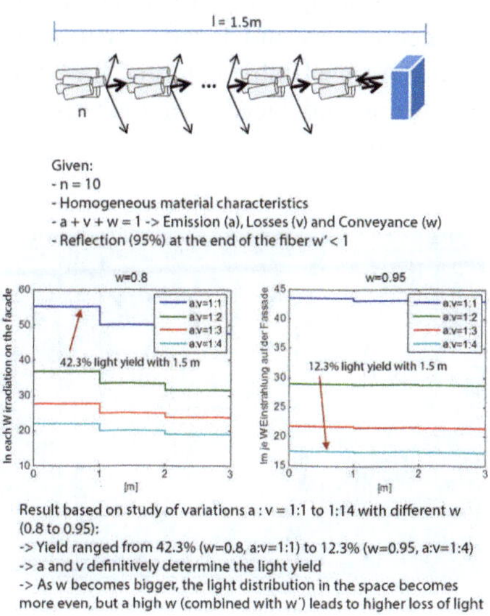

Fig. 6.115 Analysis of the radiance of fiber structures

On the basis of the biological function principles of the orange puffball sponge, a 3D knitted fabric of fiber-based material with light directing capabilities should be able to provide for an even and extensive distribution of natural light (Figs. 6.113, 6.114, 6.115). As shown in Fig. 6.114, component 1 collects daylight on the building surface. Facade integrated concentrators consisting of a combination of highly reflective surfaces and concentrated lens system can be responsible for the collection of light.

These concentrators can, when formed as a sun protection system, represent a multiuse function as well. Component 2 is the actual light leader, consisting of already developed, high-efficient, optic fibers from the textile or optics industry, which directs the daylight over the required distance. Component 3 provides for an extensive and consistent light distribution in the interior space and could even be multifunctionally constructed in a best-case scenario. Further functions like acoustic absorption and heat transfer for thermally activated building parts could be assumed by these fibers.

6.31 Shading

Self-shading for cactuses, generated by rib structures

The cactus can effectively cool itself during night time, because the large surface area of their ridges together with thorns and hairs forms a well insulated envelope. Simultaneously, a strategic balance of overall sunlight to solar-active surfaces (=thermally relevant transmission) is advantageous in the daytime. Cactuses form a variety of differently shaped ridges each according to a climatic condition (Fig. 6.116). Many are elongated; others are in contrast completely round or have leaflike branches, somewhat like the Opuntia. Hierarchically arranged structures like thorns or hairs assume functions of defense against natural enemies or protection against cooling.
A good surface area to volume ratio does not only

Fig. 6.116 Self-shading in cactuses with ridges

A good surface area to volume ratio does not only have influence on building energy efficiency, that is, the measure of the compactness of a building mass, but also on the shading of the surfaces. The overheating of a building can be countered through the optimization of this ratio and its envelope structure.

At the AIT Austrian Institute of Technology, the potential of cactuses was investigated as a biological precedent for geometric optimization of building envelopes with respect to their self-shading qualities. It was determined that the ridged shapes of cactuses function as shading devices for neighboring elements during the day and cooling ridges at night (Fig. 6.117). The thorns or hairs affect the airflows around the plant. The system of ridges, needles, and hairs provides a thermally effective boundary layer for the regulation of temperature exchange.

The studies showed that the shading and energy conversation are substantially influenced by volume geometry. Figure 6.118 visualizes the result of a variation study for self-shading of basic geometries and facades with ridge forms. A translatable potential for building forms and facades of high-rises is sought-after.

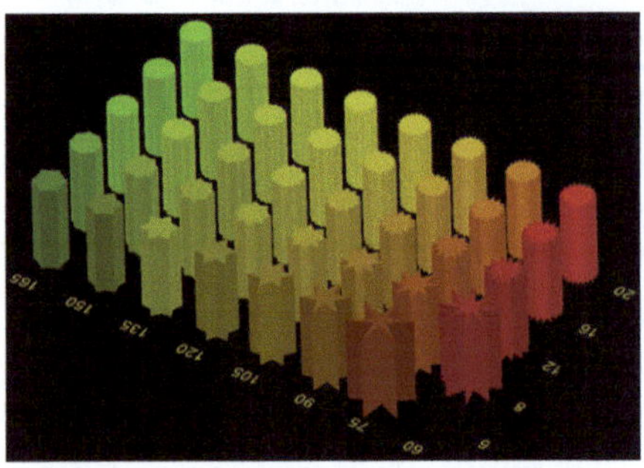

Fig. 6.117 Variations of different building geometries with ridge shapes for shading analyses

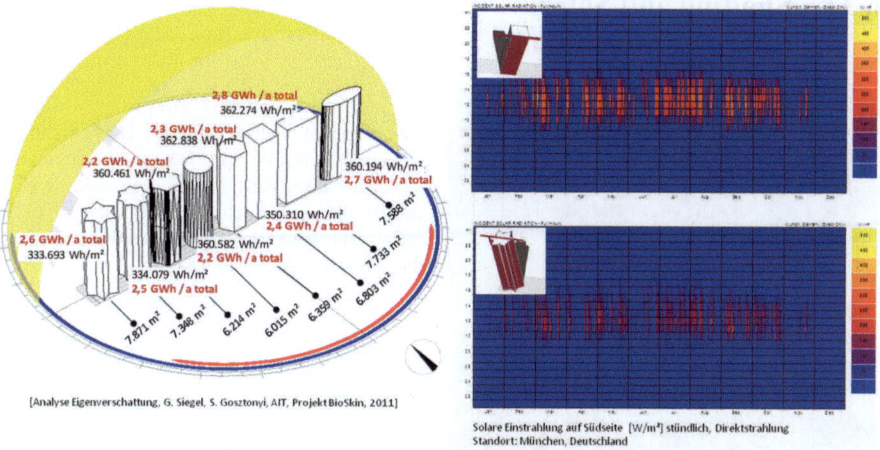

Fig. 6.118 Investigations of basic geometries and facades with "cactus geometric" ridge forms: The potential for interpretation in building forms of taller construction is apparent

The results of the calculational analyses for solar energy potential on different facade surfaces are

- Total irradiation per meter square and year (Wh/m^{2}a): a lower total irradiation is shown by cactus forms than by geometrically simple volumes (lower solar exposure per meter square)
- Yearly total irradiation (GWh/a): The yearly total irradiation is higher by cactus forms than geometrically simple and "not south-oriented" building envelopes (higher solar irradiation over the year)
- The ratio of total exposure/irradiation per meter square each year is more advantageous with a ribbed exterior surface (cactus forms) and south-oriented forms than with classical geometric forms

Potentials of the Study for Building Development

The insights from the studies for self-shading cactus forms can be effectively applied in predominantly hot climate regions.

The cooling effect at night is more efficient, based on the enlarged surface area.

Irradiation during the day is lower on a cactus because of its self-shading ridges than by common building geometries.

The studies were able to prove that geometric base forms have a major influence on solar gain/shading. Cactuses efficiently use these effects. The translation of ridge structures of cactuses to facade surfaces of buildings can lead to climatically ambitious folding facades, especially for buildings in hot and sunny climate regions.

6.32 Shading and Solar Energy Production

The leaves of plants are almost always organized for a maximal solar gain: The specific orientation of leaves effectively prevent self-shading (Fig. 6.119) and simultaneously direct themselves towards sun. Under the leaves, however the opposite is to be found: ideal shading.

Fig. 6.119 Solar energy production in a fern: the leaf arrangement avoids self-shading

---

Fig. 6.120 Shading and solar production with various panel arrangements

Researcher Lidia Badamah, from the research group of U. Knaak at the TU Delft, the Netherlands, recognized the essential organizational characteristic of this system (Table in Fig. 6.123). The adjustable shading system developed by Badamah is adaptively independent of a surface geometry and consists of individual shading panels, which are fixed with an attachment device. The leaf-like elements are arranged on a grid allowing their free movement to follow the Sun's position. The system produces a highly effective shade and at the same time can allow high to maximized solar gain.

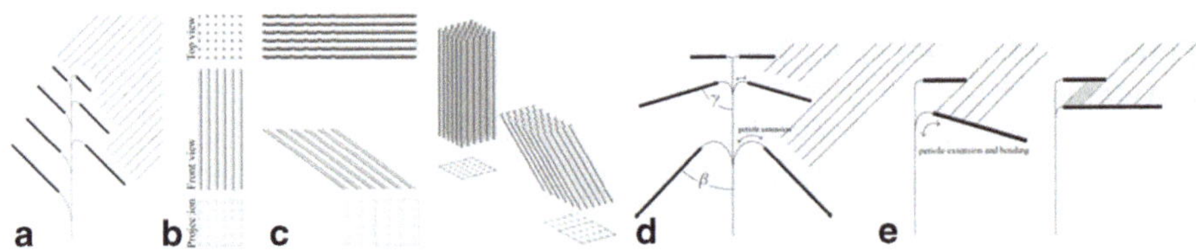

Fig. 6.121 Elevation, section, perspective

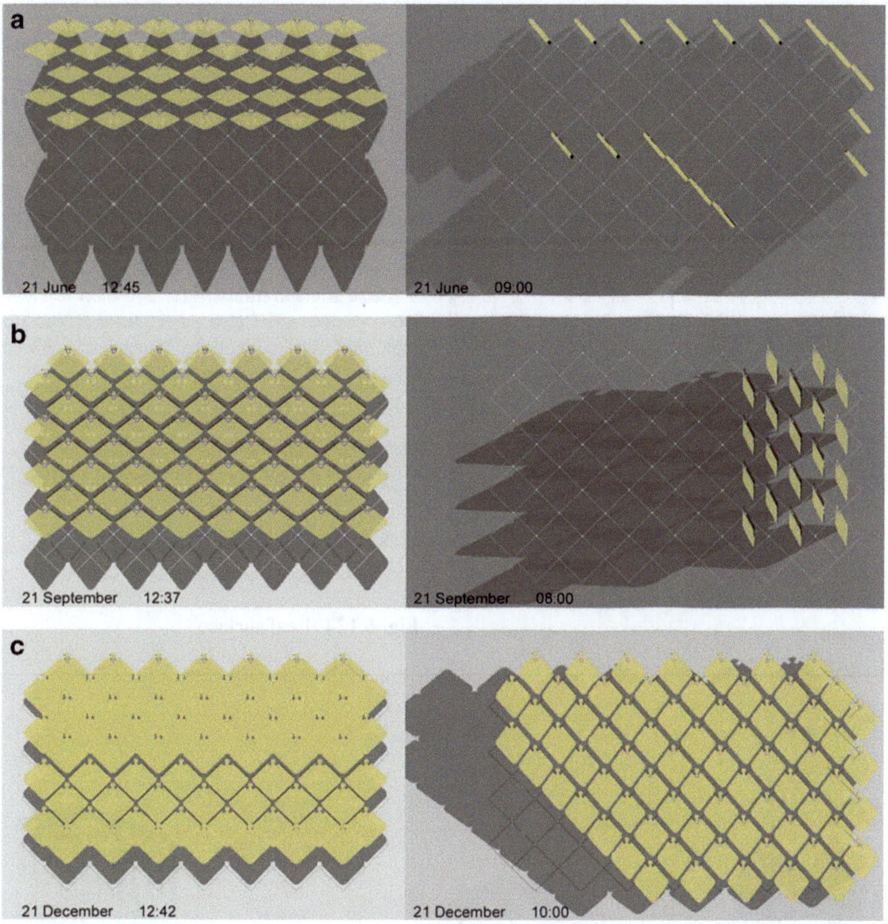

Fig. 6.122 Shading for a south-facing facade. A simulation was performed for the morning and midday Sun positions of each day

	Distribution/position	Orientation/inclination	Dynamics
Maximum light exposure	• loose distribution at multi-layer • dense distribution at mono-layer • maximum projected area • Fibonacci series for compact pattern packing • extending stem • horizontal expansion	• perpendicular to sun-rays (Diaheliotropic) • facing east for maximum exposure at morning and afternoon • facing south/north for maximum exposure at winter noon's	• increasing internode and petiole length • increasing leaf area combined with reducing mass per unit • plasticity, nastic structure • different flexibilities of the sides of a blade • special surface properties- uncoated cell clusters (for flexibility) • convex surface shape

Fig. 6.123 Position, orientation

6.33 Shading and Light Utilization 1

The leaf element of most plants is the main energy producer with the process of photosynthesis. Behind an outer protection layer, called the epidermis, lie the photosynthesis cells. For the exchange of gasses, the leaves possess pores in the epidermis layer, called stomata. This region is where transpiration (water emission in steam form) occurs. CO_2 also passes through the stomata into the intercellular space and diffuses from there into the photosynthesis cells.

Fig. 6.124 Leaf surface

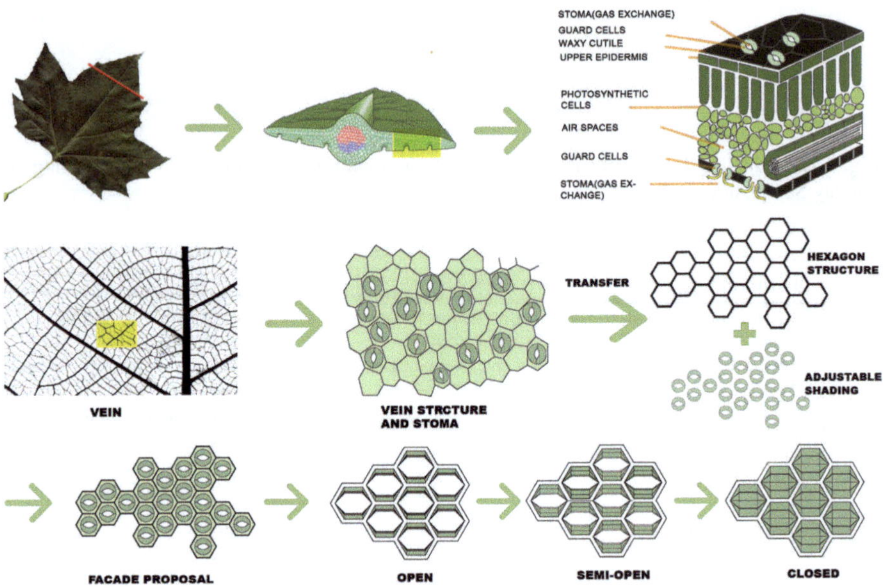

Fig. 6.125 Sketches of the function processes of leaves. Abstraction and transformation of the system

Fig. 6.126 Facade system

A double-layered facade corresponding to the stomata system of plants could be constructed, with outer layer "guard cells" and movable elements applicable for controlling light and heat transmission.

In the frame of an international student workshop "Facade Design & Performance" at the University of Melbourne, Australia, an adaptive facade concept was developed by H. Jin under the guidance of Eckhart Hertzsch and

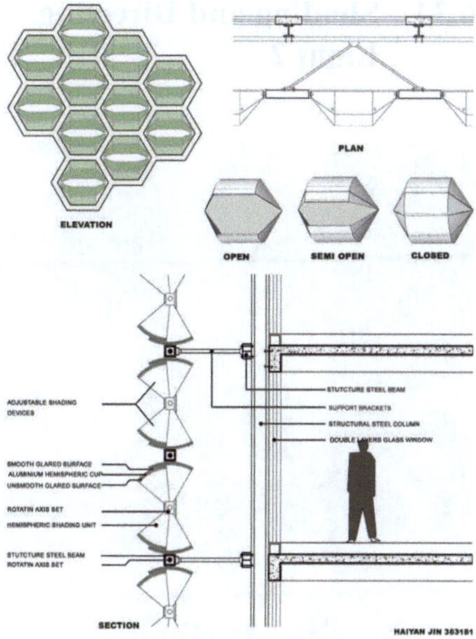

Fig. 6.127 Detail studies

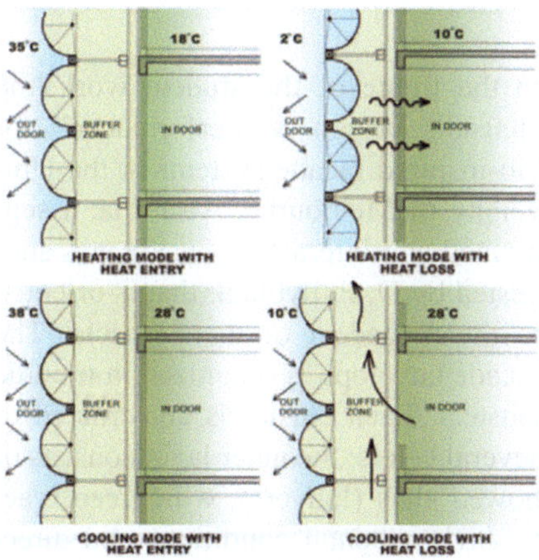

Fig. 6.128 Temperature management through adaptivity of the building envelope

Göran Pohl to depict the application potential of natural envelope structures and also to recognize their complexity (Figs. 6.125, 6.126, 6.127, 6.128).

6.34 Shading and Directing Light 2

The Balanidae are a family of barnacles that adhere themselves onto rocks. This ability is problematic for boats, as they can cement themselves onto almost any type of material. They can also be found on large sea mammals, i.e. whales. The binding substance is an extremely strong, organic cement. The barnacles live primarily in tidal waters, necessitating the capability to survive out of water as well. They produce a lime housing around themselves which can be opened or closed with movable plates (Fig. 6.129 and 6.130). With aid of this closing mechanism they can protect themselves from enemies and desiccation.

Fig. 6.129 Barnacles, *Chthamalus stellatus*

In the frame of the student workshop "Facade Design & Performance" for bio-inspired facade systems at the University of Melbourne, Australia, a segmented and interactive facade was suggested by D. Pullyblank that is oriented to the precedent of the barnacle. The facade envelope is organized in module clusters. Each cluster is constructed of several layers; the inner layer consists of bowed slats (louvers), which can react to environmental conditions and direct or prevent light into the interior space.

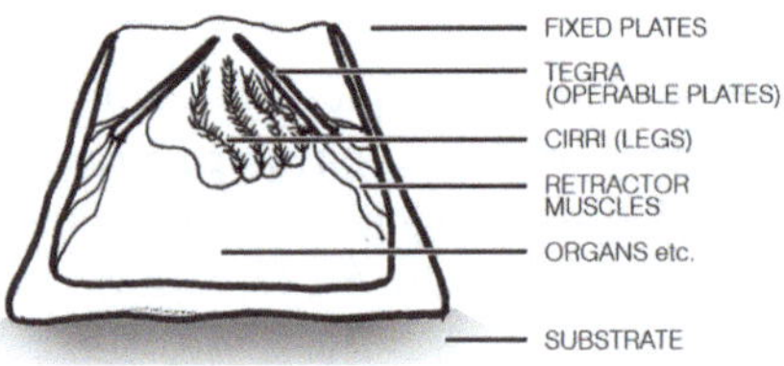

Fig. 6.130 Functioning system of barnacles

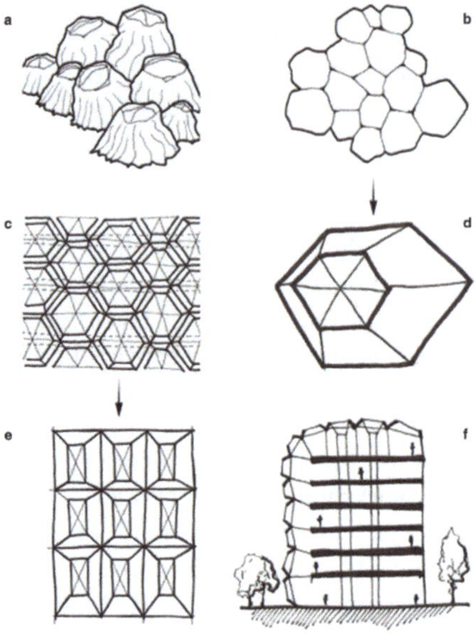

Fig. 6.131 a–f The concept of module clustering of barnacles is applied as the solution for a segmentable and reactive facade system

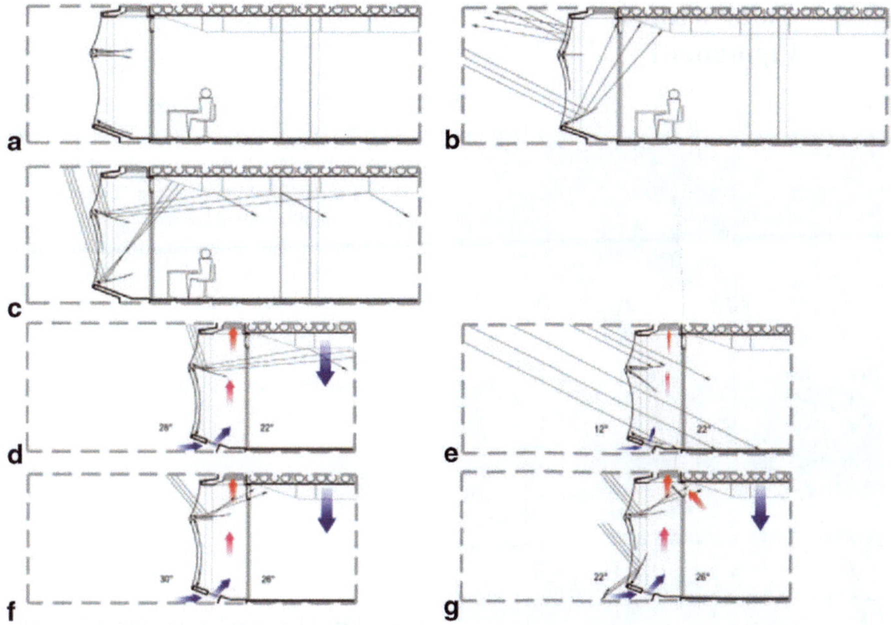

Fig. 6.132 Phases of the facade adaptation in relation to Sun's position: **a** diffuse light, **b** direct, low-angle sunlight, **c** direct, high-angle sunlight, **d** heat emission, **e** heat entry and dissipation in the fall, and **g** heat emission

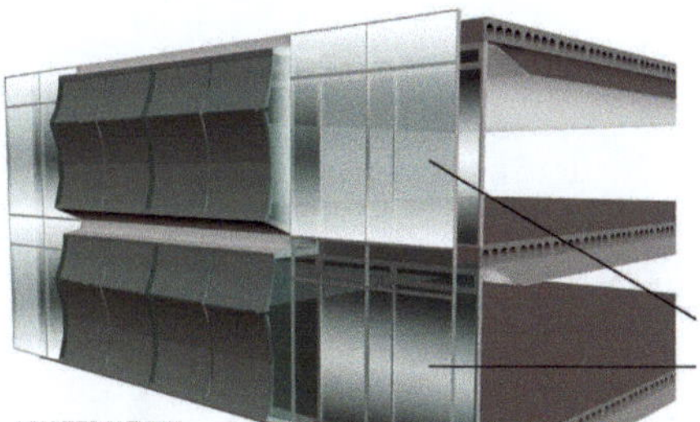

Fig. 6.133 Louvers completely closed

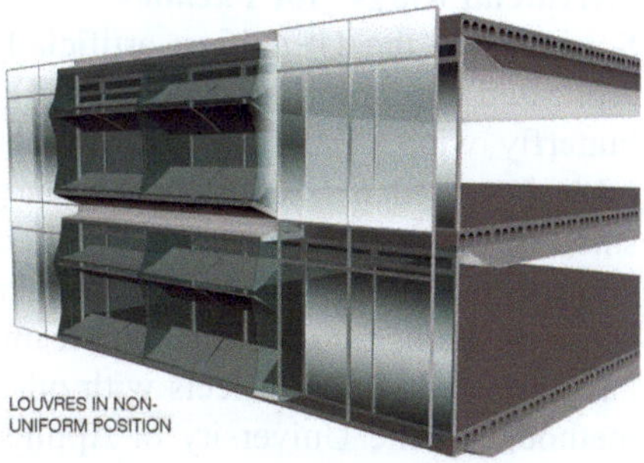

Fig. 6.134 Louvers partially opened

The depictions 6.132–6.134 visualize the phases of the facade adaptation in relation to Sun's position.

With diffuse light, (Fig. 6.132a) the louvers are completely open to allow maximum light entry. With direct, low-angle sunlight (Fig. 6.132b) the upper louvers are closed and prevent glare. The lower louvers are partially opened to let in a diffuse light. With direct, high-angle sunlight (Fig. 6.132c) both upper and lower louvers are sloped to angles in order to direct the light deep into the interior space. The middle elements are closed to reduce glare (Figs. 6.130, 6.131, 6.132, 6.133, 6.134).

6.35 Color without Pigments 1

Butterflies obtain brilliant colorations with the addition of surface pigments. The colors are therefore durable and never fade. The colors of butterfly wings result from tiny, grate-like surface structures. This so-called diffraction grating splits white light into its colored elements with reflection. Individual colors are reflected at specific angles. As opposed to conventional diffraction gratings the iridescent colors of a butterfly are perceived as a result of many different angles.

Fig. 6.135 Play of colors in a butterfly wing

"Artificial wings" for Facades

Scientists in the USA have artificially replicated the coloring structures of a butterfly wing. The copy, as with the natural precedents, consists of many small diffraction gratings, which reflect white light as blue from a particular angle position. German architect teams have attained similar effects with other methods. At the University of Applied Sciences, Cologne, Germany they developed so-called holographic-optic elements (HOE). The scientists hope in the future to be able to replace inks and pigments with more imperceptible and permanent methods.

The optics expert Mool Gupta and his colleagues at Old Diminion University in Virginia produced an artificial version of the structures of butterfly wings using electron beam lithography. In this process, an electron beam breaks down

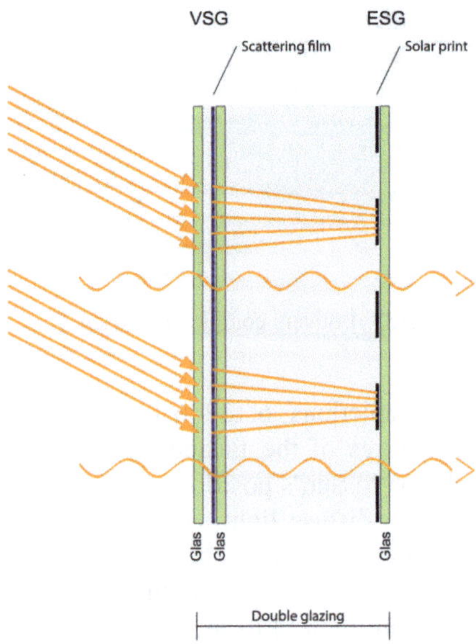

Fig. 6.136 Technology of a facade system used by holographic-optic elements (HOE)

Fig. 6.137 HOE colored facades at the media center of the Bauhaus University Weimar by Pohl Architects

the carbon bonds of an organic surface. With directed deflection of the beam, the surface can be furnished with a fine structure as wished. Each hexagon of the structure provides a different alignment of the provided diffraction pattern. The "wing," as produced in this manner by Gupta and his colleagues, consists of tiny diffraction gratings in a hexagonal honeycomb pattern. The diffraction patterns of side-by-side hexagons are additionally rotated from one another—a structure that is encountered in the wings of numerous species of butterflies. The surface structures alone are only 125 nm (millionth of a millimeter) thick and 220 nm wide. When light beams are directed at the artificial wing, the blue portion of the light is reflected back in various directions of view. (Fig. 6.136)

In 1947 an optical imaging tool was introduced by Denis Gabor, which seemingly reproduced a 3D object on a flat projection screen. This type of imagery is called a hologram. An analogous development can be seen in the technology known as "HoloSign Eyefire" developed by Michael Bleyenberg with the German research community in Bonn. The development uses holographic-optic elements (HOE) that diffract white light into its spectral components (Fig. 6.136), producing illuminated images on the facades of a building.

The developed facade system of Pohl Architects for the central building of the faculty of Media at the Bauhaus University Weimar, Germany, uses optical effects for producing color (Fig. 6.137). In this case, the same technology that provides coloration also provides solar production. Using holographic-optic elements light is scattered on the facade. Common insulating glass panes consist of two or more panes with an air- or gas-filled intermediate space. The panes developed from the precedent of the butterfly wing consist of a bound structure of two panes with a microstructured film incorporated in between, which scatters the light onto a pane behind the interspace, on which a thin-layered light absorption sheet is pressed. The light scattering elements concentrate the light on the photovoltaic elements similar to a lens to produce photovoltaically supplied energy. The diffraction grating affects the redirection of light waves, which ensues only for the determined angle. This technology provides a play of colors, enabling different variations by reflecting light in different directions with the physical effect of diffraction, comparable to other optical tools such as mirrors, lenses, and prisms. In architecture, holographic-optic elements can be used for various applications, such as light redirection, graphic and artistic facades, and shading.

6.36 Color without Pigments 2

Color adaptive capabilities of beetles

The Hercules or rhinoceros beetle Dynastes hercules (Fig. 6.138) can change its color from light to dark. At the University Namur in Belgium researchers investigated the nature of these chitin shells and discovered that the color appears black by 80% or higher humidity, otherwise the color is brown-green. They discovered that a porous structure under the wax coating of the shell yields a mesh of tiny perpendicular beams and thin fibers, which reflects the light in dry conditions. In humid conditions the porous structure fills itself which causes no more light to be reflected, resulting in a black-colored beetle.

Fig. 6.138 Hercules beetle *Dynastes hercules*

Solaradaptive Envelopes
The phenomenon of color adaptivity in insects (*Dynastes hercules,* rhinoceros beetles, tropical rainforests, Peru, Ecuador, and *Cyphochilus* beetles, Southeast Asia) has inspired research groups to study the applicability of the systems for the purposes of color changing facades. In the frame of an international research workshop for biomimetic-inspired fa-

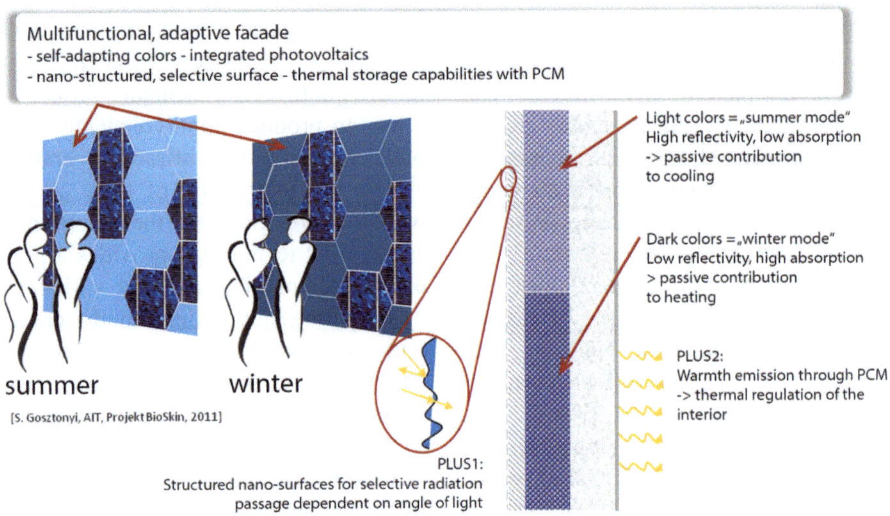

Fig. 6.139 Adaptive facade surfaces—passive thermoregulation

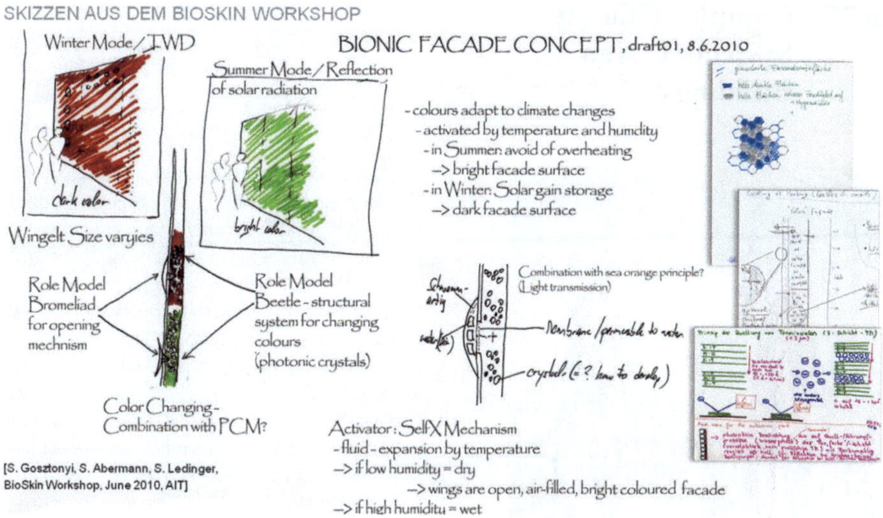

Fig. 6.140 Brainstorming in the frame of BioSkin (AIT) for the formation and application of solar adaptive envelopes

cade constructions at the University of Melbourne, students under the leadership of E. Hertzsch and G. Pohl investigated this phenomenon closer and developed different scenarios for its use.

In the project BioSkin under the leadership of Susanne Gosztonyi at AIT in Austria, the potential of color changing facades was likewise understood.

The research teams came independently to the following conclusions:

Color change in winter (dark) and in summer (light) can generate—applied to facades—different degrees of light reflectivity and absorption and are able to differently warm the materials behind with daylight: with darker colors the facades heat up faster with sunlight, with lighter colors slower. These properties can lead to the development of a solar adaptive envelope for seasonal changes, a condition to which building envelopes in large swathes of the Earth are exposed. The envelopes would not remain uniform for longer than each atmospheric condition; they could adapt themselves and therefore save energy and reduce CO_2 emissions.

Within the frame of BioSkin, as well as within the frame of the research workshop in Melbourne, they concluded

- Based on the precedents in nature, color change on facades cannot be dependent on pigment if it is to function lastingly.
- Color change can correlate to energy-saving effects with changing temperatures in winter/summer and therefore to a reduction of heating/cooling necessities (Figs. 6.139 and 6.140).

6.37 Complex Climate Systems 1: New Buildings

The location, orientation and architecture play an essential role in the thermoregulation of climate within nest structures. Many animal species living in warm regions use the wafting air at the earth's surface to ventilate the ground-level passages of their structures with a funnel effect. Termites utilize similar methods. The tunnels in the interior of their structures are connected to these lower ground-level passages. Many termite species also use a chimney effect: warm air rises and pulls cooler air with it. Termites have developed, as a result, tall nest constructions (Fig. 6.141).

Fig. 6.141 Termite construction in the Australian Outback with illustration of the chimney effect in the air passages

Fig. 6.142 Facade view, technology center in Erfurt

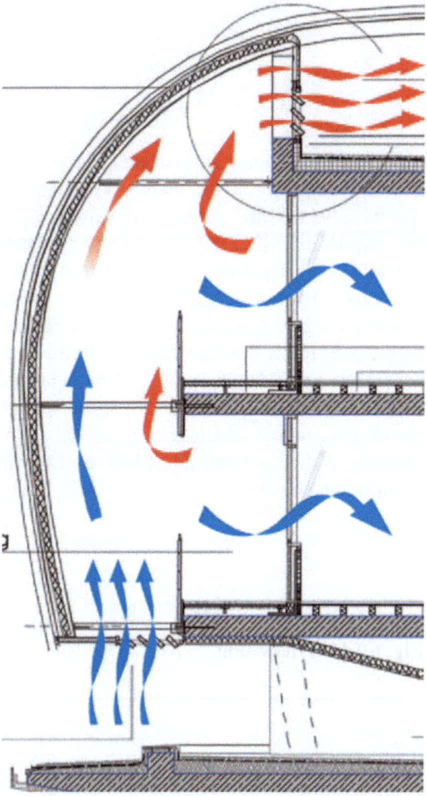

Fig. 6.143 Scheme of chimney effect (*above right*, from *above* to *below*). Air circulation, solar production, and thermal system

Analogous Technological Methods

Some termite species possess, in addition, the capability to sense the CO_2 concentration in their passages and, with too high concentrations, increase the cross-section of those passages to gain better air circulation. Many termite species are able to form chambers in the center of their nest to allow air from the ground-level openings to circulate upward. They lay leaves in these chambers, which are moistened by ground water, thereby cooling the entire structure with evaporation.

Termite hills can grow to several meters in height. They consist of a cement material that becomes rock-solid; soil and sand particles, cemented together by glandular secretions. Nonetheless, the material exhibits a certain porosity, additionally enabling the exchange of gasses.

Abstraction and technical interpretation of termite structures generate important insights for new thermoregulating components of buildings. The research results of thermoreactive structures of animals were systemized by Pohl Architects and used within the frame of an EU research program for the design of a technology center in Erfurt and subsequently evaluated for several years. (T = Termite construction, N = Application for new construction) (Figs. 6.142 and 6.143)

- Passive ventilation from a vacuum-funnel effect (T,N)
- Passive ventilation from a chimney effect (induction) (T,N)
- CO_2 and heat detection (T,N)
- Evaporation (T,N)
- Active closing of openings by rain to prevent cooling (T,N)
- Active change of the vein systems for thermoregulation to meet consistent requirements (T), application as controllable concrete core cooling with circulating fluid heat exchange system in the solid building parts (N).
- Gas exchange enabled by material selection (T) without definite application
- Use of earth storage masses for heat exchange (T,N)
- Pre-convecting ventilation with ground-level collector pipes (T,N)
- Insulation, light direction, and scattering in a transparent fiber system in the curved building envelope (cf. polar bear fur, N)

6.38 Complex Climate System 2: Building Reuse

Prairie dogs (Cynomys ludovicianus, Fig. 6.144) use the topography of the land to generate the Venturi Effect. They utilize "already present building materials" for their tunnel systems, suchs as small pre-existing hills or rises. These instincts are related to their construction of ventilation passages. The wafting air at the Earth's surface produces a Venturi Effect and flushes out the earthen tunnels with fresh air. Coincidentally, building and tunneling underground in hot regions has another advantage: the tunnel walls release heat into the surrounding soil, cooling off the air in the tunnels.

Fig. 6.144 Prairie dog

Potential Ideas from Nature

For the search of possibilities of the influence and thermal capitalization of present materials for a building, organisms in nature that connect passive currents with an activation of a "building part."

The hills of the mole *Talpa europaea* are variously designed and/or laid on a slight slope so that they lie at different heights. The moles ventilate their connecting passageways as such using the Bernoulli principle. Naturally ventilated tunnel structures are built by sea inhabiting lugworms of the family *Arenicola* with particularly designed openings that funnel air in. Around one opening of their tunnel they form a hill; on the other opening an indentation. With this configuration, they fulfill the conditions for a pressure differential, a small "tidal power plant."

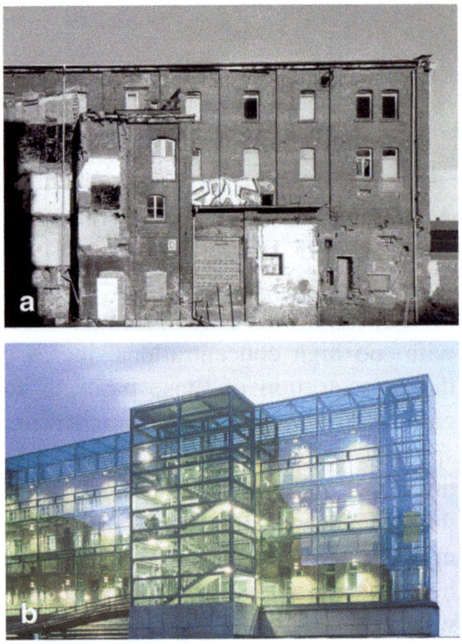

Fig. 6.145 Bauhaus University Weimar: **a** former brewery building and **b** with thermoregulating facade

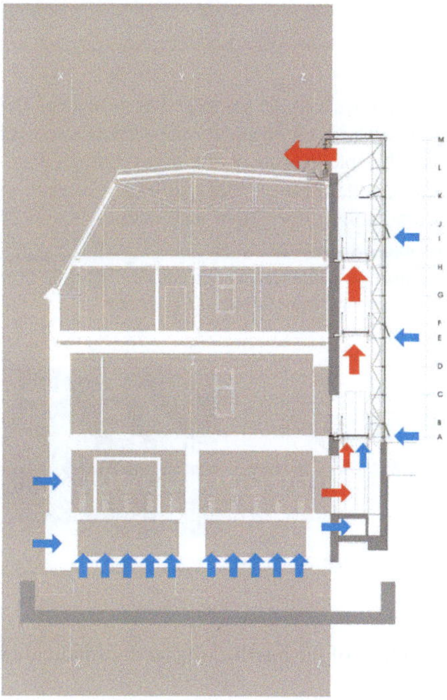

Fig. 6.146 Building section with illustration of the air and temperature flows

Ideal and natural functionalities are

- Passive ventilation with a "tidal power plant" (vacuum and funnel effect),
- Passive ventilation with a chimney effect (induction),
- Activation of thermal storage masses of present material environment, and
- Active closing of openings

The functional principles and the adaptations to specific given environments based on these precedents were adapted by Pohl Architects to a preexisting building at the Bauhaus University Weimar. For a former brewery building with thick walls in the basement level

Fig. 6.147 Detail of elevation with facade flaps

a complex thermoregulating system (Figs. 6.145, 6.146, 6.147) was developed that is functionally comparable to the natural precedents, in consideration of

- The integration of cross ventilation sluices for passive air circulation;
- The use of the vertical draft effect with a specifically constructed updraft facade: it forms a vacuum, providing for ventilation as well as serving as a warm buffer when air sluices and facade flaps are closed;
- The use of the natural topography and temperature differential of a cool (north-facing) street side to a warm (south-facing) back side to achieve cross flow for cooling (summer);
- The use of preexisting storage masses in the basement and ground levels: over 1 m thick stone walls store and buffer heat; and
- The detection of warmth and air quality, activation of thermoregulating elements.

6.39 Spatial Panels

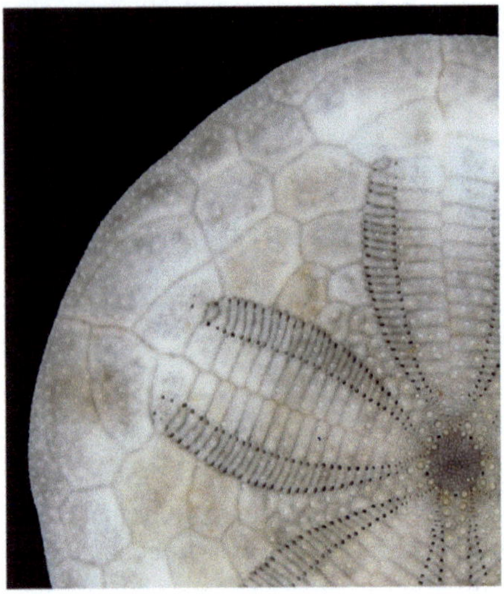

Sea urchin shells are composed of individual elements. These elements are stably interlocked yet must be able to grow on the edges through material accretion. Three plate segments always meet at one point; a principle that enables flexurally strong yet deformable structures, although only the normal and shear forces and not bending moments can be transferred at the joints.

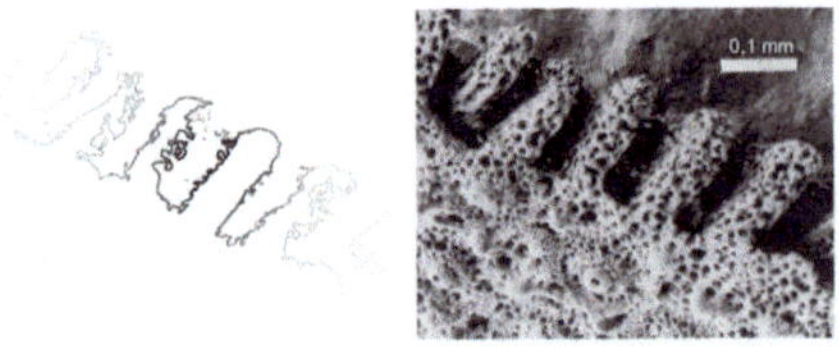

Fig. 6.148 **a** Sea urchin shell "sand dollar." **b** REM Image of the interlocking teeth of individual plates of the sea urchin shell

The plate members of the sea urchin shell are generally joined together by bracing elements in each cell. This idea was applied to the structure of a pavilion at the University of Stuttgart. Computer-based, robotic fabrication enabled

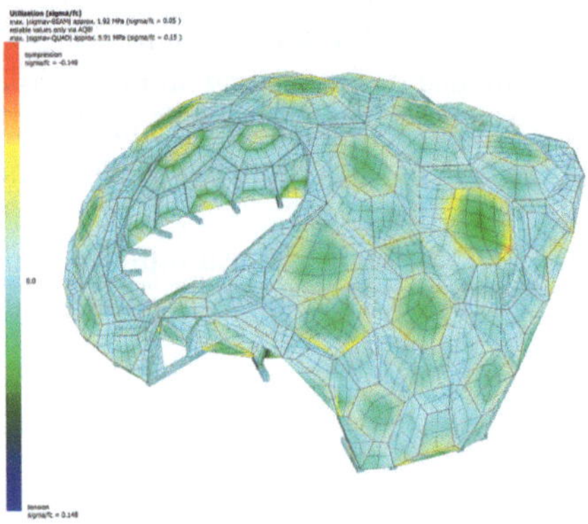

Fig. 6.149 Computer simulation and structural calculation

Fig. 6.150 Robot fabrication

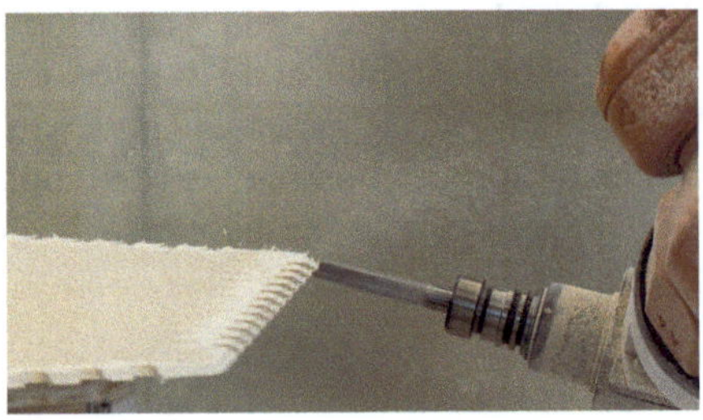

Fig. 6.151 "Finger joint" connections

Fig. 6.152 Bird's eye perspective

Fig. 6.153 Night view

Fig. 6.154 Interior view

the precise production of the individual members. Plywood panels 6.5-mm-thin were joined into spatial shell elements.

Fig. 6.155 First structure element

The research group of the ICD and ITKE institutes at the University of Stuttgart, under the leadership of A. Menges and J. Knippers respectively, managed to use only 1.6 m^3 of wood for 200 m^3 of total interior space. (Figs. 6.149, 6.150, 6.151, 6.152, 6.153, 6.154, 6.155)

6.40 Spines

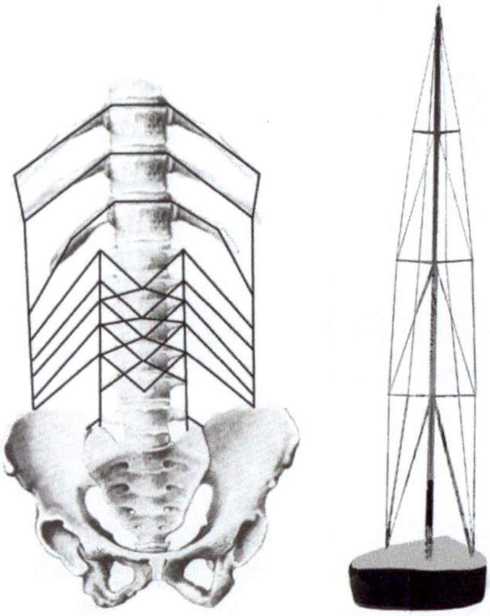

The human spine as a system of ligaments, tendons, muscles, and bones is an exceptionally flexible structure that can efficiently bear loads and distribute them evenly over the structural limbs and members. The ligaments, tendons and muscles therefore help to stabilize the spine against lateral kinks and torsional buckling. The spine is stabilized similarly to a ship's mast, though the shrouds (the mast's lateral supports), as analog to the muscle-ligament apparatus of the spine, offer no active stabilization.

Fig. 6.156 Spine in comparison to a ship's mast, which is stabilized with riggers (*cross-beams*) and shrouds (*cables*)

"Extensive studies on the architecture of the vertebrae of humans and various animals have led the author (1874) to the conclusion, that human as well as animal spines represent a framework construction. The framework was and is the only mechanically possible construction for an entity such as the spine, not only of man but also of animals, for the most varying roles: connecting a pair of extremities to the other, providing the main framework for the entire body of a vertebrate, carrying the bowels, the head, and the extremities, and can support itself on both pairs of extremities or only one pair… . The main difference between the human and animal spine resides in the fact that, with the former, the corresponding, supporting, perpendicular cross-beams come to the fore due to the predominantly upright position. The spine of the quadruped is a truss system, as it is with our modern, iron railroad bridges…." (from Real-Enzyclopädie der gesammten Heilkunde, Medicinisch-chirurgisches Handwoerterbuch für praktische Aerzte, 1893–1901)

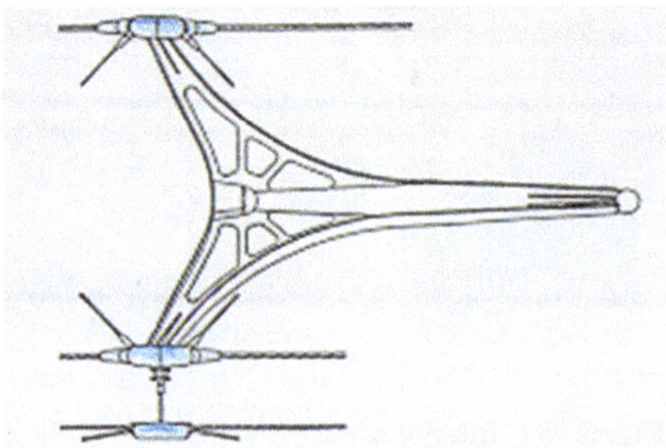

Fig. 6.157 Individual vertebrae of the structure for a 260-m-free-spanning roof

Fig. 6.158 Section from a model of the spine-supported structure

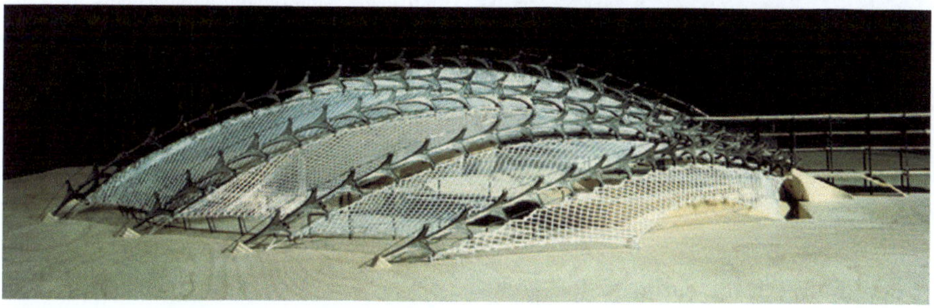

Fig. 6.159 The model shows the dominant spine supports and the delicate net structure of the roof

Frei Otto has already concerned himself with the structural system of the spine and exhibited a series of comparison studies, that relate the spine with frame structures and tensioned, freestanding masts. In his early work, on the basis of the foundational studies of Frei Otto's galloping crocodiles, Göran Pohl developed a structure of poured and cast elements that are strung together and articulated like the vertebrae (Figs. 6.157, 6.158, 6.159). Instead of the muscles, tendons, and ligaments used in anatomy, he used steel cables, which are integrated in pre-tensioning and retained in their pre-tensioned condition by an electrohydraulic tension system. The spine system, as interpreted in architecture, is tensioned in the longitudinal as well as in the radial and counter-radial directions of the arched support beams. The construction consists of altogether ten beams arranged in a fan with each beam consisting of up to 26 individually strung vertebrae. With the tensile structure strung between, it spans 260 m unsupported. Pohl further developed this structure and later implemented it in a competition entry for a new design of the natatorium and velodrome in Berlin during Berlin's application for the 2000 Olympic Games (Fig. 6.158, 6.159).

6.41 Spatial Structures of Curved Modules 1

Opuntia

The Opuntia belongs to the family of cactus plants. These plants retain stability, similar to technical fiber composites, with a fibrous sclerenchyma skeleton and surrounding tissue (parenchyma).

The structure of the sclerenchyma skeleton is of particular interest for technical application. The system consists of major and minor supports (Fig. 6.160). Parallel fiber bundles form the major supports directed towards the highest structural loads. Sinusoidally curved fibers form the subsupports and assume the role of cross bracing.

Fig. 6.160 Sclerenchyma skeleton of an *Opuntia*

Curved, free-form surfaces for architectural application: roofs, envelopes, and facades can be developed from the abstraction of the *Opuntia* structure (Figs. 6.161, 6.162, 6.163, 6.164, 6.165).

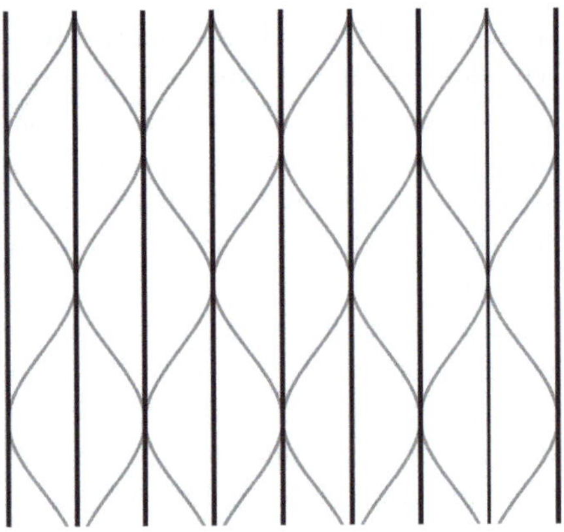

Fig. 6.161 Geometric abstraction of the sclerenchyma structure of an *Opuntie*

Fig. 6.162 Free-form shell model with spatially curved modules

Fig. 6.163 Envelope form in a 3D model

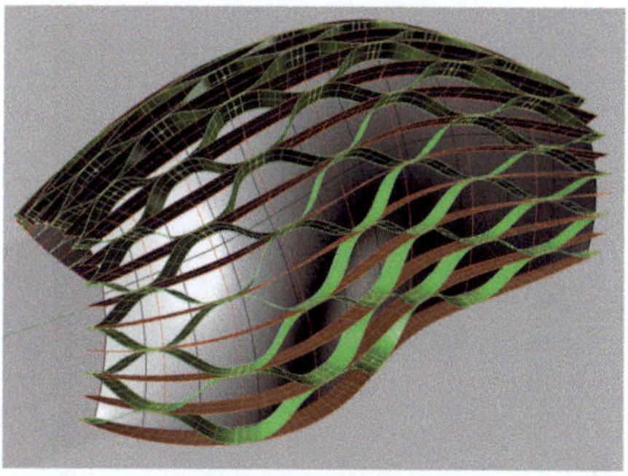

Fig. 6.164 Model attempt of a free-form shell following the precedent of the sclerenchyma of *Opuntia*

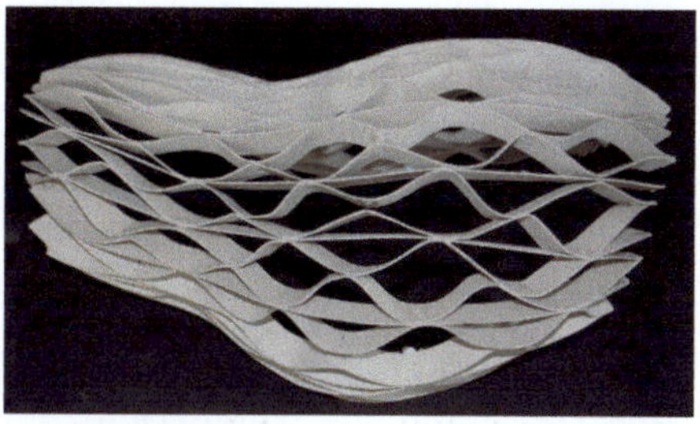

Fig. 6.165 Model

Planar elements of relatively long length and thin section assume the function of the major and minor supports. During assembly the minor supports and in turn the major supports are elastically deformed; they adjust themselves to an equal distribution of weight and stress, an effect that leads to a stable structural form of a shell. With observation of the system of forces in section, lateral forces due to the horizontal portions of stress are found to emerge at the bottom of the structure.

Malleable building materials can be joined together into spatially complex entities. At the SAS, School for Architecture of Saarland, Göran Pohl with his B2E3 Institute for Efficient Buildings developed concepts using curved wood strips that yield a free-form envelope structure with higher stability.

6.42 Spatial Structures from Curved Modules 2

Weaver birds (Ploceidae) build hanging nests (Fig. 6.166) using tightly-bound, bendable plant fibers. The nests are spherical or bottle shaped. The nest material is anchored with tension elements of straw, which the weaver bird secures with special knots and loops on branches or reed stalks. With nest construction in trees or shrubs the weaver bird initially knots the first nest fibers together and then weaves the rest of nest around these fibers.

Fig. 6.166 Curved elements yield stable structures for nest constructions of weaver birds

An analogous, architectural development uses the braiding method in a similar manner to the precedent in nature: The ICD, A. Menges and the Institute for Building Structures and Structural Design (ITKE), and J. Knippers of the University of Stuttgart realized the idea of curved building modules in a temporary research pavilion using wood.

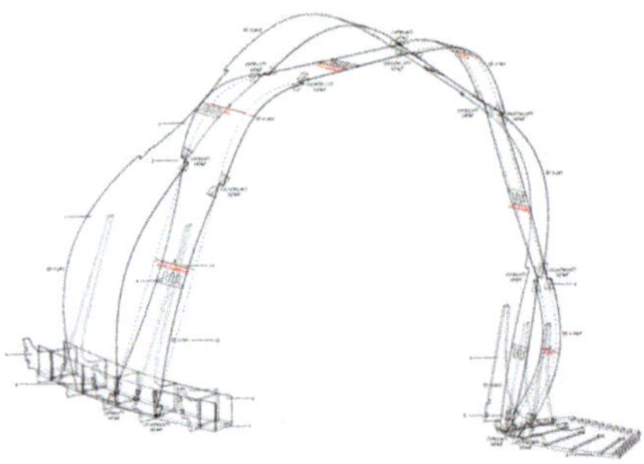

Fig. 6.167 Construction of a support of thin, curved elements

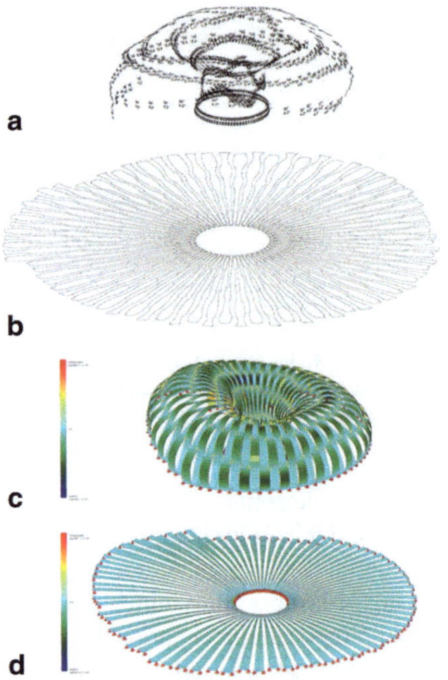

Fig. 6.168 Computer-based design and simulation results from *above to below*: data model for fabrication (**a** and **b**) and curved model (**c** and **d**)

Fig. 6.169 Top view

Fig. 6.170 Interior

Fig. 6.171 Computer-driven robotic fabrication of the individual plywood strips

Computer-based design, simulation, and production processes enabled this structure. The experimental construction consists of elastically curved plywood strips that were joined together into a complex support structure. With the curving of 10-m long, 6.5-mm-thin birch plywood strips the self-stabilizing structure is set under its own tension. The soft plywood strips then join themselves into a rigid structure (Figs. 6.167, 6.168, 6.169, 6.170, 6.171).

6.43 Layered Tissues

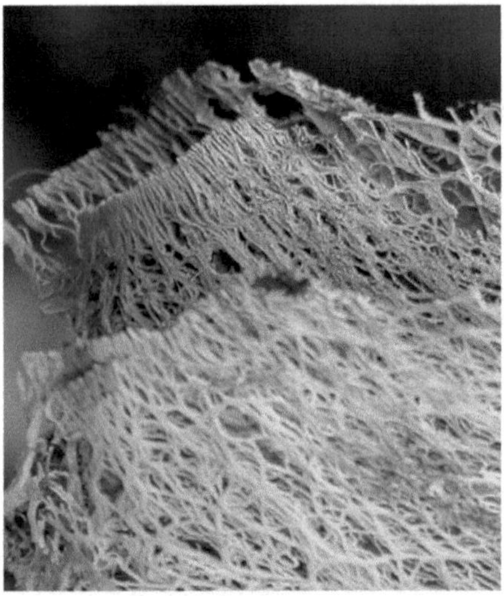

Tissues of plants are specialists for each of their respective tasks. While the enclosing layer of tissue provides for the protective envelope of the plant (primary enclosing layer: epidermis, secondary layer cork, i.e.), the various types of inner tissues have very differentiated features: the storage parenchyma forms "fleshy" storage organs (tubers, onions), the hydrenchyma serves to store water in extremely large vacuoles, i.e. in succulents, the aerenchyma is an air circulating tissue, the chlorenchyma is a chloroplast-rich photosynthesizing tissue. Structural tissues in contrast are thick tissues, whose cell walls are solidified by accretion of cellulose (collenchyma, sclerenchyma).

Fig. 6.172 Layered tissue of the sclerenchyma skeleton of an *Opuntia* leaf

The structural tissues can become pressure-resistant and rigid in their cell walls through the process of lignification (i.e., trees). Collenchyma is the name of the structural tissue of growing and herbaceous plant parts. These cells are capable of dividing and growing, therefore not lignified. In contrast, the lignified sclerenchyma consists of dead tissue, which is formed out of thick-walled, narrow cells. Sclerenchyma does not however appear in young plants, only in matured ones; sclerenchyma fibers are one example.

In a similar manner to the layered tissues of sclerenchyma, as they have been demonstrated in the natural precedents, the product designer Jens Otten developed a chair as his diploma thesis at the Kunsthochschule Kassel that focuses on lightness instead of material mass. The surfaces are extremely porous and form only an "interface" between the sitter and the chair legs (Figs. 6.173, 6.174, 6.175). The shell of the seat is constructed from three layers of airplane plywood each with 1.5 mm thickness, glued together at connection points. The construction method entails a spatialized framework in its essence, a highly resolved plywood shell, in which succeeding veneer layers change their direction per layer. The main direction of wood grain is in each case arranged in the lengthwise direction of the indi-

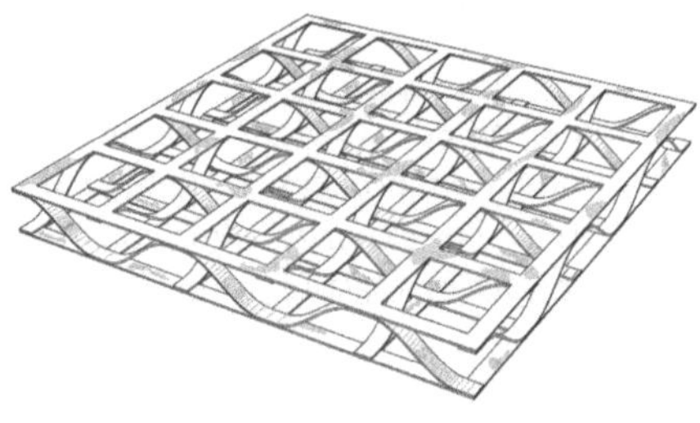

Fig. 6.173 Generation of a shell from thin, curved individual elements

Fig. 6.174 (*above*) Detail of the seat shell

Fig. 6.175 (*below*) Chair model

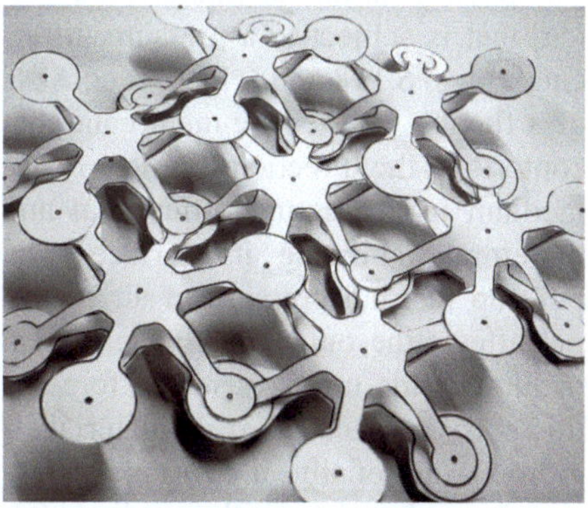

Fig. 6.176 Further development for free formed building facade parts (model)

vidual strips. The shell consists of 60 individual strips and weighs 1013 g. Jens Otten refined this build typology in the frame of his activity at the School for Architecture, Saar within the research project BOWOOSS under Göran Pohl, and translated it to free formable building facade parts (Fig. 6.176).

6.44 Pneu

Nature utilizes pneus as individual cells or cell clusters for both rigid and moving systems, such as unfolding processes. For example, the dragonfly wing unfolds itself with pneus; the poppy flower unfurls itself in magnificence from inside of a pneu with help from turgor pressure-driven hydraulic processes (Fig. 6.177). There arises so to speak a pneu inside of another pneu that then produces enough pressure so that it can spring open from the formerly protecting outer pneu.

Fig. 6.177 Unfurling of the poppy flower due to the increase of turgor pressure in the flower petals. The flower unfurls as a pneu

The pneu is an air- or liquid-filled system that is subject to a pressure difference (Figs. 6.178 and 6.179). It consists of a flexible and tensile membrane that contorts in the direction of a less dense medium in a pressure differential and therefore stabilizing its surface. The built-up internal pressure of air or liquid affects the outer membrane, which in turn builds up a resistance force to this pressure because of its material rigidity. Additionally, a resistance pressure is produced from the medium (air or water) surrounding the pneu. In a pneu there always exists a relationship between internal pressure, the geometric constraints, the stability of the membrane, external pressure, and the resulting form of the pneu. For the nonmoving parts of the pneumatic structure, the air or liquid medium becomes a support medium and support element with the absorption of the outer loads in a closed system. A consistent pressure differential is required for the stabilization of the membrane, which must be sustained by a control system that adapts to changing conditions of the environment: If one of the conditions changes, then the geometric form also changes. This necessary regulation, used as an actuator, enables wanted movements in a structure and relates to the pneumatic and hydraulic actuators of nature.

Pneus are not only used as structural elements in nature, but also as initiators

Fig. 6.178 Air structure in soap bubbles

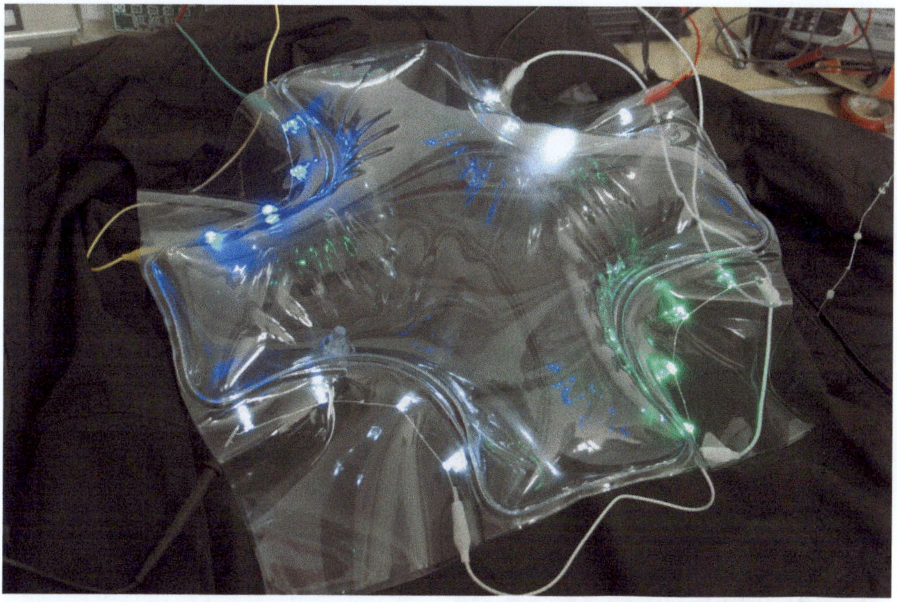

Fig. 6.179 Pneumatic system, prototype from student work developed under the direction of Göran Pohl for a media skin at the School for Architecture Saarland

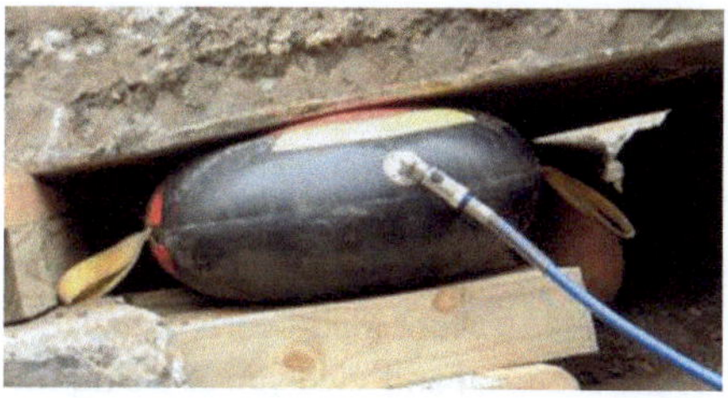

Fig. 6.180 Pneumatic lifting device

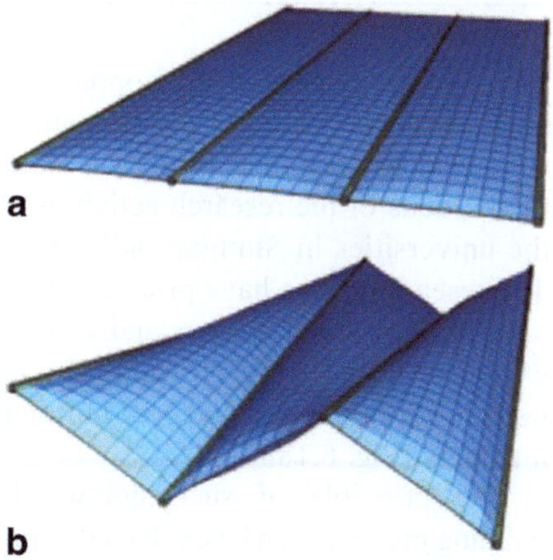

Fig. 6.181 a and **b** Utilization of a pneu for moving roof systems at the TU Berlin, Mike Schlaich

of movement. For this purpose there exist the most diverse applications (Fig. 6.180); the basic principle for movement always rests on form change because of the uptake or removal of air or fluid (fly wings, spider legs, earthworm, flower petals, *Mimosa*, …)

Pneumatically supported structures are predestined for movable and thus transformable structures because of their lightness (Fig. 6.181). These kinds of structures, which admit changes to their shapes and carry out movement from inside out and not rigidly moved as solid entities, are the goal of the research at the TU Berlin under the leadership of Mike Schlaich and Annette Bögle (HCU Hamburg).

6.45 Solid, Efficient Load-Bearing and Heat-Insulated Lightweight Structures

The inner structure of bone arises from certain regulations. By 1870 the german doctor Hermann von Meyer had already supposed this: he concluded that there is a functional ordering scheme at the basis of the apparent chaos of cavities and bone trabeculae. The bone trabeculae are correspondingly accorded the most stress; to save material, natural processes produce interstitial cavities at specific locations where structural support at a microscopic level is not necessary. This "building tradition" interpreted as material saving is to be found in many places in nature, such as in bones, sponges, but also in the biosilicate shells of i.e. algae.

Fig. 6.182 Supporting skeleton of the sea sponge as precedent for gradient optimized building materials

The biological principle of optimized, stress-bearing, solid lightweight structures and optimal heat insulation stand at the focus of the research activities of the universities in Stuttgart and Berlin. The research teams have produced concrete of a particular mixture and set with expanded clay aggregate and can be poured into light wall and roof building members (Fig. 6.183).

The principle of such lightweight building methods can be explained in nature with the construction of bone. In a bone, the areas with higher stress receive more strength and exhibit increased production of spongy bone (*Spongiosa*), whereas areas with less stress exhibit in relation many pores and cavities. With the increase of rigidity in bones, the organic mass also increases. Similarly the other way around, with decreasing structural bone material the rigidity decreases.

In concrete it was attempted to induce focused strength by varying the amount of porosity throughout a form. The reduced weight of more porous concrete also reduces the load-bearing capacity of the material and vice versa. Using this characteristic, monolithic building parts were successfully produced according to the direction of loads with improved heat-insulating properties. To achieve the improved insulation, highly porous aggregates are added, resulting in so-called infra-lightweight concrete.

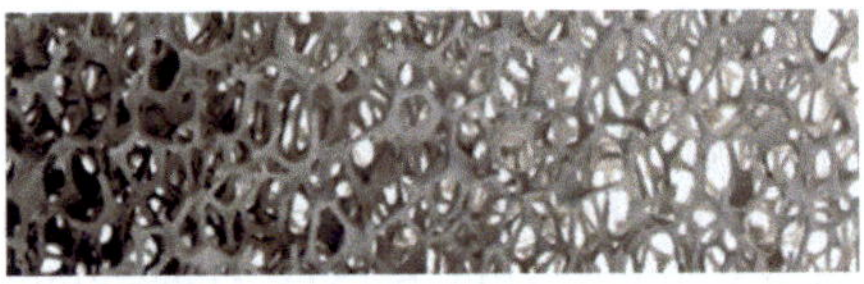

Fig. 6.183 Support structure, detail of a test body for gradient concrete, produced at ILEK, University of Stuttgart

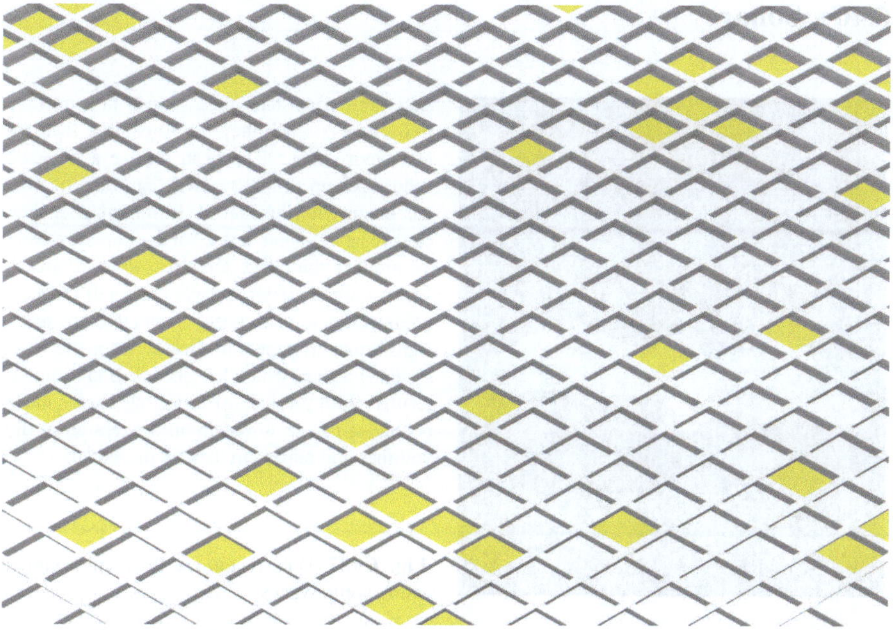

Fig. 6.184 Gradient concrete floor with differently treatable fields, developed by Pohl Architects and Lightweight Construction Institute (Leichtbauinstitut) Jena

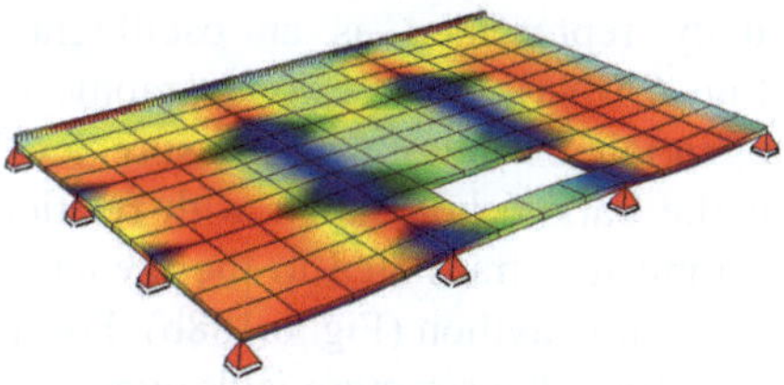

Fig. 6.185 Stress distribution of FE model of a floor deck under its own weight

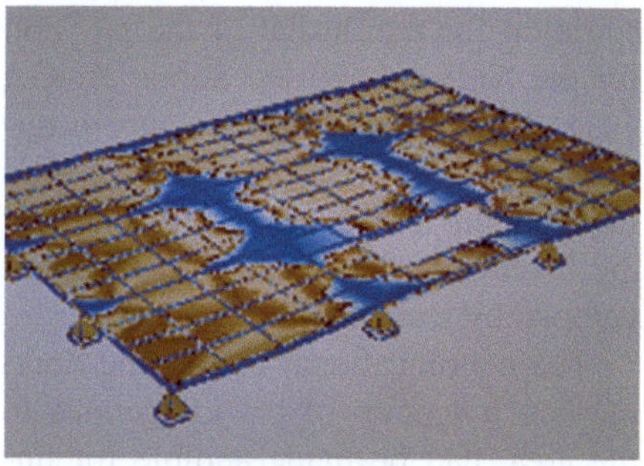

Fig. 6.186 Gradient concrete, implemented in a floor deck with varying distribution of the infra-lightweight concrete

A research team of Mike Schlaich at the University Berlin and specialists of the ILEK, Institute for Lightweight Structures and Conceptual Design at the University of Stuttgart under Werner Sobek, developed the infra-lightweight concrete further under the label "Gradient Concrete" and were able to vary the solidity and heat-insulating ability over a section of concrete (Fig. 6.183). High-stressed zones of structures can be located with computer-supported calculations and, with precision, structurally strengthened. Therefore, less structurally important regions can be completed with infra-lightweight concrete, leading to other positive effects (Fig. 6.184 and 6.185): with gradient concrete, well-insulated exterior walls can be finished without additional insulating layers, resulting in building parts that are reduced in overall weight, easier to transport, and more efficient in their raw material consumption (Figs. 6.185 and 6.186). The decreased raw material usage correlates to a reduced carbon footprint of the building material.

6.46 Sonar

Bats emit consistent click sounds in ultrasonic frequencies, which they use to explore their direct environment. They can receive a reflected echo from up to 300 m away and then be able recognize what lies ahead. In this manner they can create a "sound image" of their environment. To produce the clicks some bats emit ultrasonic waves with the larynx, others flick their tongues. Meanwhile it has been attempted to teach sight-impaired people to flick their tongues in a similar manner, which has already lead to initial results. With the help of reflecting click sounds people can learn to obtain an impression of the volumes and masses in their environment.

Fig. 6.187 Bats recognize their environment using sonic waves

The echolocation of bats provided inspiration for the design of the pavilion for the National Garden Show (BUGA 2011) in Koblenz. The structure originated under the leadership of Mandfred Feyerabend and Markus Holzbach of the Fachhochschule Koblenz and codeveloped by students.

The echolocational call of the noctule bat, a species of bat indigenous to the area, can be made audible for humans with the aid of sound technology. Using a music editing program the sound waves of the calls were visually represented as an oscillogram (Fig. 6.188a). The resulting graphic illustration of the sound pressure level of the bats' echolocation with relation to time was translated to the layout of the future pavilion (Fig. 6.188b). For its basic form the structure is designed according to naturally occurring catenary curves, for example in spiderwebs, interpreted as supporting arches. In order to finish the structure using small wood members, the surfaces had to be divided

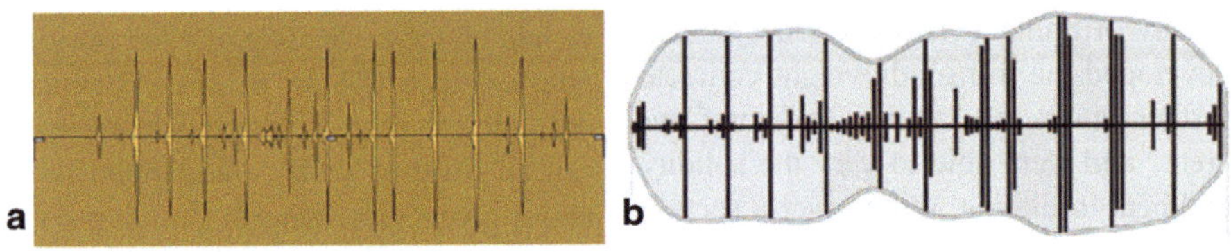

Fig. 6.188 **a** and **b** Echo calls of bats are translated into a structure layout

Fig. 6.189 Sonar pavilion at the National Garden Show in Koblenz 2011

Fig. 6.190 Detailed images

into parallel sections of three planes. These three planes stand each at 60° to one another, so that a spatially stable triangle and hexagonal grid is produced, as it occurs with beehive honeycombs in nature. The spatial network of small wooden rods, which consists of an overtruss, undertruss, and diagonal connecting members, was implemented following a continuous digital work process. The 3D basis data of the complete frame for the pavilion described ca. 6000 members and all their connections. The oscillogram of the noctule bat's echolocation was then projected on the floor of the pavilion structure with the use of LED light strips; the human-audible echolocation calls were emitted over loudspeakers (Figs. 6.189, 6.190).

6.47 Fiber Composite Sensors

For the production of high-stressed materials, the plant and animal world follows the principle of minimal energy expenditure, which essentially correlates to the least possible application of material substances. A further strategy of nature is the flexible adaptation of material characteristics and forms to each different stress and environmental demand. Structural materials of nature are fiber-reinforced materials (Fig. 6.191) that consist of a matrix, inside of which stress-suitable strengthening fibers are embedded. Further capabilities are embedded within the composites of living beings: specialized receptors that serve the recognition of chemical substances, of cold, or the perception of light.

Fig. 6.191 Fiber bundle and composite of a bamboo root

In relation to the seeing, hearing, and tasting capabilities of natural organisms, engineers are trying to produce "perceptions" in our technology that could be achieved using new techniques, such as building sensors into lightweight-fiber composite constructions. At the Technical University (TU) Chemnitz new methods are being pursued for the steering of complex systems that are comparable to natural systems both in their generation and in their function.

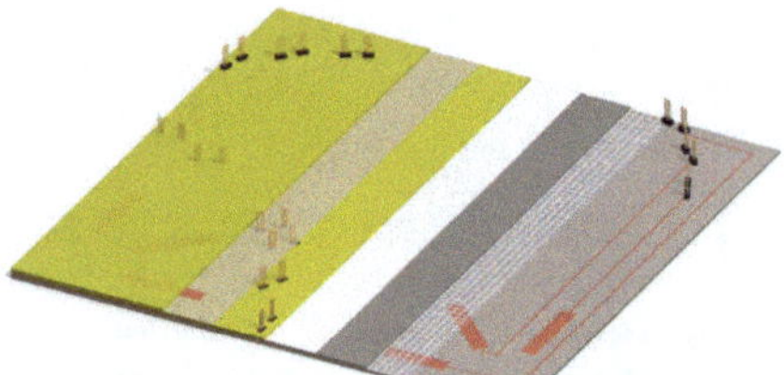

Fig. 6.192 Schematic construction for the integration of sensors

Composite structures, which exhibit an optimum of efficiency, functionality, precision, adaptivity, capability for self-repair, and lifespan, stand at the focus of this development. Fiber and textile strengthened plastics offer particular advantages because of the high variability in the adjustments of a desired characteristic profile and their high potential for lightweight structures. An interdisciplinary cooperation of scientists, of

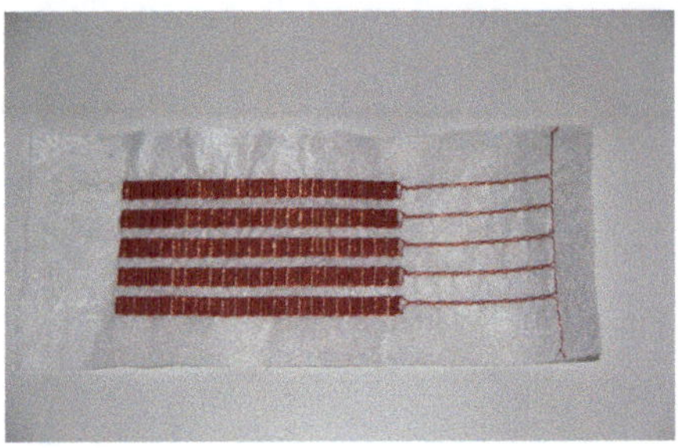

Fig. 6.193 Wire sensor stitched into a textile

Fig. 6.194 Sensor embedded in cement mortar

Fig. 6.195 Integrated measurement and data acquisition

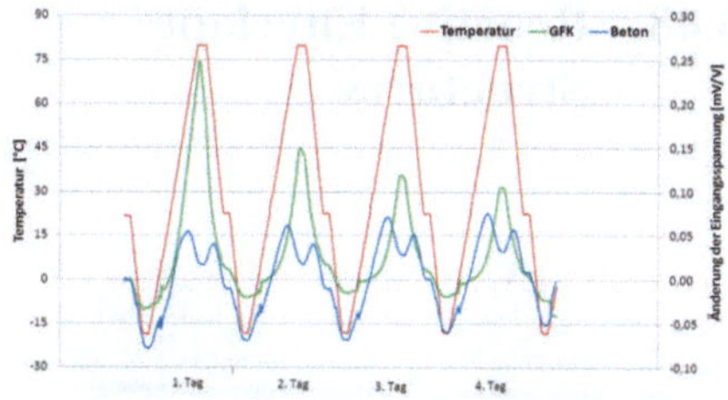

Fig. 6.196 Results of the integration of sensors in fiber-strengthened concrete: Both the strengthening fibers and the concrete expand and contract with rising and falling temperatures

the Competence Center of Lightweight Structures (Kompetenzzentrum Strukturleichtbau, SLB) e.V. at the TU Chemnitz with the professorship of lightweight construction and plastic manipulation under the leadership of Lothar Kroll and the professorship of circuitry and system design, successfully developed so-called direct material control (DMC) system regulation (Figs. 6.192, 6.193, 6.194, 6.195, 6.196). Active structure concepts have the ability to adapt their behavior and characteristics to a multitude of outside influences. The high flexibility in the structural design and technological execution of fiber–plastic composites primarily allows active structures to be outfitted with integrated sensors and actuators and to connect to intelligent and complex systems with an appropriate control strategy in combination with a capable signal manipulation.

The first stitch sensors developed at the SLB e.V. were given their first application in the innovative DMC system. The textile-like fabricated sensor is an element of a variably constructed lightweight composite structure and serves to generate the controlling signal, that is, for a robot. The system commands self-adapting and evaluating electronics, which on the one hand allows adaptation to outside requirements and on the other realizes a standardized output of data. The further development of intelligent fiber composite materials leads, based on the integration of information, sensory, and actuator technology, to complex function-oriented systems. Up to now the most diverse materials have been available, that is, piezoelectric, fiber-optic fibers (so-called fiber Bragg grating), also shape memory alloys and prefabricated information elements, that is, strain gauge strips. From these components active composite materials or "smart composites" can be produced with targeted characteristics that are particularly suited for application in stressed building parts consisting of fiber composites. An example of an application in architecture is the ability to determine the amount of moisture in cement-bound systems with means of stitch sensors. Further applications occur for the measurement of strain in fiber–plastic composites or in fiber-based probes.

6.48 Reactive Envelope Structures

Cones of conifers, i.e. the cones of pines and spruces, possess hygroscopic form-changing abilities, which are evoked by anisotropic behavior of the wood fibers. The cones open and close themselves as a reaction to the environmental conditions of "humidity" and "aridity" and as a result release their seeds. The opening mechanism requires no mechanical systems, but is a result of passive elongation of the wood fibers, by which the movement is evoked. The material physically changes with water uptake (adsorption) or release (desorption); the cones open in dry conditions and close in moist conditions.

Fig. 6.197 Spruce cones, *right opened* (dry conditions) and *left closed* (wet conditions)

Technical Application

Reactive envelopes following the precedent of conifer cones have incited researchers of the ICD under the leadership of Achim Menges at the University of Stuttgart to develop systems that can react to weather conditions without motors. A. Menges and Steffen Reichert executed studies for this purpose: "This anisotropic elongation was used to develop an air humidity-driven veneer composite. A thin cut of maple wood veneer was utilized, as it exhibits a relatively high tangential elongation with comparably low modulus of elasticity. A change in the relative humidity from i.e. 40–70 % leads to a quick change in size of the veneer, which is translated to a notable change in shape: from an originally flat form to a highly warped one. The veneer composite element uses the reactive material characteristics in surprisingly simple building part that is at once an integrated sensor, energy-less

Fig. 6.198 Reaction of the veneer elements to humidity

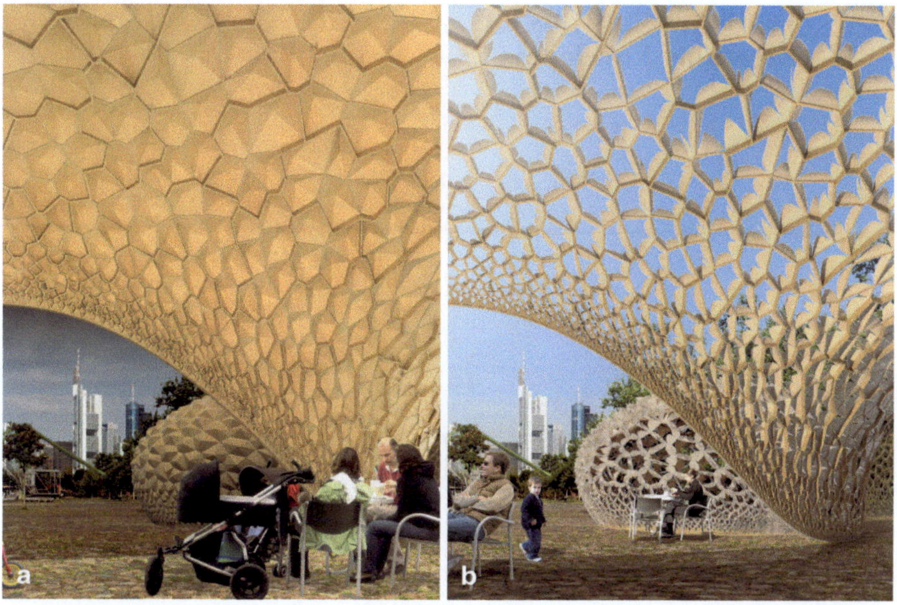

Fig. 6.199 Opening mechanism of a roof structure: *left closed* (wet conditions) and *right opened* (dry conditions)

actuator, and modulating flap. An integrated functionality of this type on the material level allows complex, decentralized behavior patterns without any control units. Each veneer composite element reacts to its specific location, functions completely independent from the others and forms in combination a highly robust, decentrally driven, adaptive system" (Figs. 6.198, 6.199, and 6.200)

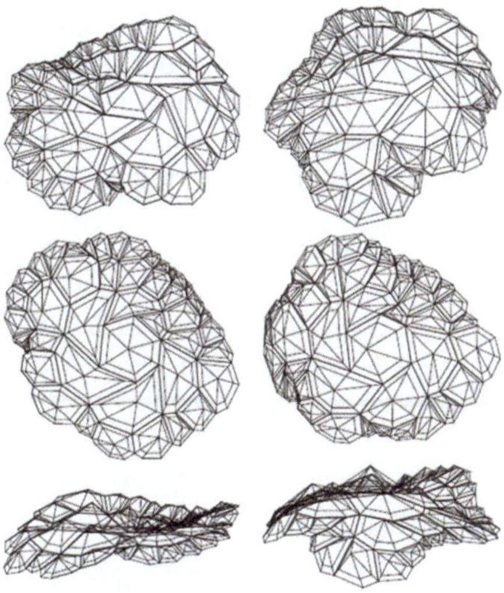

Fig. 6.200 Computer-based generation for the translation to CNC cut patterns of the roof surfaces

6.49 Ventilation Systems for Breathing Envelopes

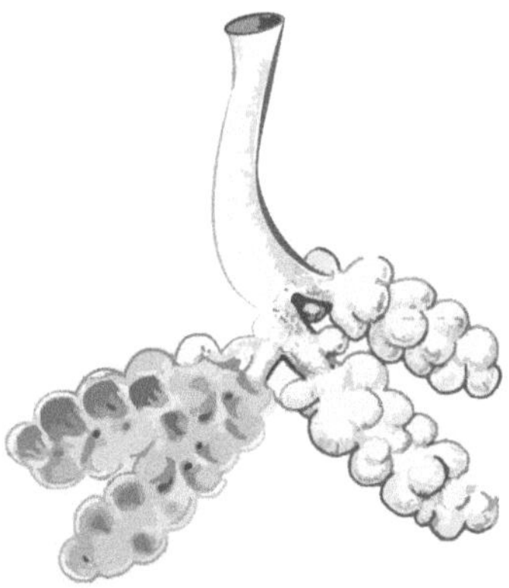

Living organisms have developed particularly effective technologies and geometric arrangements that allow them to distinctly condition themselves to different media. They can sense local changes in global movements and react to them. The matter exchange in biological systems, the ventilation of respiratory organs and circulation of substances are particularly inspiring. The sponge Ascinoide possesses an interesting geometric regularity that allows water to be circulated through and around its skin. The sea sponge is exemplary for stability and water passage. Precedents for respiratory systems are: gills, lungs, and spiracles.

Fig. 6.201 Lung—bronchial anatomy

Technical Interpretation

The technology of breathing components developed by Lidia Badarnah at the TU Delft (Badarnah and Knaak 2007) translates the active principles and methods of natural respiratory and circulation systems to elements for building envelopes (Figs. 6.202 and 6.203). The biomimetic-inspired breathing envelope for buildings functions on the following principles:

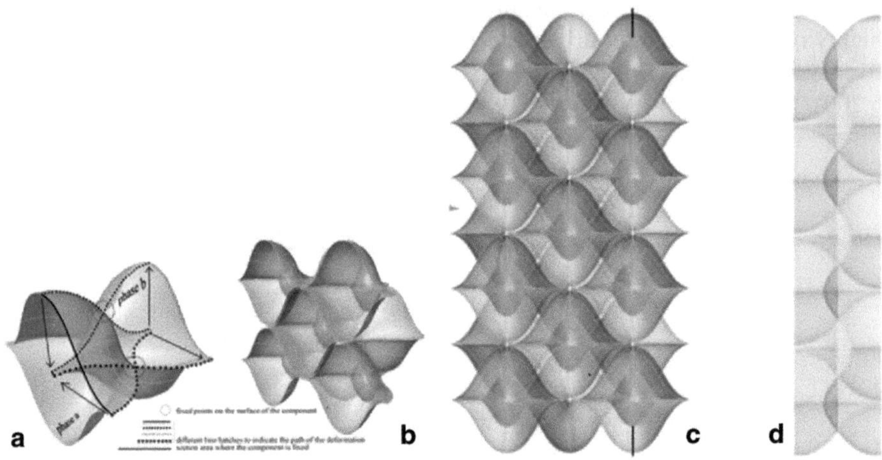

Fig. 6.202 a–d Construction of the basic components of a breathing envelope

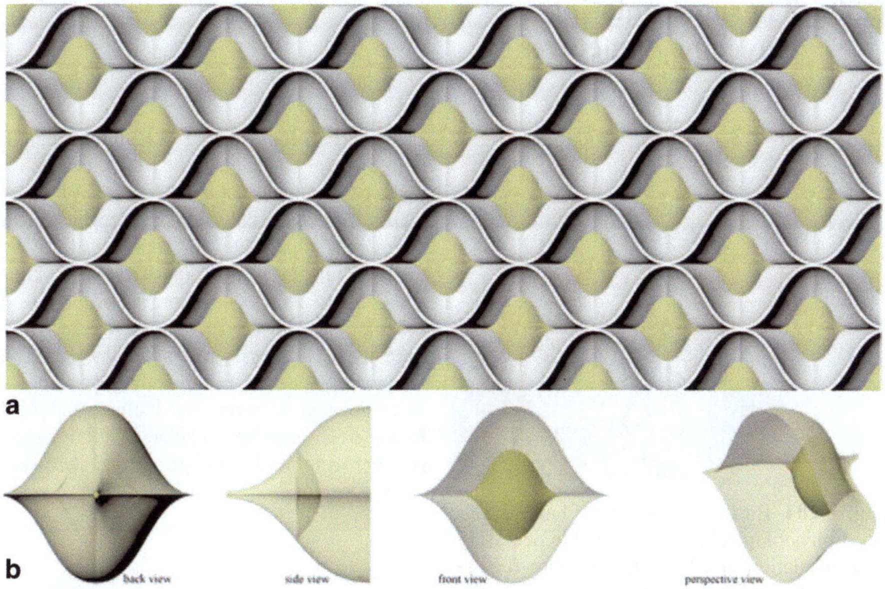

Fig. 6.203 **a** Arrangement in an envelope system. **b** Elevations and section

1. Generation of pressure gradients with the movement of building parts
2. Extension and contraction of volumes for the generation of "suction" and "exhaust"
3. The system is hierarchically membered
4. The air exchange is controlled by the shape of the surface form.

The respiratory organ, and the ventilating system as a whole, is an active system and forms the protective envelope for the building. The envelope consists of singular active components that are arranged similar to cell walls. The macro-arrangement of the cells as well as the microsystem within the cells react dynamically.

The cells are constructed of membranes of different porosity and result in gas exchange. They are partially gas permeable, half-permeable to impermeable. The inner membrane elements form, using a double membrane system, lung-like chambers as central respiratory organs and are similar to air supported expanding structures. These chambers are coupled to sensors and can elongate themselves with piezoelectric signals and change in volume. They are outfitted with openings and simple flap mechanisms that intake air with expansion and remove air with contraction. The complex system can simultaneously "inhale" and "exhale": While certain chambers allow for the intake of air, others provide for the outflow. Air then cannot flow in the opposite direction. With development of new materials, a self-adaptive facade technology is proposed that can adjust itself intelligently to changes.

6.50 Thermoregulating Envelope Structures

Fig. 6.204 Birds and mammals have developed physiological traits that enable necessary heat insulation and body temperature regulation for thermoregulative function

Living organisms require a certain temperature ratio in order to maintain their biological functions. Many organisms have developed physiological methods or behavioral patterns that yield thermoregulation. The activators that affect the thermoregulating system are affected by radiation, convection, conduction, absorption, and evaporation. The organisms have respectively developed specific systems in each instance. The thermoregulation of ectothermic organisms is determined by behavior and biologic function. Endothermic organisms regulate their body temperature with the aid of physiological characteristics.

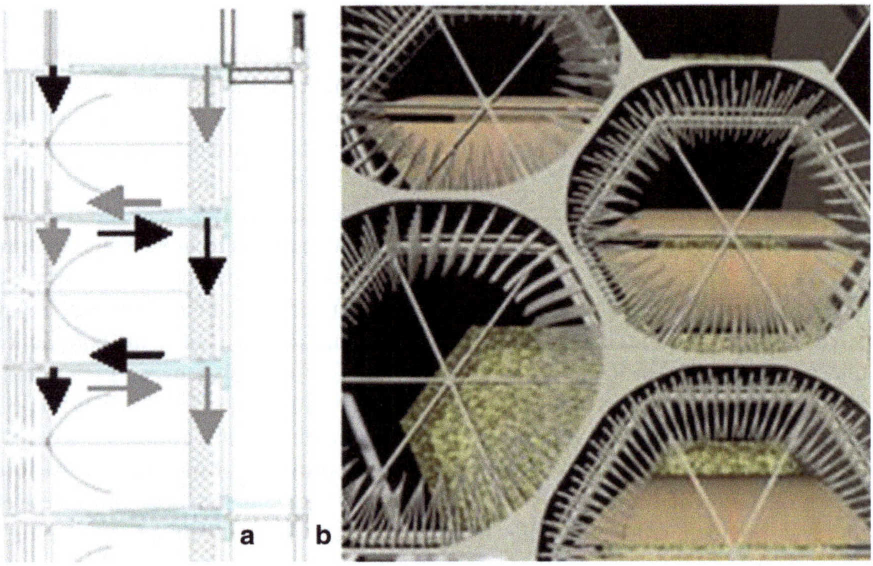

Fig. 6.205 **a** and **b** Cooling facade: integrated irrigation system for water evaporation in the building envelope

Natural system	Deep principle
Stoma of plants	*Osmotic pressure changes control openings for evaporation*
Pine cone	*Relative humidity changes cause material deformation*
Hair around eyes	*Protection against small particles (e.g. dust and sand)*
Human skin	*Latent heat transfer - Cooling trough evaporation*

Fig. 6.206 Functioning principles that were used for the development of the cooling facade

Fig. 6.207 View of a cooling facade for an arid region

Technical Interpretation
Living nature contains an uncountably large number of organisms that could act as the model for the technological development of thermoregulatory processes. The unique strategies of natural thermoregulation enable an adaptation behavior of the "envelope" as the answer to variable temperatures in the environment and inside the organisms. The various species are individually adapted to different climate regions with certain temperature ranges within which the organism can survive. The examples in nature selected here for observation were chosen due to their having developed strategies and mechanisms for a constant body temperature that it can adhere to despite variable climates (Figs. 6.205 and 6.206).

Thermoregulating systems of living nature were more closely investigated at the TU Delft by Badarnah et al. They focused themselves particularly on the following systems and functions:

- Termite hills—passive ventilation
- Tuna fish—heat exchange
- Human skin—transpiration
- Birds—Cooling by means of larynx vibration

The group at the TU Delft (Badarnah et al. 2010) compiled a classification (Fig. 6.206) of possible applications for building envelopes and discussed primary interpretations of these systems. These interpretations included an evaporation-cooled wall for application in arid regions (Fig. 6.205). Figure 6.207 illustrates a system for warm, arid, or humid regions using an envelope with an enclosure system that, when moist, allows air to flow inward and during dry and hot weather conditions allows air to flow outward. The system consists of four integrated modules (Fig. 6.206, courtesy of Lidia Badarnah). (1). "Stoma Brick": the functional part of the thermoregulation system consists of an outer filter with filter hairs (filter for debris) and a "venous" enclosing flap that enables opening and closing in relation to air moisture. A large part of the inner layer is spongelike to absorb moisture for evaporation. (2). "Mono-brick": contains the irrigation mechanism. (3). Steel frame. (4). Inner layer: HEPA filter for air purification or acrylic glass panels for window openings.

6.51 Modifiable Surface Elements 1

Active and passive movements are of particular interest for scientists. The folding pollination mechanism of the bird of paradise flower Strelitzia reginae is a reversible mechanism that is not autonomously evoked. The strelitzia offers birds a perch of two flower petals grown together that bends downward from the weight of the bird. With the accompanying horizontal movement the flat lamina open outwards which otherwise enclose the stamens. When the stamens are exposed the pollen clings onto the body of the bird and is transmitted by by the bird to the stigmata of the next plant.

Fig. 6.208 *Strelitzia*

The movement of the *Strelitzia* flower does not occur autonomously but dependent on an outside influence. The reversible elastic deformations require no additional "mechanics" and can function with a nearly endless number of uses. The Plant Biomechanics Group at the University Freiburg (Prof. T. Speck) investigated the function morphology of the *Strelitzia* and confirmed that the flap mechanism retains its reversible functionality even after over 3000 uses (Fig. 6.209).

At the University of Stuttgart through the Institute for Building Structures (ITKE; J. Knippers) and the Institute for Textile Technology (ITV) Denkend-

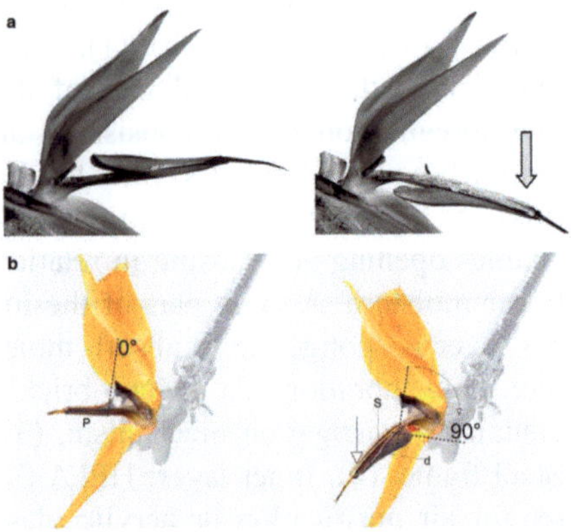

Fig. 6.209 **a** and **b** Flap mechanism of the flower and **c** measured displacement

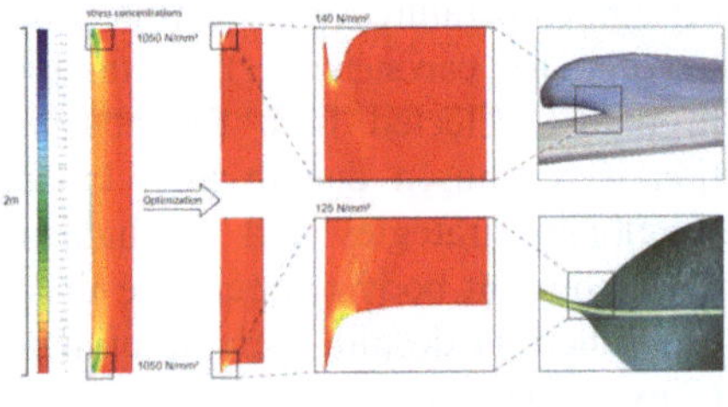

Fig. 6.210 Reduction of the stress at the fold through biomimetic optimization of contour lines

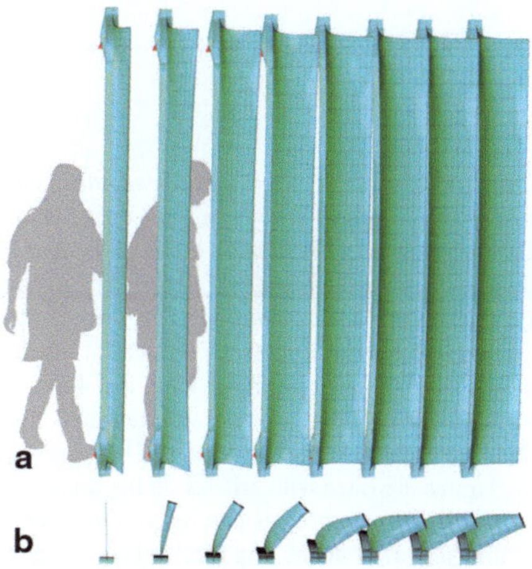

Fig. 6.211 **a** and **b** Finite element analysis of facade shading

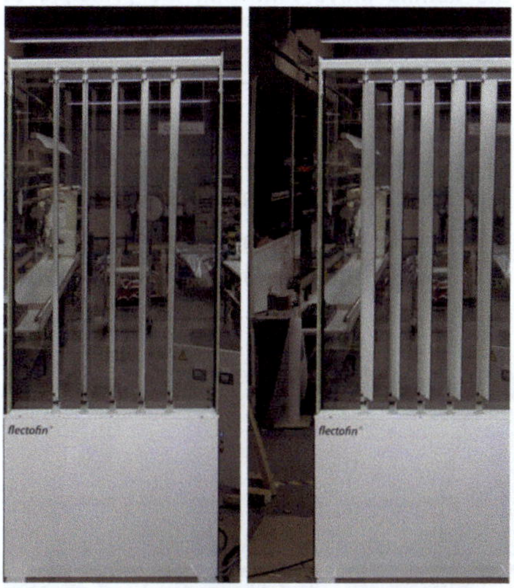

Fig. 6.212 Abstraction of a section: the laterally displaced spine leads to lateral torsional buckling of the entire shell element

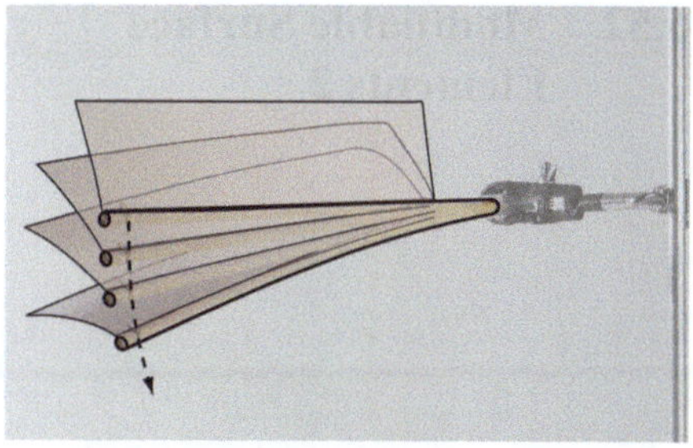

Fig. 6.213 **a** and **b** Model illustration of the functioning concept

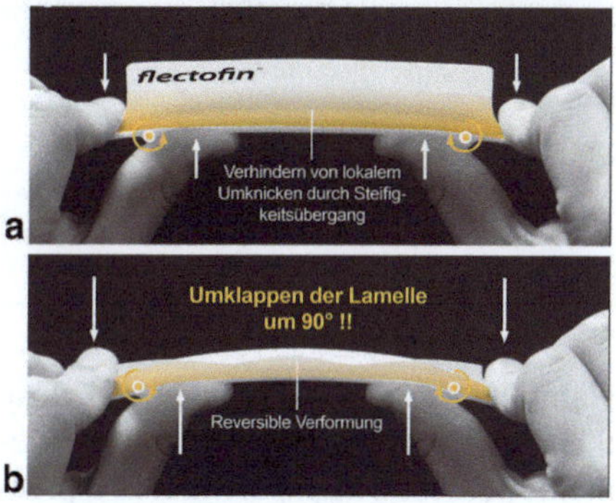

Fig. 6.214 Flectofin© in the first mock-up

orf, this function was then translated to a shading lamella that reacts to an external force with a lateral bend due to lateral torsional buckling. The research results concluded in a patented technology for a shading facade, Flectofin© (Figs. 6.210, 6.211, 6.212, 6.213, 6.214)

6.52 Modifiable Surface Elements 2

Carnivorous plants have developed different mechanisms to catch their prey. The venus flytrap Dionaea muscipula is a carnivorous plant of the family Droseraceae (sundew family). Their root system serves for the anchoring of the plant in the ground and for water intake. For nutrient intake they utilize several centimeter long, leaf-formed traps (lamina). These are red colored and lure their prey with secretions of nectar. The edges of the lamina are lined with bristles. In the center of lamina sit sensory bristles that react to the touch of prey and activate the trap mechanism.

Fig. 6.215 Venus flytrap *Dionaea muscipula*; the lamina move following the principle of thigmonasty

Natural apparatuses that open and close themselves without mechanical elements possess a high potential as a precedent for application in the building industry. Light reflecting and shading systems for buildings that essentially draw on material–property changes for their mobility and thus simplify the mechani-

Fig. 6.216 a Movement principle

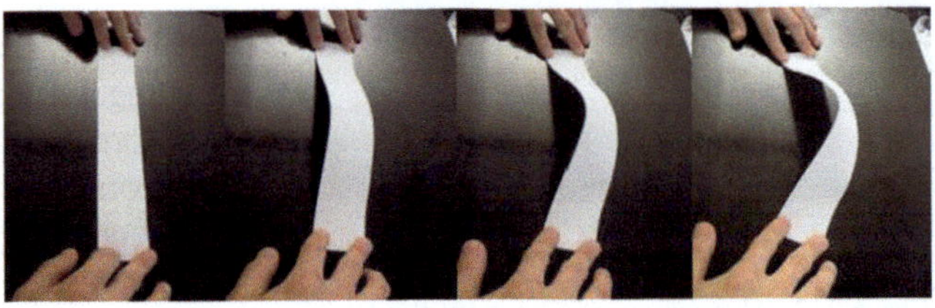

Fig. 6.217 b and **c** Calculation study of elastically deformed strips

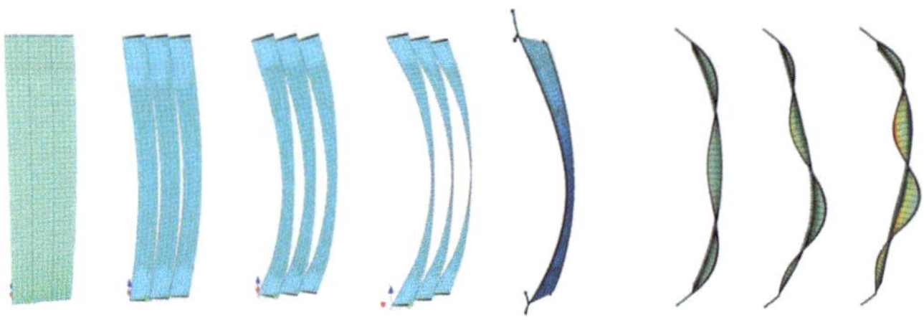

Fig. 6.218 **a** and **b** "One Ocean" EXPO Pavilion, Korea, SOMA architects

Fig. 6.219 **a** and **b** Facade segment of "One Ocean" EXPO Pavilion, Korea

cal building parts could be drawn from these natural precedents. (Figs. 6.216, 6.217, 6.218, 6.219).

The opening elements in the facade of the Theme Pavilion of EXPO 2012 in Yeosou, Korea by SOMA architects achieve movements with elastic deformation. Inspired by the research on Flectofin©, Knippers Helbig Advanced Engineering developed the technical concept of the moving elements.

The up to 15 m-tall-lamellae with strengthening ribs on both sides consist of only 8-mm-thick glass fiber-reinforced plastic. They are elastically deformed by an extrinsic force initiated from above and below.

6.53 Multiaxially Modifiable Surface Elements

The Mimosa species belonging to family of Leguminosae are indigenous to the American continent. They possess oval, longitudinally-formed pinnae arranged in pairs that when touched fold together within about a second. Additionally, the entire frond folds downward with the already folded pinnae. The frond and the individual pinnae gradually return to their original position after approximately a half hour. During the night in "sleep position" the leaves of the Mimosa assume the same folded position. This phenomenon of plant movement activated by touch is called seismonasty.

Fig. 6.220 The *Mimosa* plant in opened condition

Mimosa conveys a touch stimulation inside of the plant so that neighboring fronds react. This reaction is not autonomous but is caused by change in turgor pressure, triggered by a chemical messenger substance or electric impulses. Movement studies on plants have inspired the research teams of Pohl Architects and Knippers Helbig Advanced Engineering to the most diverse technical interpretations independently from one another (Figs. 6.221 and 6.222). For a covering of a courtyard in a former monastery, the teams cooperatively developed a multiaxial moving envelope (Figs. 6.223 and 6.224). For the protection of spectators of the local festivals in Feuchtwangen, a new type of roof envelope is being planned that can react to rain, sun, and heat and, like the

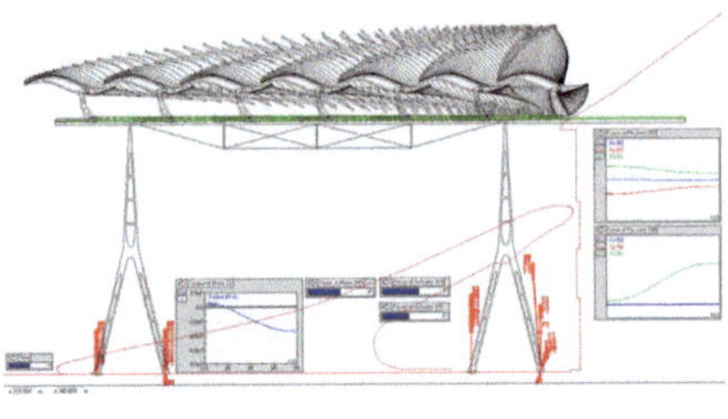

Fig. 6.221 Movement simulation with structural analysis

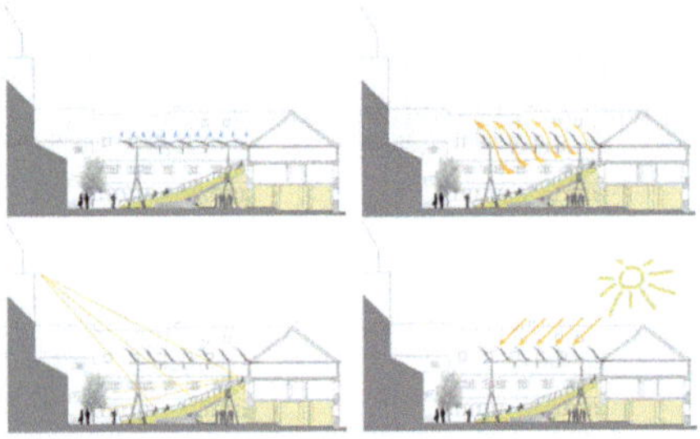

Fig. 6.222 Movement in different weather conditions: above the amplitude of movement, middle in sunny conditions, below in rainy conditions

Fig. 6.223 The roof over the monastery courtyard in closed position with light staging

Mimosa frond, consists of individually linked, adjustable panels. This adjustable roof developed analogous to the nastic movements (not autonomous) of plants consists of a leaf plumage with seven individual "pinnae" that span the entire breadth of the courtyard. In the opened position the panels are driven to the back. With the onset of rain detected by sensors the roof closes itself within 2 min. For the air circulation in the audience space the panels of the roof can smoothly position themselves into a slanted position. The vaulted shape of the panels and their shading structures provide for thermal wind ventilation and cooling. In the closed position the rain on the roof is directed into an integrated gutter system. The panels consist of specially finished ultra-lightweight parts entirely of glass fiber-composite construction.

Fig. 6.224 Various open roof positions. The complete retracted position of panels is not represented

6.54 Reactive Contraction Systems

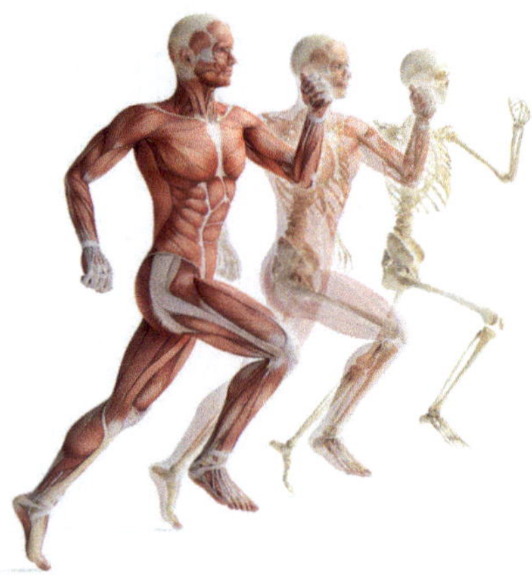

Muscles

The muscle (lat. Musculus) is a contractile organ, consisting of tissue that can tighten or slacken itself. The particular coordinations of muscles enable movements, i.e. running, but also the ability to stand upright and the abatement of burdens. Along with ring muscles (eye, mouth, anus), vertebrates possess cavity muscles (esophagus, stomach, heart) and multi-headed muscles, like the bicep muscle (M. biceps brachii).

Fig. 6.225 Muscle biology

Lightweight structures, like the stressed ribbon bridge, vibrate heavily under the weight of pedestrian traffic. At the TU Berlin under Mike Schlaich and Achim Bleicher, an active vibration control system was developed and tested using artificial muscles to reduce the exceptionally high susceptibility to vibrations in stressed ribbon bridges using carbon fiber-reinforced plastic bands (Figs. 6.227, 6.228, 6.229, 6.230). The concept of reactively contracting systems is based on the controlled input of induced forces in the handrail structure. For the generation of these forces industrially manufactured pneumatic muscles of the firm FESTO (Fig. 6.226) were used. These artificial muscles expand themselves radially with an increase of internal pressure causing them to

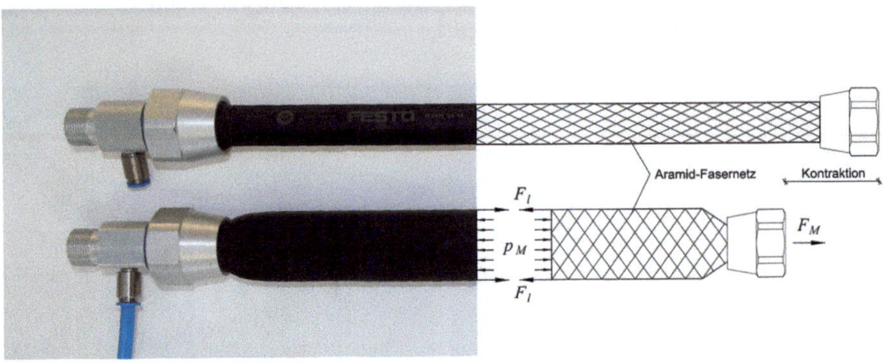

Fig. 6.226 FESTO pneumatic muscle

Fig. 6.227 Actively regulated stressed ribbon bridge with sensors, actuators, and controllers

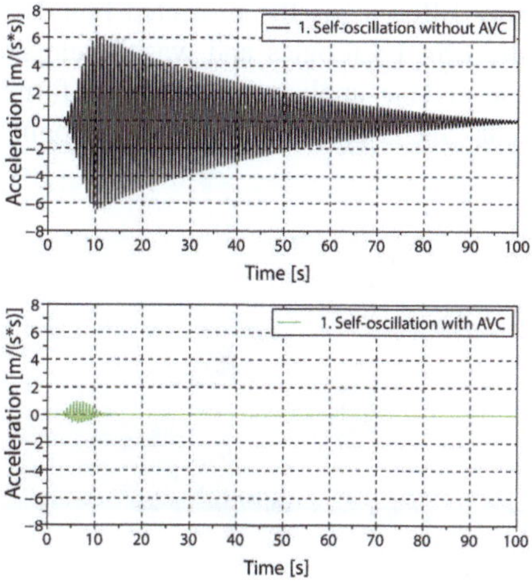

Fig. 6.228 Pedestrian induced accelerations in the natural vibration frequency with and without active vibration controls

Fig. 6.229 Load-bearing/vibration damping simulation, "walking" on the prototype, TU Berlin

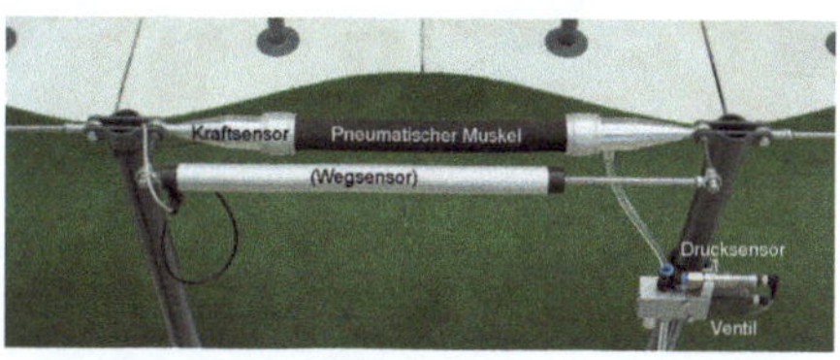

Fig. 6.230 Pneumatic muscle with related equipment

contract in length. With especially developed algorithms, the contractions and the induced forces are regulated in relationship to the occurrence of bridge vibrations, thereby stabilizing the bridge with "muscle strength."

6.55 Self-responsive Movements, Fin Ray Effect®

The tail fins of some fishes (Fig. 6.231) consist of fin rays that always run nearly parallel to one another and are bound with connective tissue. These rays then are the structural elements of the tail fin and each are formed with different hardness according to the species of fish. The noteworthy aspect of this simple and robust structure is that the fins concavely deform to lateral pressure while the connected tips of the rays experience roughly no displacement, as one would expect from a normal cantilever.

Fig. 6.231 Tail fins of two supporting fin rays

The Fin Ray Effect® discovered with the movement patterns of fish fins depicts a function principle that is interesting for various technical applications. The "Fin Ray Effect®" is the protected brand of the firm EvoLogics and was developed for diverse applications, such as form-adapting gripping elements for gripping

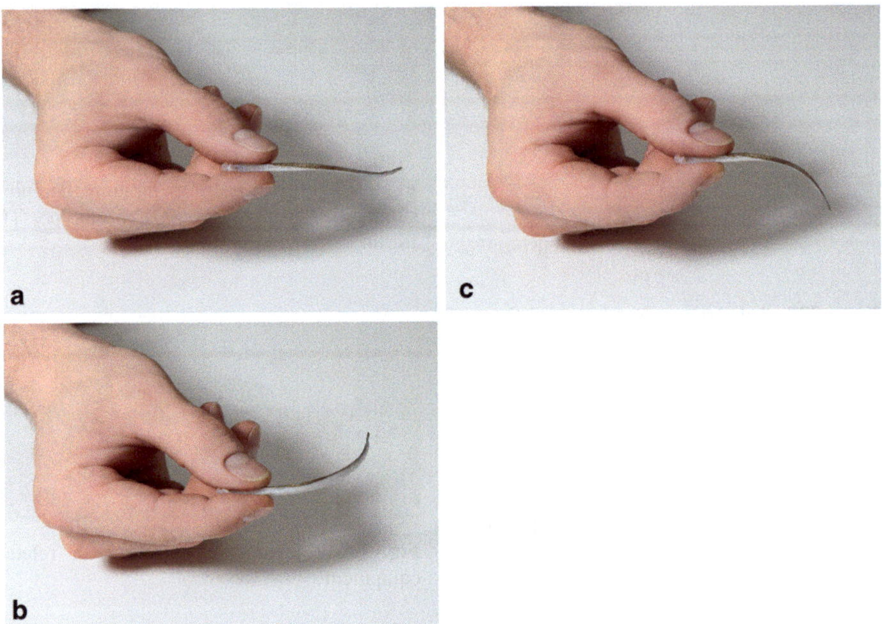

Fig. 6.232 A subtle shift of the fingers moves the fish fin

Fig. 6.233 (*above*) Fin Ray Effect® on a plaice

Fig. 6.234 Fin Ray Effect®: Different movement patterns illustrated in model. The number of cross braces is not important for the bending behavior of the entire system

devices or demonstrations at exhibitions. With this naturally occurring effect the elastically coupled element "fish fin" reacts to pressure with a movement in the direction of the pressure (Figs. 6.232 and 6.233). The same happens when bands running in the direction of the fin rays are energized. With a deformable tail fin of this type the fish can propel themselves from the alternating eddies of the vortex streets they produce. Researchers at the TU Berlin with Mike Schlaich and Annette Bögle of the HCU Hamburg are developing applications for construction that can use the Fin Ray Effect® (Fig. 6.234). An openable joint solution for textile canopies and membrane roofs according to the fin ray principle does not behave like traditional joints to the principle of squishing the material, but instead uses the geometric deformability and the system-given nestling of the structure.

6.56 Flexible Shells

High flexibity despite a hard shell

Evolution has enabled some species of isopods, for example the pill bug (Armadillidium vulgare (Fig. 6.235) to roll themselves up into a stable, closed ball in situations of danger. Isopods are completely covered on the surface with a carapace that is formed by an exoskeleton, which gives the body its shape and stability. This skin represents a product of secretions from the epidermis cells and is what enables the arthropods to live on land at all. They can either be rigid, as in the case of plate-shaped spicules, or soft, as with bendable membranes, connected by spicules.

Fig. 6.235 Pill bug *Armadillidium vulgare*

Technical Application
Interesting possibilities of design development are offered by these relocating shells, present for example in pill bugs. Similar principles have been known for quite some time: In the Middle Ages knights' armor was designed to offer a certain freedom of movement despite the rigidness of the material. Joint pieces at the knee and elbow were especially outfitted with elements that overlap each other in the manner of the pill bug.

At the School for Architecture Saar of the HTW Saar, studies for a segmented bridge following the precedent of pill bugs have occurred under the leadership of Göran Pohl (Figs. 6.236, 6.237). The roll bridge is composed of individual elements that can be rolled together like the shells of the pill bug. The elements are connected to one another with a hinge joint that hinders lateral torsion and enables it to be rolled up. The bridge unrolls all of the segments to a

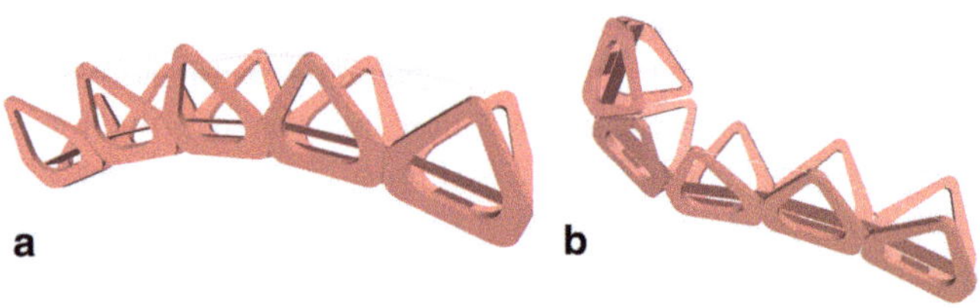

Fig. 6.236 Roll bridge in the process of movement (**a** and **b**). The bridge rolling together (**c** and **d**)

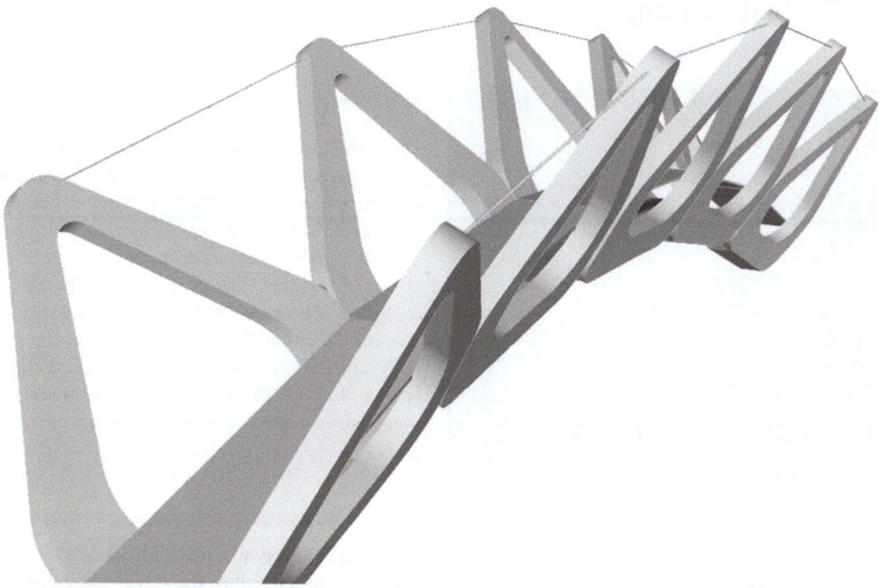

Fig. 6.237 Model of a roll bridge

slightly overextended, curved position yielding a structurally sound arch-shape form. The bridge is then operational in the completely unrolled position. Connecting the individual segments with the means of cables enables the bridge to be rolled up to a compact bundle like the pill bug. A further connection is considered for the underside of the roll bridge between each neighboring element, which would distribute the loads occurring in the unrolled position to the rigid frame pieces of the bridge. With this construction method a small roll bridge for pedestrians for short spans is imaginable; one that is flexibly opened and closed.

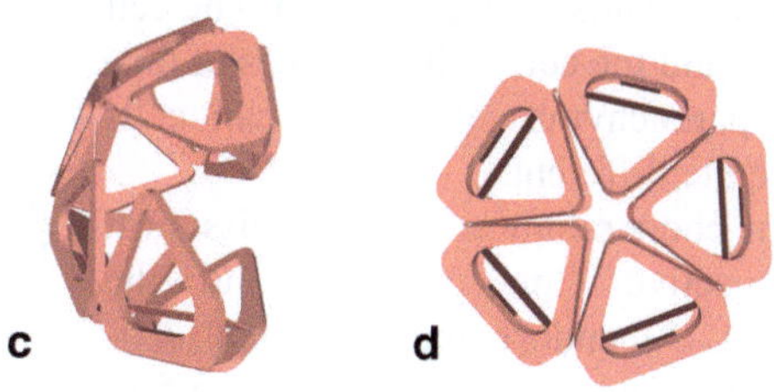

6.57 Self-healing

Tree-bark that encloses wounds

The bark closes off injured regions with the help of a self-repairing procress. Fig. 6.238 illustrates a large protective enclosure around a branch. Such branch stubs enclose themselves after being damaged by storms or trimming and are often completely sealed. The probability of complete enclosure depends essentially on the intensity of damage. A circle-shaped injury is correspondingly circularly enclosed, a linear injury linear or oval. In a beech tree the complete closure of saucer-sized wound lasts about 10 years.

Fig. 6.238 Tree wound with clearly visible enclosure as a result of the self-healing process

The vine *Aristolochia macrophylla* was investigated in the botanical garden at the University Freiburg as study object for the effects of self-healing. The process is carried out by the closure of a wound in several phases. In the first phase parenchyma cells swell in the wound and seal it (Figs. 6.239, 6.240, 6.241).

The sealing presumably occurs by a viscoelastic–plastic expansion of cells, driven by cell-turgor pressure. Subsequently, parenchyma repair cells form that grow into the wound and thicken their cell walls (Fig. 6.242b, c). The cell shape and the cell wall thickness of normal parenchyma cells and repair cells are quite different.

In cooperation with the Swiss firm Prospective Concepts AG and the EMPA Dübendorf, the biological self-healing process was translated to a fabricated membrane using the Tensairity® concept. The technology consists of an air-filled membrane, pre-stressed with a slight internal pressure of 50–500 mbar, that is stabilized by cable elements and pressure bars. In the instance of damage

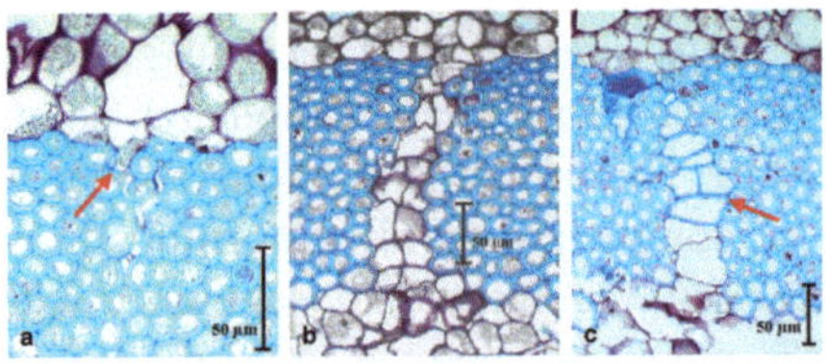

Fig. 6.239 a–c Self-healing in *Aristolochia macrophylla*. Parenchyma cells close the wound

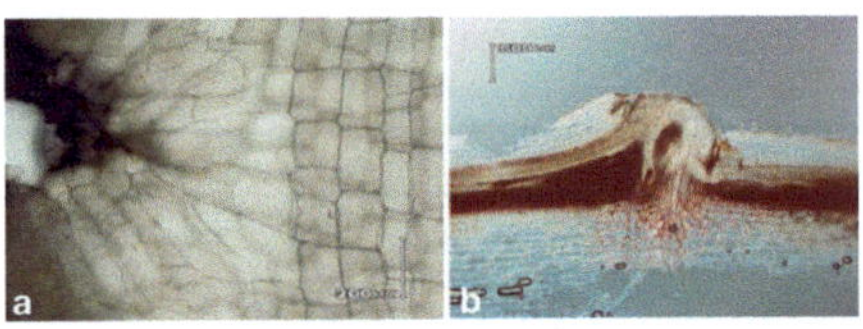

Fig. 6.240 Self-healing after an outer wound on the bean plant *Phaseolus vulgaris* (**a**) early phase of self-repair: parenchyma cells fill the wound, (**b**) later phase with swelling and wound closure

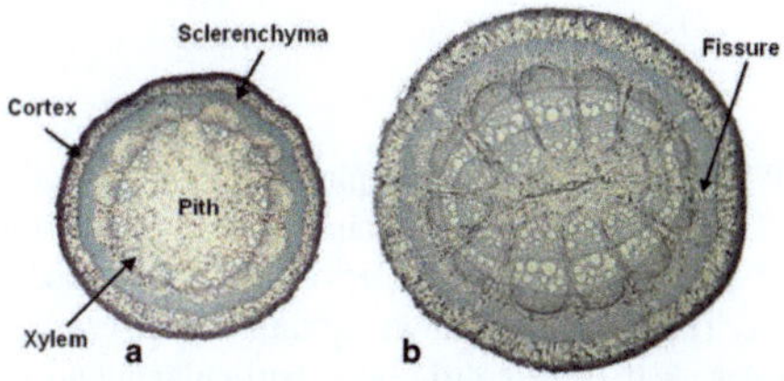

Fig. 6.241 Section of the vine *Aristolochia macrophylla* (**a**) 1-year-old trunk with closed ring of sclerenchyma fibers. (**b**) As a consequence of the yearly growth, a 2-year-old-trunk displays a growth of the xylem and a segmenting of the sclerenchyma ring

Fig. 6.243 (*above right*) Tensairity® system

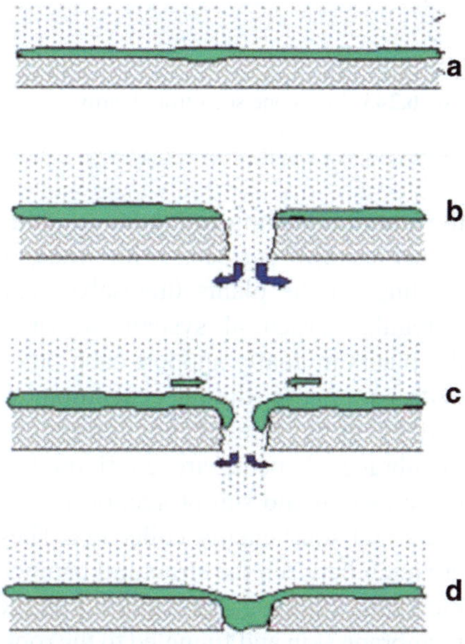

Fig. 6.242 **a** Layer construction from above to below: Air with high pressure. *Green*=Active repair layer. Membrane. **b–d** A hole in the air chamber is subsequently sealed with the foam until the hole is completely filled and the air loss stopped

Fig. 6.244 (*middle*) Components of the Tensairity® system are a long pressure bar, an inflatable membrane for the pneumatic base element, cables for radial and counter-radial tensioning, anchoring parts

to the membrane air begins to flow out of the Tensairity® element. As the air is introduced into the system with little pressure, it escapes slower than usual. In the studies the biological principle was translated to a self-healing membrane. The technological development is based on an additional foamy, "cellular" membrane layer, which can reseal the membrane in case of damage. The repairing layer is located on the inside of the membrane. The repair process functions in a similar manner to the natural precedent: the injury is sealed by a closed pored foam layer (Fig. 6.242). The possibility of self-repair depends on the amount of damage. Layers with a polyurethane basis have already yielded promising results. Initial uses for lightweight bridge structures and pneumatic roofs are currently being tested (Figs. 6.243, 6.244).

6.58 Bambootanics

Within the field of botany, phyllotaxis is the study of leaf arrangements of plants. The coordination of outer leaf organs like leaves, branches, seeds, etc. (Fig. 6.245) follows certain nature-created rules; leaf organs grow at a particular angle to one another. The structures of plants often exhibit spiral patterns that are determined by the Fibonacci sequence. With this arrangement a large number of leaves or seeds can be fit into a small space, and also arranged so that the most amount of sunlight can be distributed. This arrangement allows no instances where one leaf shades another leaf.

Fig. 6.245 Pine cone seen from below

Technical Application—Bambo(o)tanics

Bambo(o)tanics is a self-growing structural system using bamboo, which was developed by Niko Feth at the School for Architecture with G. Pohl and L. Bergrath, HTW Saarbrücken. Following the insights of phyllotaxis of pine cones, the structure consists of individual, modular canopies formed from bamboo stalks and strung together (Figs. 6.246, 6.247, 6.248, and 6.249). For the construction of the individual canopies, the bamboo stalks are bent into the canopy form while being grown. The curving of the stalks according to Fibonacci spiral yields canopy surfaces with minimal shading on the plants themselves and a regular structural system in which the plants mutually support each other. With parametric, digital generation according to the rules of phyllotaxis, the membrane elements are configured to serve as rain and sun protection for the users and solar energy collectors. They are hung between the structural bamboo members. Rainwater is captured on the membrane skin and funneled to the roots of the plants. The photovoltaic film-coated, semi-permeable membranes are

Fig. 6.246 Bambo(o)tanic during its growth process

Fig. 6.247 Fully developed bamboo canopy for an outdoor market in a tropical region. The solar membranes serve as Sun and rain protection, lead water to the plant roots, and integrate photovoltaic (PV) modules for solar energy production

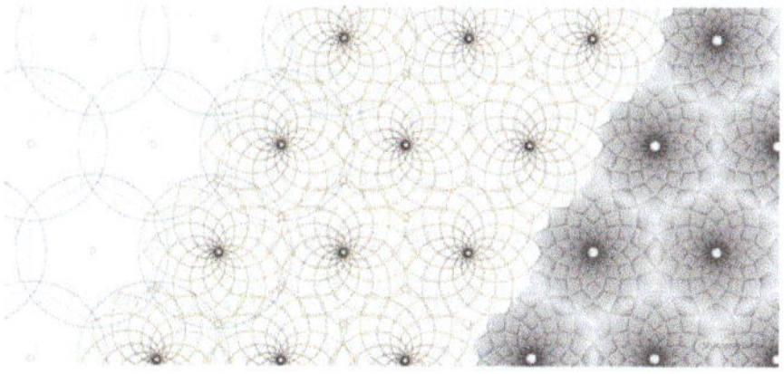

Fig. 6.248 Plan view of the system: Hexagonal arrangement of the supports, arrangement of the bamboo stalks, and arrangement of the solar membranes

aligned following the principles of phyllotaxis arrangement, so that the growth of the bamboo branches and twigs is not hindered in their natural tendency to fill light gaps. This manner of construction is particularly suitable for the creation of weather protected shelters and offers a system that integrates technology and nature for a canopy in tropical regions.

Fig. 6.249 Elevation

6.59 Floating Volumes

The Portuguese man o' war (Physalia physalis) is actually a conglomeration of smaller organisms which is buoyed on the surface of the ocean by means of an air-filled volume. Indeed, the appearance of the man oí war strongly resembles that of a jellyfish; in reality, it consists rather of a colony of polyps that are completely independent from one another. These individuals specialized themselves to certain tasks in the course of evolution, such as defense or reproduction. The Japanese water chestnut (Trapa natans) is an annual water plant that appears in standing waters. Floating elements on the fan-shaped leaves allow them to float in a rosette form on the water's surface.

Fig. 6.250 **a** Portuguese man o' war *Physalia physalis* **b** Japanese chestnut

Floating Habitats

The system inspired by the Portuguese man o' war *Physalia physalis* is a concept for an urban habitat structure for inland bodies of water, which was developed by Claudia Pommer in the frame of her university thesis at the Institute for Industrial Design in the subject area of engineering and industrial design of the Hochschule Magdeburg-Stendal under Ulrich Wohlgemuth (Figs. 6.251, 6.252, and 6.253). Inspired by the polyp colony organism *Physalia physalis,* the idea emerged for a "camping site philosophie" on water. A complex, urban habitat structure yielded itself as a combination of differently sized public spaces, housing units, and connecting footbridges following the precedent of the Japanese water chestnut. Every platform offers five docking positions for the ca. 32 m^2 large, modularly constructed living units. Each unit is movable with the use of an electric engine so that the arrangement on the water can always be differently reconstructed. Therefore the units are able to attach and detach at different docking positions as they wish. A pontoon forms the core of living unit. For protection while docking, a surface of flexible material that can adjust to the movement of water surrounds the core and offers a place for relaxation. A pneumatic structure with a place for sleeping is attached to the top of the pontoon. The "cocoon" dwelling is protected by

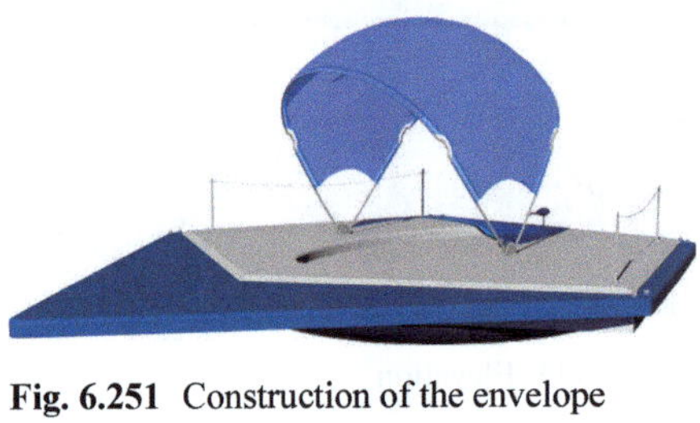

Fig. 6.251 Construction of the envelope

Fig. 6.252 Visualization of habitat structure

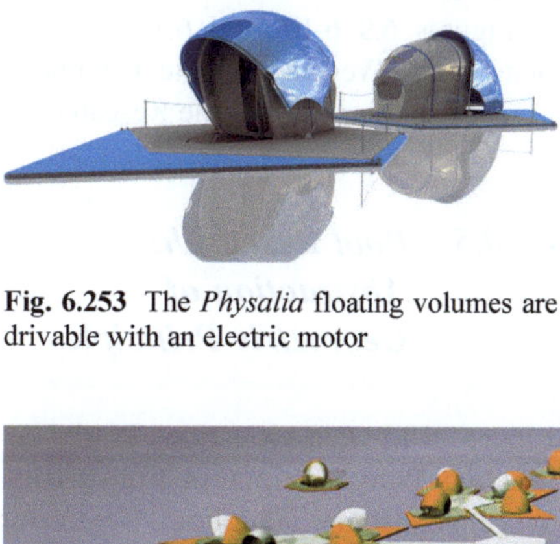

Fig. 6.253 The *Physalia* floating volumes are drivable with an electric motor

an adjustable covering with integrated solar panels that provide energy for the cocoon. To lend the individual units a certain level of individuality and recognizability, different colors, patterns, and lighting can be used (Fig. 6.254).

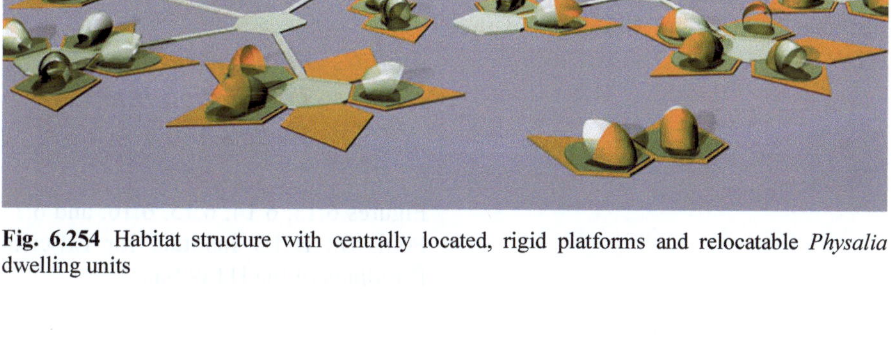

Fig. 6.254 Habitat structure with centrally located, rigid platforms and relocatable *Physalia* dwelling units

6.60 Sources, Figure Index, Authors and Project Contributors in Chap. 6

Further information and advice for the subchapters in Chap. 6. If not written separately, the institutions are based in Germany.

6.60.1 Biomimetics on the Basis of Algae, a Biological Example

Hamm, C. 2005, Kieselalgen als Muster für technische Konstruktionen, BIOSpektrum 1/05, 41–43

Figure 6.1 Hustedt Collection, Alfred Wegener Institute Bremerhaven, Photo: Hinz/Crawford

Figure 6.2 L Friedrichs, Alfred Wegener Insitute, Helmholtz Centre for Polar and Marine Research

Figure 6.3 after IL38 Diatomeen,2, S. 45

6.60.2 Pool Research as Biomimetic Method in Application

Figure 6.4 Construction scheme of a diatom shell. Image: Pohl, G., Otten, J., Research group BOWOOSS, B2E3 Institute for Efficient Buildings of the HTW Saar

6.60.3 Pool Research: Abstraction through the Classification of Biological Precedents

Figure 6.5 Excerpt from the classification of diatoms, Pohl, G.

Figure 6.6 Basic forms of diatoms, Pohl, G.

6.60.4 Pool Research: Analysis and Evaluation

Figure 6.7 C. Hamm, Alfred Wegener Insitute, Helmholtz Centre for Polar and Marine Research

Figure 6.8 Alfred Wegener Insitute, Helmholtz Centre for Polar and Marine Research

Figures 6.9 6.10, and 6.11 L Friedrichs, Alfred Wegener Insitute, Helmholtz Centre for Polar and Marine Research

6.60.5 Pool Research: Abstraction of Geometric Principles

Figure 6.12 Classification Pohl, G., graphics Pohl, G., images from Alfred Wegener Insitute, Helmholtz Centre for Polar and Marine Research

6.60.6 Pool Research: Translation into CAD Models

Figures 6.13, 6.14, 6.15, 6.16, and 6.17 Pohl, G., B2E3 Institute for Efficient Buildings of the HTW Saar

6.60.7 From Pool Research to Applied Research

Figure 6.18 Pohl, G., B2E3 Institute for Efficient Buildings of the HTW Saar

6.60.8 Generative Design

Figure 6.19 Pohl G.

Figure 6.20 Bartenbach Light Laboratory; Project team Behnisch Achitects, Pohl Architects

Figure 6.21 Project team Behnisch Achitects, Pohl Architects

Figures 6.122, 6.123, 6.124, and 6.125 Pohl Architects

6.60.9 Physical Models

Figure 6.26 Pohl, G., B2E3 Institute for Efficient Buildings of the HTW Saar

Figure 6.27a, b Pohl, G., B2E3 Institute for Efficient Buildings of the HTW Saar

Figure 6.27c, d Pohl, G.

Figures 6.28 and 6.29 Pohl, G., B2E3 Institute for Efficient Buildings of the HTW Saar

6.60.10 Biomimetic Potentials: Ribs and Frameworks

Figure 6.30 Feth, N., Pohl, G., Research group BOWOOSS, B2E3 Institute for Efficient Buildings of the HTW Saar

Figure 6.31 L Friedrichs, Alfred Wegener Insitute, Helmholtz Centre for Polar and Marine Research

Figure 6.32 Feth, N., Pohl, G., Research group BOWOOSS, B2E3 Institute for Efficient Buildings of the HTW Saar

6.60.11 Biomimetic Potentials: Rectangular Frames

Figure 6.33 Pohl, G., B2E3 Institute for Efficient Buildings of the HTW Saar

Figure 6.34 Pohl Architects

Figure 6.35 Pohl, G., Otten, J., Research group BOWOOSS, B2E3 Institute for Efficient Buildings of the HTW Saar

Figure 6.36 Pohl, G., B2E3 Institute for Efficient Buildings of the HTW Saar

6.60.12 Biomimetic Potentials: Layered Structure

Figure 6.37 L Friedrichs, Alfred Wegener Insitute, Helmholtz Centre for Polar and Marine Research

Figure 6.38 Pohl, G., Otten, J., Research group BOWOOSS, B2E3 Institute for Efficient Buildings of the HTW Saar

6.60.13 Biomimetic Potential: Offset Beams

Figure 6.39 Image N. Abarca, Botanical Garden and Botanical Museum Berlin-Dahlem, Free University Berlin

Figure 6.40 Pohl, G., Stolz, F., Research group BOWOOSS, B2E3 Institute for Efficient Buildings of the HTW Saar

6.60.14 Biomimetic Potentials: Incisions and Curvature

Figure 6.41 Image P. Höbel, M. Eng.

Figures 6.42, 6.43, and 6.44 Pohl, G., Stolz, F., Research group BOWOOSS, B2E3 Institute for Efficient Buildings of the HTW Saar

6.60.15 Biomimetic Potentials: Curvature

Figure 6.45 Pohl, G.

Figure 6.46 Pohl, G., Otten, J., Research group BOWOOSS, B2E3 Institute for Efficient Buildings of the HTW Saar

6.60.16 Biomimetic Potentials: Hierarchical Structures

Figure 6.47 L Friedrichs, Alfred Wegener Insitute, Helmholtz Centre for Polar and Marine Research

Figure 6.48a Cuma, Alfred Wegener Insitute, Helmholtz Centre for Polar and Marine Research

Figure 6.48b Pohl Architects

Figure 6.49 Pohl, G.

Figure 6.50 Knippers Helbig Advanced Engineering

6.60.17 Biomimetic Potentials: Fold Systems

Figure 6.51 Pohl, G.

Figure 6.52 Pohl, G., Otten, J., Research group BOWOOSS, B2E3 Institute for Efficient Buildings of the HTW Saar

6.60.18 Translation and Technological Implementation using the example of the BOWOOSS Research Pavilion

Figures 6.53 and6.54 C. Hamm, Alfred Wegener Insitute, Helmholtz Centre for Polar and Marine Research

Figure 6.55 L Friedrichs, Alfred Wegener Insitute, Helmholtz Centre for Polar and Marine Research

Figure 6.56 Pohl Architects, dept. Institute for Lightweight Structures Jena

Figure 6.57, 6.58, and 6.59 Pohl, J., Pohl Architects, dept. Institute for Lightweight Structures (Leichtbauinstitut) Jena

Figures 6.60 and6.61 Pohl, G., Otten, J., Research group BOWOOSS, B2E3 Institute for Efficient Buildings of the HTW Saar

6.60.19 *BOWOOSS Research Pavilion: Methods and Results of Building-Biomimetics*

Figures 6.62, 6.63, 6.64, and 6.65 G. Pohl, N. Feth, Research group BOWOOSS, B2E3 Institute for Efficient Buildings of the HTW Saar

Figure 6.66 M. Martin, Saarbrücken

Figure 6.67 and 6.68 Pohl, G.

Figures 6.69, 6.70, and 6.71 Halbe, R., Roland Halbe Architecture Photography

Figure 6.72 Pohl, G.

6.60.20 *Building Biomimetics in Examples: Biomimetics and Analogous Developments*

Figure 6.73 Pohl, G.

6.60.21 *Structural Optimization*

Karlsruhe Institute of Technology, KIT, Mattheck, C.

Sauer A. 2008, Untersuchungen zur Vereinfachung biommetrisch inspiriertер Strukturoptimierung,

Diss., FZKA 7406

Hochschule Magdeburg-Stendal, department of Engineering Sciences and Industrial Design, Biller, S., Mühlenbehrend, A. „Die Jahr100 Kurve“

Figure 6.74 Pohl, G.

Figure 6.75 Mattheck C., Sauer A, KIT Karlsruhe

Figure 6.76 Mattheck C., “Stupsi erklärt den Baum,” Publisher KIT Karlsruhe, 4. revised printing 2010, p. 44 and ”Mechanik am Baum“ Publisher Forschungszentrum Karlsruhe, 2002, p. 64

Figure 6.77 Biller, S., Hochschule Magdeburg

Figure 6.78 Biller, S., Hochschule Magdeburg

6.60.22 *Self-organization*

Dr. Mirtsch GmbH, Mirtsch, F.

www.woelbstruktur.de

Figure 6.79 Pohl, G.

Figures 6.80, 6.81, 6.82, 6.83 Dr. Mirtsch GmbH

6.60.23 *Evolutionary Design*

University of Stuttgart, Institute for Computer-based Design ICD, Menges, A.

Menges A., 2011, Morphogenetic Design Experiments, Institute for Computer-based Design, University of Stuttgart

Figure 6.84 Pohl, G.

Figures 6.85, 6.86, and 6.87 Menges, A., University of Stuttgart, ICD—Institute for Computer-based Design

6.60.24 *Morphogenetic Design*

Pohl Architects, dept. Institute for Lightweight Structures (Leichtbauinstitut) Jena

Kooistra W. and Pohl G. (2015), Diatom Frustule Morphology and its Biomimetic Applications in Architecture and Industrial design. In: Hamm, C. Evolution of Lightweight Structures—Biomechanic Adaption and Biodiversity of Plankton Shells: Analyses and Technical Applications, Springer Berlin

Pohl G. (2015), Fibre Reinforced Architecture Inspired by Nature: COCOON_FS. In: Hamm, C. Evolution of Lightweight Structures—Biomechanic Adaption and Biodiversity of Plankton Shells: Analyses and Technical Applications, Springer Berlin

Figure 6.88 Lars Friedrichs, Alfred Wegener Insitute, Helmholtz Centre for Polar and Marine Research

Figure 6.89 Christian Hamm, Alfred Wegener Insitute, Helmholtz Centre for Polar and Marine Research

Figure 6.90 Pohl, J., Pohl, G.

Figures 6.91 and 6.92 Pohl J., Pohl G., Pohl Architects, dept. Institute for Lightweight Structures (Leichtbauinstitut) Jena

6.60.25 Geometric Optimizations: Sectional Optimization

Technical University (TU) Berlin, Institute for Civil Engineering, Chair of Conceptual and Structural Design, Schlaich, M., Gaulke, A.

Figure 6.93 Pohl, G.

Figure 6.94 Schlaich, M.

Figure 6.95 Gaulke, A.

Figure 6.96 Pohl, G.

6.60.26 Hierarchical Structures

Alfred Wegener Institute Bremerhaven (AWI)

Hamm, C. 2005, Kieselalgen als Muster für technische Konstruktionen, BIOSpektrum 1/05, 41–43

Pohl Architects, dept. Institute for Lightweight Structures (Leichtbauinstitut) Jena

Figures 6.97 and 6.98 L Friedrichs, Alfred Wegener Insitute, Helmholtz Centre for Polar and Marine Research

Figure 6.99 Pohl, G., Pohl Architects

Figure 6.100 Pohl Architects

Figure 6.101 Knippers Helbig Advanced Engineering

6.60.27 Evolutionary Urban Planning

Institute for Computer-based Design ICD, University of Stuttgart

Krampe F., Voss C., Ahlquist S., Menges A. 2011, Integrated Urban Morphologies. Entwicklung eines evolutionären, klimaorientierten Entwurfsprozesses auf Maßstabsebene des städtischen Blocks Institute for Computer-based Design ICD, University of Stuttgart

Figures 6.102, 6.103, 6.104 Krampe F., Voss C., Ahlquist S., Menges A. 2011, Integrated Urban Morphologies, ICD Uni Stuttgart

6.60.28 Exterior Surface Effects

http://www.spektrum.de/alias/materialwissenschaft/selbstreinigung-ohne-lotoseffekt/1126247

http://www.bionik.tu-berlin.de/institut/s2skink.html

http://www.bionikvitrine.de/mediapool/99/996537/data/PDFs/Haihaut/Haihauteffekt.pdf

Figure 6.105 Maren Beßler_pixelio.de/ www.pixelio.de

Figure 6.106 Cornerstone/pixelio.de/ www.pixelio.de

Figure 6.107 Bionic StreamForm Frank Wedekind, Saarbrücken

6.60.29 Foundations of Resource-Efficient Facade Technologies

Gosztonyi S., Judex F., Brychta M., Gruber P., Richter S., 2012, BioSkin—Bionische Fassaden, Potentialstudie über bionische Konzepte für adaptive energieeffiziente Fassaden, AIT Austrian Institute of Technology, foundation study in frame of the Austrian promotial program "House of the Future Plus," promoted by the Ministry for Transportation, Innovation, and Technology

Gruber P., Gosztonyi S., 2010, Skin in architecture: towards bioinspired facades. In: Brebbia, C.A. & Carpi, A. (eds.), Design and Nature V, Comparing Design in Nature with Science and Engineering, Volume 138, WIT press, Southampton, ISBN: 978-1-84564-454-3

Figure 6.108 Pohl, G.

Figure 6.109 Gosztonyi S., 2011, BioSkin, AIT Austrian Institute of Technology

Figure 6.110 BioSkin Workshop Team, 2011, BioSkin, AIT Austrian Institute of Technology

6.60.30 Daylight Usage

Gosztonyi S., Judex F., Brychta M., Gruber P., Richter S., 2011, BioSkin—Bionische Fassaden, Potentialstudie über bionische Konzepte für adaptive energieeffiziente Fassaden, AIT Austrian Institute of Technology, foundation study in frame of the Austrian promotial program "House of the Future Plus." promoted by the Ministry for Transportation, Innovation, and Technology

Figure 6.111 "Licht im Schwamm." 17.11.2008 Uni Stuttgart

Figure 6.112b, c Richter S., 2011, BioSkin, AIT Austrian Institute of Technology, http://idw-online.de/de/news289131

Figure 6.113 Richter S., 2010, BioSkin, AIT Austrian Institute of Technology

Figure 6.114 Gosztonyi S., 2011, BioSkin, AIT Austrian Institute of Technology

Figure 6.115 Judex F., 2011, BioSkin, AIT Austrian Institute of Technology

6.60.31 Shading

Gosztonyi S., Judex F., Brychta M., Gruber P., Richter S., 2011, BioSkin—Bionische Fassaden, Potentialstudie über bionische Konzepte für adaptive energieeffiziente Fassaden, AIT Austrian Institute of Technology, foundation study in frame of the Austrian promotial program "House of the Future Plus." promoted by the Ministry for Transportation, Innovation, and Technology.

Figure 6.116 Pohl, G.

Figure 6.117 Siegel G., 2010, BioSkin, AIT Austrian Institute of Technology

Figure 6.118 Siegel G., Gosztonyi S., 2010, BioSkin, AIT Austrian Institute of Technology

6.60.32 *Shading and Solar Energy Production*

Badarnah, L., Knaack, U., 2008, Shading/energy generating skin inspired from natural systems. Proc. of the 2008 World Sustainable Building Conf. SB08, Eds G. Floiente and P. Paevere, pp 305–312

Figure 6.119 Pohl, G.

Figures 6.120, 6.121, 6.122, 6.123 Badarnah, L., Knaack, U., 2008, Shading/energy generating skin inspired from natural systems

6.60.33 *Shading and Directing Light 1*

Hertzsch, E., Pohl, G 2011, international Student Workshop on Façade Design & Performance, University of Melbourne, Australien.

Jin, H., 2011, Second Skin Façade inspired from the epidermal stoma of leaves. Design proposals, Bio-Inspired Façade Systems.

http://de.wikipedia.org/wiki/Stoma_%28Botanik%29

Figure 6.124 Pohl

Figures 6.125, 6.126, 6.127, 6.128 Jin, H., 2011, Second Skin Facade inspired from the epidermal stoma of leaves. Design proposals, Bio-Inspired Façade Systems.

6.60.34 *Shading and Directing Light 2*

Hertzsch, E., Pohl, G 2011, international Student Workshop on Façade Design & Performance, University of Melbourne, Australien.

Pullyblank, D., 2011, Modular Façade inspired by Barnacles. Design proposals.

http://de.wikipedia.org/wiki/Seepocken

Figure 6.129 Sea barnacles, Kiser, K., sxc.hu

Figures 6.130, 6.131, 6.132, 6.133, 6.134 Pullyblank, D. 2011, Modular Façade inspired by Barnacles. Design proposals.

6.60.35 Colors without Pigments 1

Pohl Architects, www.pohlarchitekten.de

Figure 6.135 Pohl G.

Figures 6.136–6.137 Wilhelm J., Pohl Architects

6.60.35 *Color without Pigments 2*

Gosztonyi S., Judex F., Brychta M., Gruber P., Richter S., 2011, BioSkin—Bionische Fassaden, Potentialstudie über bionische Konzepte für adaptive energieeffiziente Fassaden, AIT Austrian Institute of Technology, foundation study in frame of the Austrian promotial program “House of the Future Plus,” promoted by the Ministry for Transportation, Innovation, and Technology.

Figure 6.138 Kirsanov, V., fotolia.de #30191504

Figure 6.139 Gosztonyi S., 2010, based on results from BioSkin Creative Workshop, AIT Austrian Institute of Technology

Figure 6.140 Gosztonyi S., Ledinger S., Abermann S, Haslinger E., 2010, BioSkin Creative Workshop, AIT Austrian Institute of Technology

6.60.36 Complex Climate Systems 1: New Construction

Pohl, G., Technology and Media Centre Erfurt, Tensinet Symposium: Designing Tensile Architecture, September 2003, Brussels, Belgium

Pohl, G., Fabric Architecture march/april 2004,USA

Figure 6.141a, b Pohl, G.

Figures 6.142–6.143 Pohl Architects

6.60.37 Complex Climate Systems 2: Building Reuse

Pohl Architects, Media Center, Bauhaus University Weimar, Erfurt, Germany; info@pohlarchitekten.de

Hochschul- und Forschungsbauten, 2003, Stiftung Baukultur Thüringen

Figure 6.144 Post, K., http://www.klauspost.com

Figures 6.145, 6.146, 6.147 Pohl Architects

6.60.38 Spatial Panels

Institute for Computerbased Design ICD, Menges, A., University of Stuttgart, Institute of Building Structures and Structural Design (ITKE), Knippers J., University of Stuttgart

http://www.itke.uni-stuttgart.de/img/background/95-110829_Above-web.jpg

Figure 6.148aBas van der Steld, F., Hendriklaan 259 A, NL-2582 Gravenhage

Figure 6.148b Seilacher, A. Engelsfriedhalde 25, D-72076 Tübingen

Figure 6.149 Waimer, F., La Mangna, R., Knippers, J., Institute of Building Structures and Structural Design (ITKE), University of Stuttgart

Figures 6.150, 6.151, 6.152, 6.153, 6.154, 6.155 Menges A., ICD University of Stuttgart

6.60.39 Spines

Pohl, G., Pohl Architects: Competition for the Olympic Games 2000 in Berlin

Figure 6.156 André, A. B2E3 Institute for Efficient Buildings at HTW Saar.

Figures 6.157, 6.158, 6.159 Pohl, G.

6.60.40 Spatial Structures with Curved Modules 1

Research Group BOWOOSS, Pohl, G., B2E3 Institute for Efficient Buildings at HTW Saar

Figure 6.160 Pohl, G.

Figures 6.161, 6.162, 6.163, 6.164, 6.165 Pohl, G., Otten, J., Research Group BOWOOSS, B2E3 Institute for Efficient Buildings at HTW Saar

6.60.41 Spatial Structures with Curved Modules 2

Menges, A., Knippers, J., (2010) ICD/ ITKE Research Pavilion 2010, Institute for Computer-based Design (ICD), J. Knippers and Institute of Building Structures and Structural Design (ITKE), A. Menges at University of Stuttgart

Figure 6.166 Sias van Schalkwyk, http://sxc.hu

Figure 6.167 Menges A., Eisenhardt, Vollrat, Waechter, ICD University of Stuttgart

Figure 6.168a-b Menges A., Eisenhardt, Vollrat, Waechter, ICD University of Stuttgart

Figure 6.168c-d Knippers, J., Lienhard, J., ITKE University of Stuttgart

Figure 6.169 Halbe, R., Roland Halbe Architecture Photography

Figure 6.170–6.171 Menges A., ICD University of Stuttgart

6.60.42 Layered Tissues

Otten, J. Research Group BOWOOSS, B2E3 Institute for Efficient Buildings at HTW Saar

Figures 6.173, 6.174, 6.175 Otten, J

Figure 6.176 Pohl, G., Otten, J., Research Group BOWOOSS, B2E3 Institute for Efficient Buildings at HTW Saar

6.60.43 Expandable Structures

TU Berlin, Institut für Bauingenieurwesen, Chair of Conceptual and Structural Design, Schlaich, M., Bögle, A., Hartz, C.

Figures 6.177–6.178 Pohl, G.

Figure 6.179 Pohl, G., B2E3 Institute for Efficient Buildings at HTW Saar

Figure 6.180 Schlaich, M., Bögle, A., Hartz, C., Technical University (TU) Berlin, Faculty of Engineering

Figure 6.181 Schlaich, M., Bögle, A., Hartz, C., Technical University (TU) Berlin, Faculty of Engineering

6.60.44 Solid, Efficient, Load-bearing and Heat-Insulated Lightweight Structures

TU Berlin Institut für Bauingenieurwesen, Chair of Conceptual and Structural Design, Schlaich, M., Hückler, A.

Figure 6.182 Pohl, G.

Figure 6.183 ILEK, University of Stuttgart

Figure 6.184 Pohl Architects, dept. Institute for Lightweight Structures (Leichtbauinstitut) Jena

Figures 6.185–6.186 Sofistik Skript, Technical University (TU) Berlin, Institut für Bauingenieurwesen, Chair of Conceptual and Structural Design

6.60.45 Sonar

http://www.fh-koblenz.de/Echolot-Eine-Bionische-Struk.4211.0.html

Fachhochschule Koblenz

Objekt- und Tragwerksplanung: ilcom—Institutue for Lightweight Constructions and Materials (Institut für leichte Konstruktionen und Material), Fachhochschule Koblenz, Faculty of Architecture and Engineering, Feyerabend, M., Holzbach, M.

Planning of Light and Sound: Faculty of Mathematics und Technology, Bongartz, J.

Figure 6.187 fotolia.de #29581760

Figure 6.188a Feyerabend, M.

Figure 6.188b Fachhochschule Koblenz, Faculty of Architecture and Engineering

Figures 6.189–6.190 Feyerabend, M.

6.60.46 Fiber Composite Sensors

Technical University of Chemnitz, Department of Lightweight Structures and Polymer Technology, Kroll, L., Gelbrich, S., Elsner, H.

Technical University Chemnitz, Professorship Circuit and System Design

Kompetenzzentrum Strukturleichtbau e.V., Chemnitz

Figure 6.191 Pohl, G.

Figures 6.192, 6.193, 6.194, 6.195, 6.196 Technical UniversityChemnitz, Kroll, L., Gelbrich, S., Elsner, H.

6.60.47 Reactive Envelope Structures

Menges, A., Reichert, S., 2011, Responsive Surface Structure, Institute for Computer-based Design (ICD), University of Stuttgart

Figures 6.197, 6.198, 6.199, 6.200 Menges, A., Reichert, S.

6.60.48 Ventilation Systems for Breathing Envelopes

Badarnah, L., Knaack, U., 2007, Bio-Inspired System for Building Envelopes. Proc. of the Int. Conf. of twenty-first century: Building Stock Aviation, Ed. Kitsutaka, Y., TIPEI: Tokyo, pp. 431–438

Figure 6.201 Pohl, G.

Figures 6.202–6.203 Badarnah, L., Knaack, U., 2007, Bio-Inspired System for Building Envelopes

6.60.49 Thermoregulating Envelope Structures

Badarnah, L., Nachman Farchi, Y., Knaack, U., 2010, Solutions from Nature for building envelope thermoregulation. Proc. of the 5th Design&Nature Conf., Comparing Design and Nature with Science and Engineering, Eds. Carpi, A., Brebbia,C., WIT press, Southampton

Biomimicry taxonomy: www.AskNature.org

Figure 6.204 Pohl, G.

Figures 6.205a-c Badarnah, L., Nachman Farchi, Y., Knaack, U., 2010, Solutions from Nature for building envelope thermoregulation.

Figure 6.206 Tab. Badarnah, L., Nachman Farchi, Y., Knaack, U., 2010, Solutions from Nature for building envelope thermoregulation.

Figure 6.207 Tab. Badarnah, L., Nachman Farchi, Y., Knaack, U., 2010, Solutions from Nature for building envelope thermoregulation.

6.60.50 *Modifiable Surface Elements 1*

Poppinga, S., Lienhard, J., Schleicher, S., Masselter, T., Knippers, J., Speck, T. (2010) Gelenkfreie Klappen bei Strelitzia reginae. Conference Proceedings of the 5. Bremer Bionik Kongress 'Patente aus der Natur', Bremen, Germany, 320–326.

Lienhard, J., Schleicher, S., Poppinga, S., Walter, A., Sartori, J., Milwich, M., Stegmaier, T., Masselter, T., Speck, T., Knippers, J. (2010) Optimierung und Weiterentwicklung des Flectofin®. Conference Proceedings of the 5. Bremer Bionik Kongress 'Patente aus der Natur', Bremen, Germany, 36–45

J. Lienhard, S. Schleicher, S. Poppinga, T. Masselter, M. Milwich, T. Speck & J. Knip-pers (2011): Flectofin: a nature based hinge-less flapping mechanism. – Bioinspiration and Biomimetics, 6: DOI:10.1088/1748–3182/6/4/045001

J. Knippers & T. Speck (2012): Design and construction principles in Nature and Architecture. – Bioinspiration and Biomimetics, 7. DOI:10.1088/1748–3182/7/1/015002

S. Poppinga, J. Lienhard, S. Schleicher, T. Masselter, M. Milwich, T. Stegmaier, J. Sartori, A. Walter, H.-F. Schur, K. Vogg, T. Speck & J. Knippers (2010): Architektur und Bionik – Wandelbarkeit ohne Gelenke. – ibr RWK Informationen Bau-Rationalisierung, 38/4: 24 – 25.

S. Poppinga, T. Masselter, J. Lienhard, S. Schleicher, J. Knippers & T. Speck (2010): Plant movements as concept generators for deployable systems in architecture. – In: Brebbia, C.A. & Carpi, A. (eds.), Design and Nature V, 403 – 410. WIT Press, Southampton.

Institut für Tragkonstruktionen und Konstruktives Entwerfen ITKE, Knippers, J., Universität Stuttgart

Figures 6.208, 6.209, 6.211, 6.212 Lienhard, J., ITKE

Figure 6.213 Lienhard, J., Schleicher, S., ITKE

Figure 6.214 Schleicher, S., ITKE

6.60.51 *Modifiable Surface Elements 2*

soma. Analoge Effects. Thematic Pavillon 2012 Yeosu, South Korea, www.soma-architecture.com

Soma Architects, www.soma-architecture.com

Knippers Helbig Advanced Engineering, www.knippershelbig.com

Knippers, J.,Scheible, F., Oppe, M., Jungjohann, H. (2012) "Kinetic Media Façade Consisting of GFRP Louvers," Conference Proceedings of CICE 2012, Rome

Knippers, J.,Scheible, F., Oppe, M., Jungjohann, H. (2012) "Bio-inspirierte Kinetische Fassade für den Themenpavillon EXPO 2012 in Yeosu, Korea," VDI-Wissensforum conference proceedings ′Bauen mit Innovativen Werkstoffen′, Stuttgart

Schinegger, K., Rutzinger, S., Oberascher, M., Weber, G. (2012) "Theme Pavilion Expo Yeosu One Ocean," Residenz Publishers, Austria

Figure 6.215 Kriss Szkurlat, sxc.hu

Figures 6.216–6.217 Knippers Helbig Advanced Engineering

Figures 6.218–6.219 soma Architects

6.60.52 Multiaxially Modifiable Surface Elements

Pohl Architects, www.pohlarchitekten.de

Knippers Helbig Advanced Engineering, www.knippershelbig.com

Pohl, G., Pfalz, M., (2010), pp. 420–470, Innovative composite-fibre components, in: Textiles, Polymers and Composites for Buildings, Woodhead Publishing Limited, Oxford

http://www.diplom-biologe.de/samen/Tropische_und_subtropische_Pflanzensamen_3_0/artikel5.html

http://de.wikipedia.org/wiki/Pflanzenbewegung

Figure 6.220 Pohl, G.

Figure 6.221 Pohl Architects, dept. Institute for Lightweight Structures (Leichtbauinstitut) Jena

Figure 6.222 Pohl Architects

Figure 6.223 Spiekermann, C., Pohl Architects

Figure 6.224 Fischer, J., Pohl Architects

6.60.53 Reactive Construction Systems

TU Berlin, Institut für Bauingenieurwesen, Chair of Conceptual and Structural Design, Schlaich, M.,

Bleicher, A.: Aktive Schwingungskontrolle einer Spannbandbrücke mit pneumatischen

Aktuatoren, Bautechnik 89, Nr. 2, pp. 89–101, 2012

Figure 6.225 fotolia.de, #42890795

Figure 6.226, 6.227, 6.228: Bleicher, A.

Figure 6.229: Pohl, G.

Figure 6.230: Bleicher, A.

6.60.54 Self-responsive Movements, Fin Ray Effect®

TU Berlin, Institut für Bauingenieurwesen, Chair of Conceptual and Structural Design, Schlaich, M., Bögle, A., Hartz, C.

EvoLogics GmbH, Berlin, Bannasch, R., Kniese, L.

Massivbau kPlan AG, Abensberg, Kirchmann, H-P, Kersting, A.

LEICHT GmbH, Rosenheim, Schöne, L., Arndt, J.

Figure 6.231a: Pohl, G.

Figure 6.231b: Behnke, R.

Figures 6.232, 6.233, 6.234: Guignand, S.

6.60.55 Relocating Shells

School for Architecture (Schule für Architektur), Pohl, G., HTW Saar

Figure 6.235 André, A.

Figures 6.236–6.237 Pohl, G., Feth, N. Ghinita, I., HTW Saar

6.60.56 Self-healing

Nachtigall, W. Bau-Bionik, (2003) Springer Publishers Berlin, Heidelberg, New York, p. 215

Speck, T. et al. (2006) Self-healing processes in nature and engineering: self-repairing biomimetic membranes for pneumatic structures. Brebbia, C.A. (eds), Design and Nature III, WIT Press, Southampton, pp. 105–114

Busch, S., Seidel, R., Speck, O. & Speck, T. (2010): Morphological aspects of self-repair of lesions caused by internal growth stresses in stems of Aristolochia macrophylla and Aristolochia ringens—Proceedings of the Royal Society London B, 277: 2113–2120.

Rampf, M., Speck, O., Speck, T. & Luchsinger, R. (2011): Self-repairing membranes for inflatable structures inspired by a rapid wound sealing process of climbing plants—Journal of Bionic Engineering, 8: 242–250.

M. Rampf, O. Speck, T. Speck & R. Luchsinger (2012): Structural and mechanical properties of flexible polyurethane foams cured under pressure. – Journal of Cellular Plastics, 48: 49 – 65.

Figure 6.238 Pohl, G.

Figures 6.239, 6.240, 6.241, 6.242, Plant Biomechanics Group Freiburg

Figures 6.243–6.244 Luchsinger, R.

6.60.57 Bambootanic

Diploma thesis at the School of Architecture (Schule für Architektur) HTW Saar, Feth, N.

Figure 6.245 Feth, N.

Figures 6.246, 6.247, 6.248, 6.249 Feth, N.

6.60.58 Floating Volumes

Magdeburg-Stendal University of Applied Sciences, Department of Engineering and Industrial Design, Wohlgemuth, U., Pommer, C.

Figure 6.250a Santiago, I., sxc.hu.

Figure 6.250b University of Karlsruhe (KIT), Botanical Garden.

Figures 6.251, 6.252, 6.253, 6.254 Pommer, C.

Chapter 7
Brief Information to Biological Structures

The abstracted and compiled information in the following 50 sections originates in textbooks, original publications, and reference articles. Some data originate in the data collection of Flindt (1986). Much information is also taken from v. Frisch (1974), Freude (1982), and the Finnish collaboration "Animal Architecture" (1995).

7.1 Biological Building Materials (Outline)

1 Endogenous Materials
1.1 Secretions
1.1.1 Threads without Foreign Materials
1.1.2 Threads with Foreign Materials
1.1.3 Not Thread-like, without Foreign Materials
1.1.4 Not Thread-like, with Foreign Materials
1.2 Excretions
1.3 Skin Formations
2 Exogenous Materials
2.1 Plant Origin
2.2 Animal Origin
2.3 Inorganic Materials
2.4 Anthropogenic Materials
3 Substrates for Hollowed Structures
3.1 Organic
3.1.1 Plant Origin
3.1.2 Animal Origin
3.2 Inorganic
3.2.1 Stone, Earth
3.2.2 Ice, Snow
(Outline according to Freude (1982). The author gives examples for each on pages 177–179 for his outline points.)

G. Pohl, W. Nachtigall, *Biomimetics for Architecture & Design*,
DOI 10.1007/978-3-319-19120-1_7

7.2 Beaver Structures

Water lodges of beavers (*Castor fiber*) are built from up to 4-m-long sticks and twigs that are woven over and between one another. The structure is then hollowed out from the inside so that an inner den emerges. This den is completely sealed with mud, stones, and fine plant fibers, except for the uppermost portion, which serves as air ventilation. The beaver prefers trees with thicknesses of about 12 cm, which it chops into pieces and transports away. The dams are generally not taller than 1.5 m.

7.3 Beaver Dams

Beaver dams stall the water. The water level is controlled by the removal and addition of twigs. The longest known beaver dam is 1200 m long. "In the Voronezh region in Russia, the largest dam is 120 m long, 1 m tall, and 60–100 cm wide. In the USA the beavers build dams in the swamps of the Mississippi of several hundred meters in length. On the Jefferson River (Montana, USA) lies possibly the largest of all dams. One can walk along it for 700 m. A horseman could not break in" (v. Frisch).

7.4 Badger Structures

The structures of the badger *Meles meles* have the diameter of about 10–30 m and reach up to 5 m in depth. The chambers are laid out in up to three levels on top of one another and connected with passageways that lead to several exits. Large structures can be up to 100 m in total tunnel length with 40–50 openings.

7.5 Tunnel Systems of Steppe Marmots

The passage system reaches depths of 2–3 m, occasionally 7 m, and has one or two exits, a den, and a chamber for excrement. The den is particularly soft padded in very cold regions (most clearly with the Siberian black-capped marmot that winters in permafrost soils). Entire families remain there, curled up and snuggled next to one another, for their hibernation. The body temperature amounts then to only 5 °C. To hinder further sinking of the temperature in the den, individuals will occasionally wake up and generate metabolic heat.

7.6 Scrubfowl Mounds

The scrubfowl *Megapodius freycinet*, despite its partridge size, can build nest mounds with a diameter of up to 12 m and a height of up to 5 m, the dimensions of largest structures that have been observed with this species of fowl. Smaller scrubfowls that live on volcanic islands use geothermal warmth by building mounds with loose, warm volcanic soil.

7.7 Storage Chambers of Moles

Moles, *Talpa europaea*, gather stockpiles of partially eaten and therefore immobilized earthworms; in one instance, 1200 earthworms with a total mass of over 2 kg were counted in one storage chamber.

7.8 Storage Chambers of Hamsters

The female European hamster, *Cricetus cricetus*, gathers up to 15 kg of grain supplies for winter, in certain cases actually up to 50 kg.

7.9 Spherical Structures of the Ovenbird

Ovenbirds of the family Furnarius build up to 10-kg heavy nests from around 2000 mud clumps, with each individually weighing up to 5 g. The diameter amounts to about 25 cm; the diameter-to-wall thickness ratio is about 7.5:1.

7.10 Mortar Structures of the Potter Wasp

Potter wasps of the species *Polybia singularis* finish their thick-walled "ceramic" nests with slits on the sides as entrances and can reach up to 30 cm in length and 1.5 kg in mass.

7.11 Weaver Bird Nests

The weaver bird *Philetairus socius* completes communal nests that can be up to 9 m wide and around 2 m thick.

7.12 Tallest Ant Mounds

In Finland, an above ground structure of the red wood ant *Formica rufa* was observed. Its height was 2 m and base diameter was close to 6 m.

7.13 Stockpiles of the Harvester Ant

Harvester ants of the genus *Messor* can fetch 20,000 grains for a nest in a single day. The nest can reach up to 3 m deep and up to 50 m in extent. The nests can contain thousands of storage chambers, sometimes with several kilograms of grains.

7.14 Structures of Compass Termites

The "tower" structures of compass termites reach a height of 3.7 m, a length of 3 m, and a width of about 1 m, whose direction is exactly north–south.

7.15 Elongated Termite Structures

The South African termites of the genus *Odontotermes* form regular, wave-like structures of 2 m height and up to 11 m length, which run in distances of around 50 m through the landscape.

7.16 Earth Mounds of Less Organized Termites

The termite *Cornitermes cumulans* of South Africa builds approximately egg-shaped, underground nests with diameters up to 40 cm, which stands on stilts and is thereby thermally insulated. Later an earth dome is built on top of the nest, inside of which the nest is gradually relocated. Eventually, an overground termite structure that can measure up to 1.6 m tall with a base of 1 m emerges.

7.17 Largest Termite Structures

The maximum height measurements are around 9 m. A large termite mound weighs around 12 tons. Termite passages to groundwater sources can be up to 40 m long. In the Karakum Desert of Central Asia, termites can build shafts to groundwater of up to 200 m in length. "Geologists can take advantage of this by studying the excavation materials for the search of earth minerals deep underground" (Freude).

7.18 Nest of the Goldcrest

The exterior of the nest of the goldcrest *Regulus regulus* consists of weaving materials, moss, and lichen, about 7 g of weaving materials and 4 g of moss or lichen per nest. The goldcrest additionally collects the egg cocoons of spiders (along with the young spiders) as well as the cocoons of certain wasps and caterpillars, and builds an outer layer with them. The middle layer contains loosely packed moss stems with or without the addition of lichens. The inner cushion layer consists of small feathers or animal hairs. In one particular case, 2818 moss stems (total 3.1 g), 1422 lichen pieces (3.5 g), and 2674 feathers (1.8 g) were counted in one nest. The three-layered nest connects structural stability to thermoinsulation.

7.19 Tree Frog Nests

The Brazilian tree frog *Hyla faber* builds a 10-cm tall and 30-cm-diameter nest with his large forelimbs, in which it lays its eggs.

7.20 Foam Nest of the Green Flying Frog

This frog, *Rhacophorus reinwardtii*, is on the one hand well known due to its broad webbed feet, which allow it to more or less glide at length from tree tops to the forest floor, and on the other hand, for its foam nest structures. The several centimeter thick nest dries on the outer surface, which causes it to become brown and unnoticeable. The interior is made damp so that a small pond forms for the eggs and eventual tadpoles.

7.21 Egg Raft of the Purple Snail

Sea snails of the genus *Janthina* build foam rafts, on the underside of which they secure their egg cocoons. The air bubbles are adhered to a spiral band of with a length of 12 cm and width of 2 cm. Up to 500 cocoons with a total of 250,000 eggs can be adhered to the underside. The foam nests are also known from other snails, insects (praying mantises), some fishes, as well as tree frogs.

7.22 Honeycombs of the Honeybee

A honeycomb with an area of 37 cm × 22.5 cm can be built from merely 40 g wax; however it can contain no less than 1.8 kg honey.

7.23 Precise Constructions of the Honeybee

Wax glands of the bee workers secrete wax flakes of about 0.5 mm thick and 1.5 mm long, and each weighs 0.25 mg. The bees can build about 80,000 cells with 1 kg wax. The cell depth is 12 mm and breadth is 5.2 mm; the diameters of the cells have a margin of error of only 0.05 mm. The space between two parallel honeycomb strands amounts to only 9.5 mm; nevertheless the bees still have good mobility on the honeycomb. No less than 8.6 honeycombs are situated on 1 cm^2. The thickness of the honeycomb walls amounts to an average 0.073 ± 0.002 mm for workers and 0.092 for queens. The necessary sensors lie in the antennae and at the ends of the mandibles.

7.24 Temperature Differential in Bee Colonies

Bee wax is most workable at temperatures of 34–35 °C; however, the larvae can tolerate nest temperatures of only 37 °C, and at 45 °C the mature bees also die. The temperature differential, inside of which their lives are possible, amounts therefore to barely 2 °C. In too hot weather, the bees ventilate the hive at the entrance hole ("fanning") and spray water around for evaporative cooling. In too cold weather and longer frost periods, the bees crowd themselves closely together and generate then no less than 0.1 kW of warmth per kilogram of bee mass.

7.25 Spider Webs

Communal nests, fabricated by thousands of species of spider, for example *Araneus sermoniferous* and *Ulocerus republicanus*, reach 100 m in width.

7.26 Thickness of Spider Silk

A large, mature spider can produce silk which is 0.010–0.012 mm in diameter. Cribellate spiders weave around 50,000 silk threads of only 0.002 mm in diameter into one single thread.

7.27 Egg Containers of the Sac Spider

With silk threads the sac spider *Agroeca brunnea* builds an egg container of about 0.6 cm width and with a bell-shaped form, in which around 50 eggs can be accommodated. It is covered with earth and clay clumps that dry to form a solid outer layer.

7.28 Silkworm Cocoons

Silkworms have been used by humans for at least 5000 years. For the construction of a pupa cocoon, the larvae handle up to 4 km of self-produced silk in single threads of around 1 km in length, of which about 70 % is usable.

7.29 Nest Structures of the Swift

Swifts of the species *Panyptila cayennensis* from Central and South America build a tubular nest of about 60 cm in length with an opening on the bottom. The nests are built in 6 months with animal hairs, plant fibers, and feathers, mixed with a saliva secretion, onto an overhanging bluff or rock formation.

7.30 Dung Balls of the Scarab Beetle

The Egyptian scarab beetle *Scarabaeus sacer* weighs only 2 g, but can roll dung balls with the size of a fist and mass of 40 g.

7.31 Coral Reefs

The largest structure built by any animal is represented by the Barrier Reef in northeastern Australia, with a total length of over 2000 km. The coral colonies there produce no less than 4 tons of limestone material per square kilometer in 1 day.

7.32 Sand Coral Reefs

Bristle worms of genus *Sabellaria* build closely snuggled together, organ-like tubes known as "sand coral reefs" on the North Sea shore. On the island of Norderney, a 60 m long reef with about a half meter height emerged within 2 years, in which one breakwater was sheeted with around 75 million tubes. Similarly, tall reefs are built by the tropical genus *Phragmatopoma* with heights of up to 1 m.

7.33 Fishing Nets

Many South American caddisflies produce nets with a tiny mesh dimension of 3–20 μm. A net with 1.5 cm diameter contains around 2,000,000 meshes.

7.34 Storage Hideaways

The eastern European house mouse species, known as the steppe mouse *Mus musculus spicilegus,* constructs hideaways inside of which two-to-six mice together collect 5–7 kg of seeds and these are then covered with earth. Below these hideaways they build their nests of 60–120 cm in diameter and up to 50 cm in height.

7.35 Path Constructions

Leaf-cutter ants of species *Atta sexdens* maintain up to 200-m-long path free of vegetation from the nest. The width of the path is up to 7 cm and can lead through formerly dense grass areas.

7.36 Bowers of the Bowerbird

In the course of a year, the bowerbird *Prionodura newtoniana*, although only blackbird sized, builds "two stems with giant, bristly towers, one 2 m and the other up to 2.70 m tall; the just under one meter space in between is transformed into a dance floor" (Animal Architecture).

7.37 Regulating Humidity

Termites of genus *Macrotermes* maintain a humidity level within 89–99 %. To reach water in dry regions, they build up to 40-m-long passages to groundwater, in some cases possibly up to 200 m in length.

7.38 Gas Exchange

Turtles leave air holes for the exchange of gases for their eggs that they bury in the sand. Many water-dwelling worms and caddis fly larvae drive water through their structures by either a muscle-driven vibration of their appendages or the pump-like movement of their abdomen. Nest-building fishes, for example, the stickleback *Gasterosteus aculeatus*, accomplish circulation through their nests with fin movements. African lungfish of the genus *Protopterus* build for the summer a mud cocoon that is dried out on the interior and whose upper lid is kept porous for gas circulation. For high tide, the Malaysian crab *Mictyris longicarpus* builds a sand-castle structure that contains an air chamber in which it resides. The prairie dog *Cynomys ludovicianus* uses, as stated previously, pressure differences "according to the Bernoulli principle" for the forced air circulation of its structures.

7.39 Vertebrate Temperature Regulation

Occupied nests of the weaver bird of the species *P. socius* maintain an internal temperature of 20 °C over exterior temperatures in environments such as the Kalahari Desert of South Africa, where during winter the temperatures at night can fall up to − 10 °C. Snow hares and redpolls build snow dens in igloo style, inside of which the temperature is 7–8 °C higher than the external temperature. Australian "thermometer birds" of the species *Leipoa ocellata* build large nests out of decomposing plant material whose warmth incubates the eggs. By heaping an earthen mound on top of the eggs, the interior temperature remains at a constant 34 ± 1 °C independent from the exterior climate.

7.40 Temperature Regulation by Insects

Wasps can maintain 30 °C in the breeding chamber of their nests. When it becomes too cold, the worker bees generate warmth with muscle vibrations; when it becomes too hot, evaporating water is released in the nest. Honeybees maintain a temperature of 35 °C within their honeycomb structures, as the wax is easier to work with at this temperature. Red forest ants of the species *F. rufa* provide numerous ventilation openings for their nests, which they seal shut during nights and cold weather by plugging them with their heads. The slope of the mounds is regulated by the worker ants so that it acts as an ideal collector of sunlight. In springtime, the ants warm themselves on the exterior of their mounds and then return to the interior where they "digitally" radiate their warmth.

7.41 Sizes of Populations of Colony-Forming Insects

Sizes of populations of colony-forming insects are as follows: paper wasps 140 individuals, bumblebees up to 2000, hornets up to 1500, yellow meadow ants up to 20,000, honeybees up to 80,000, army ants and leaf-cutter ants 100,000–600,000, red wood ants up to 800,000, and termites of the genus *Bellicositermes* up to 3,000,000.

7.42 Leaf Surfaces of Plants

An average apple tree has around 20,000 leaves with a median surface area of each leaf of 18 cm^2 and therefore 32 m^2 of total surface area. A large beech possesses 450 m^2 of total leaf surface area.

7.43 Maximum Heights of Trees

Maximum heights of trees are: sycamore 40 m, ash 50 m, pine 48 m, coconut palm 32 m, silver fir 75 m, sequoia 132 m, and giant eucalyptus 152 m.

7.44 Maximum Trunk Diameters of Trees

Maximum trunk diameters of trees are: field maple 0.7 m, pine 1.0 m, spruce 2.0 m, red beech 2.0 m, fir 3.0 m, sequoia 11.0 m, and baobab 15.0 m.

7.45 Slenderness of Plants

Slenderness is described here as the quotient of the height and the base diameter.

Baobab (20 m): 2.5
Giant sequoia (135 m): 11
Fir (70 m): 42
Spruce (60 m): 60
Sunflower (4 m): 100
Bamboo (40 m): 133
Sugarcane (6 m): 200
Rye stalk (1.5 m): 500

For comparison: The TV tower in Stuttgart (200 m): 19. (The "thickening" of taller structural system results from the Barba-Kick law of "proportional resistance", where the diameter d is not directly proportional to a height h, but to the product $h\sqrt{h}$).

7.46 Specific Masses of Wood

As the lightest wood variety, balsa wood weighs 0.18 g cm^{-3}, pine 0.49 g cm^{-3}, buckeye 0.57 g cm^{-3}, pear tree 0.72 g cm^{-3}, red beech 0.74 g cm^{-3}, rosewood 0.82 g cm^{-3}, and the heaviest wood, *Guaiacum* weighs 1.23 g cm^{-3}.

7.47 Elasticity Moduli of Biological Building Materials

The "most elastic" material of the animal kingdom, the protein Resilin, possesses the lowest E-Module of merely 0.002 GPa whereas spruce wood possesses the E-Module of 10 GPa. For comparison the E-Module of silicone rubber is 0.01 GPa whereas that of V2A steel is 200 GPa.

7.48 Elastic Efficiencies of Biological Stretching Elements

Resilin has the highest elastic efficiency of 96%. In comparison, the elasticity efficiency of sheep tendons is 90%.

7.49 Tensile Strength of Biological Building Materials

Coniferous wood, Class III, possesses a tensile strength of 90 N mm^{-2}; spider silk possesses a tensile strength of 500 N mm^{-2}. For comparison the tensile strength of hard PVC is 75 N mm^{-2}, that of structural steel ST 33 is 310 N mm^{-2}, and that of special spring steel is up to 3090 N mm^{-2}.

7.50 Root Depths of Plants

The depths of dandelion roots reach around 30 cm, over 1 m in case of silver thistle, just under 3 m in case of wheat and rapeseed, up to 10 m in case of forest trees, and up to 20 m in case of desert plants.

Additional Literature

Adobe-Fragen (1998) http://www.nmia.com/~eaci/eaci-FAQ.html. Accessed 14 Mar 1998

Aldous S (2001) How solar cells work. http://www.howstuffworks.com/solar-cell.htm

Anonymous (ed) (2000) Grow your own house. Zeri Fundation, Vitra Design-Museum

Bachmann T (1922) Jb Wiss Bot 61:372

Bahadori M (1978) Passive cooling systems in Iranian architecture. Sci Am 238(2):144–154 (Feb 1978)

Balz M (2002) Organische Bauformen. Vortragssammlung zum AIV-Symposium: "Das Stadtquartier der Zukunft", Stuttgart, 25.–27.04.02

Becker P-R, Braun H (eds) (2001) Architektur und Lebewesen. Berichtsband des Übersee-Museums Bremen, nestWerk, Isensee, Oldenburg

Becker S, Brenner K, Frankhauser P, Humpert K, Kull U, Schaur E (1994) Selbstorganisation urbaner Strukturen. Ungeplante Siedlungen und großstädtische Agglomerationen. Zeitschrift für Architektur und Städtebau, März 1994

Behling S, Behling S (1996) Sol power. Die Evolution der solaren Architektur. Mit einem Vorwort von Sir Norman Foster. READ-Publikation, Prestel

Berichte IL (1975a) Netze in Natur und Technik—Nets in Nature and Technics. Institut für leichte Flächentragwerke, Bericht Nr. 8, pp 430

Berichte IL (1975b) Wandelbare Pneus—Convertible Pneus. Institut für leichte Flächentragwerke, Bericht Nr. 12, pp 186

Berichte IL (1986) Bambus—bamboo. Institut für leichte Flächentragwerke, Bericht Nr. 31, pp 432

Berichte IL (1992) Ungeplante Siedlungen—non-planned Settlements. Institut für leichte Flächentragwerke, Bericht Nr. 39, pp 256

Bernreuter J (2001) Der Zeitdruck ist enorm: die Photovoltaikindustrie braucht dringend eine neue Quelle für Silizium. Photon 9:10–21

Bertelsmann Lexikon-Institut (1979) Biologie. Daten und Fakten zum Nachschlagen. Gütersloh

Betz G, Tributsch H (1988) Light-induced proton transfer reactions at polymer electrolyte interface. Solid State Ion 28–30:1197–1200

Birke P (2001) Bionische Strategie zur Entwicklung eines neuen audiovisuellen Raumkonzepts für die Oper in Oslo (Teil II). In: Wisser A, Nachtigall W (eds) BIONA-report 15, Akad Wiss Lit. Fischer, Mainz, pp 52–57

Braun DH (2008) Bionisch inspirierte Gebäudehüllen: Konzeption einer Bionisch inspirierter Gebäudehülle nach dem Vorbild natürlicher Hüllen und Häute. PhD, University Stuttgart, Institut für Baukonstruktion und Entwerfen Lehrstuhl 2

Burgess JW, Witt PN (1978) Spinnennetze: Plan- und Baukunst. Naturwiss Rundschau 31:269–282

Camazine S, Deneubourg J-L, Franks NR, Sneyd J, Theraulaz G, Bonabeau E (2001) Self-organization in biological systems. Princeton University Press, New Jersy

Christian Doppler Laboratory for Plastic Solar cells und Quantum Solar Energy (Project Leader: Sariciftci NS), Linz (2001) http://www.ipc.uni-linz.ac.at/os/plastic/index.html

G. Pohl, W. Nachtigall, *Biomimetics for Architecture & Design*,
DOI 10.1007/978-3-319-19120-1

Collias NE, Collias EC (1964) Evolution in Nest-Building in the Weaverbirds (Ploceidae). University of California Press
Collias NE, Collias EC (1976) External construction by animal. Stroudsburg
Crome W (1951) Die Wasserspinne. NBB 44, Wittenberg
Curtis WJR, Doshi B (1985) An architecture for India (photographs by Balkrishna Doshi). Rizzoli International Publications, New York
Czihak G, Langer H, Ziegler H (eds) (1976) Biologie. Springer, Berlin
Dietelmeier S (2000) Seifenhautmodelle. Facharbeit im Leistungskurs Biologie, Kollegstufen-Jahrgang 1998/2000, eingereicht am 1.2.2000, Gymnasium Unterhaching (Kursleiter Helmut Birkner), unpubl
Dimčić M (2011) Structural optimization of grid shells based on genetic algorithms. PhD, Inst. für Tragkonstruktionen und Konstruktives Entwerfen, Universität. Stuttgart, Stuttgart
Dubos R (1960) Pasteur und die moderne Wissenschaft. [p 28: Dubos bezieht das Zitat aus Pasteurs EXPERIMENTALWISSENSCHAFT]. Verlag Kurt Desch, München
Easton DC (1996) The rammed earth house. pp 224 and (2002) http://www.napavalleyonline.com/directory/wshomegarden.html
Eibl-Eibesfeldt I (1984/1995) Die Biologie des menschlichen Verhaltens (Grundriss der Humanethologie). In: Piper R (ed), 3rd edn. München
Emerson AE (1938) Termite nests—a studie of the phylogenie of behavior. Ecol Monogr 8:249–284
Emmerich DG (1964) Structures autotendantes. L'Architecture d'Aujourd'hui 115:XLII–XLV
Etheridge D (2001) Chill out—there's a way to cool down your office without warming up the planet. New Sci Mag 171(2298):22 (07/07/2001)
Flagge I et al (eds) (2001) Thomas Herzog: Architektur und Technologie/Architecture and Technology. Zweisprachige Ausgabe: Deutsch/Englisch, Prestel, München
Flindt M (1986) Biologie in Zahlen, 2nd edn. Fischer, Stuttgart
Frisch vK (1974) Tiere als Baumeister. Ullstein-Verlag, Berlin
Frith HJ (1956) Temperature regulation in the Nesting mounds of the Mallee-fowl, *Leipoa ocellata*. C.S.I.R.O. Wildl Res 1:79–95
Geßner F (1942) Die Leistungen des pflanzlichen Organismus. In: Bertalanffy (ed) Handbuch der Biologie, vol IV. Darmstadt Akademische Verlagsgesellschaft Athenaioin, Potsdam
Giesenhagen K (1912) Handwörterbuch der. Naturwissenschaften 2:1–35
Grätzel M (1993) Low-cost solar cells. The World & I. 228–235
Grätzel M, Liska P (1992) Photoelectrochemical cells and process for making same. U.S. Patent 5(084):365
Grojean R, Sousa J, Henry M (1980) Utilization of solar radiation by polar animals: an optical model for pelts. Appl Opt 19(3):339–347
Gruner D (1990) Die Lehm-Moschee am Niger. In: Haberland E (ed) Reihe: Studien zur Kulturkunde. Frank Steiner, Stuttgart
Hall D (1960) Ein Interview mit Henry Moore. Horizont 3(2)
Hedinger H (ed) (1967) Die Straßen der Tiere. Vieweg Vogel, Braunschweig
Herzog T (1992) Bauten 1978–1992. Hatje, Stuttgart
Herzog T (ed) (1996) Solar energy in architecture and urban planning. Prestel, München
Hesse R, Doflein F (1914) Tierbau und Tierleben, 1st edn, vol 2. Teubner, Leipzig
Horstmann K, Schmid H (1986) Temperature regulation in nests of the wood ant, *Formica polyctena* (Hymenoptera: Formicidae). Entomol Gen 11:22–236
Howse PE (1966) Air movement and termite behaviour. Nature 210:967–968
Ilg L (1998) Natürliche Belüftung von Gebäuden. http//www.hl-technik.de
Ingber DE (1993) Cellular tensegrity: defining new rules of biological design that govern the cytoskeleton. J Cell Sci 104:613–627
Ingber DE (Jan 1998) The architecture of life. Sci Am 278:49–57
Ishay JS, Shmuelson M (1996): Thermoelectric properties of the hornet comb: a device for producing transforming and storing electrical energy for the entire colony. Physiol Chem Phys Med NMR 28:41–54

Kaiser P (1962) Über Netzbau und Strömungssinn bei den Larven der Gattung Hydropsyche. Int Revue Ges Hydrobiol 50:169–224

Kaiser N (1966) Maximen für solares Bauen—auf dem Weg zu solaren Standards. In: Herzog T (ed) Solar energy in architecture and urban planning. Prestel, München, pp 20–45

Kalyanasundaram K, Grätzel M (1999) Sensitised solar cells (DYSC), based on nanocrystalline oxide semiconductor films. http://www.epfl.ch/icp/Icp-2/solarcellE.html. (Hierin Angabe von Review-Artikeln)

Kleineidam C, Ernst R, Roces F (2001) Wind-induced ventilation of the giant nests of the leaf-cutting ant *Atta vollenweideri*. Naturwissenschaften 88:301–305

Kloot WG, van der Williams CM (1953/1954) Cocon construction by the cecropia silkworm 1, 2, 3. Behaviour 5 141–174(6):233–255

Kooistra W, Pohl G (2015) Diatom frustule morphology and its biomimetic applications in architecture and industrial design. In: Hamm C (ed) Evolution of lightweight structures—biomechanic adaption and biodiversity of plankton shells: analyses and technical applications. Springer, Berlin

Koon DW (1998) Is polar bear hair fibre optic? Appl Opt 37(5):3198–3200

Kresling B (1997) Folded and unfolded nature orgami science and art. In: Miura K (ed) Proceedings of 2nd international meet of origami science and science of origami, Otsu, Japan, 29 Nov–2 Dec 1994, pp 93–106

Kühnelt W (1959/1960) Ökologische Vorlesungen. Universität Wien, pers.Mitt.u.B.Lötsch

Kurth-Zelle http://www.habito.de/HAPHO.htm und http://www.wissenschaft.de/sixcms/detail.php?id=41028

Lasarew AJ (1973) Dissertation über pflanzliche Stängelstrukturen und Turmbauwerke. Zitat Nr. [21] in Lebedew JS (1983)

Lebedew JS (1983) Architektur und Bionik. Verlag für Bauwesen, Berlin

Lengerken vH (1939) Brutfürsorge- und Brutpflegeinstinkte der Käfer. Akademische Verlagsgesellschaft, Leipzig

Lichtblau FW (2000) "Taugt für die Welt von morgen?!" Tagungsband 10. Symposium "Thermische Solarenergie", Staffelstein, Kloster Banz

Lippsmeier G (1991) Das kleine Tropenhospital als Modell der Zukunft. Vielfältige Einflüsse auf die Architektur. forschung—Mitteilungen der DFG 2/91:24–26

Loos A (1908) Ornament und Verbrechen. (Reprinted in: A. Loos "TROTZDEM"; Neudruck (1982)). Georg Prachner Verlag, Wien

Lötsch B (1980) Ökologische Begründung des regionalen Bauens. Bericht des Instituts für Leichte Flächentragwerke, IL 27:244–251

Luchsinger RH, Crettol R (2006) Adaptable tensairity. In: Scheublin F, Pronk A, Borgard A, Houtman R (eds) In: Adaptables'06, Proceedings of the joint CIB, Tensinet and AISS international conference on adaptability in design and construction, Eindhoven

Luchsinger RH, Pedretti A, Steingruber P, Pedretti M (2004a) The new structural concept tensairity: basic principles. In: Zingoni A (ed) Proceedings of the second international conference on structural engineering, mechanics and computation. A.A. Balkema/Swets Zeitlinger, Lisse

Luchsinger RH, Pedretti A, Steingruber P, Pedretti M (2004b) Light weight structures with Tensairity. In: Motro R (ed) Shell and spatial structures from models to realizations. Editions de l'Espérou, Montpellier

Luchsinger RH, Pedretti M, Reinhard A (2004c) Pressure induced stability: from pneumatic structures to tensairity. J Bionics Eng 1(3):141–148 (Article No. JBE-2004-025)

Luchsinger RH, Sydow A, Crettol R (2011) Thin-walled structures, structural behavior of asymmetric spindle-shaped tensairity girders under bending loads (Empa, Switzerland). Elsevier, Switzerland. http://www.elsevier.com/locate/tws

Lüscher M (1955) Der Sauerstoffverbrauch bei Termiten und die Ventilation des Nestes bei Macrotermes nataliensis (Haviland). Acta Trop 12:289–307

Maak N (2001) Die Natur der Dinge. FAZ 01.08.2001 und Hessen. DAB 10/2001:8–10

Martin H, Lindauer M (1966) Sinnesphysiologische Leistungen beim Wabenbau der Honigbiene. Z Vergl Physiol 53:372–404

Meissner D (o. J.) Solar technology. In: Elvers B et al (eds) Ullmanns encyclopedia of industrial chemistry. VCH, Weinheim

Mies van der Rohe L (2002) http://www.greatbuildings.com/architects/Ludwig_Mies_van_der_Rohe.html

Miura K (1989) Folding a plane-scenes from nature technology and art symmetry of structure. In: Proceedings Interdisc. Symp. Budapest 1989, p 391

Moore H, Hedgecoe J (1968) Henry Spencer Moore. Simon and Schuster, New York

Nachtigall W (1977) Biostatik. In: Hoppe W et al (eds) Biophysik. Springer, Heidelberg

Nachtigall W (1986) Der Pneu-Begriff in der Botanik des 19. und 20. Jahrhunderts. Jahrbuch Wissenschaftskolleg 1985/1986:313–327

Nachtigall W (1996) Buildings in nature. Impulses for technology. In: Proceedings 4th European conference on solar energy in architecture and urban planning

Nachtigall W (1997) Vorbild Natur. Bionik-Design für funktionelles Gestalten. Springer, Berlin

Nachtigall W (2002) Bionik. Grundlagen und Beispiele für Ingenieure und Naturwissenschaftler, 2nd edn. Springer, Berlin

Nachtigall W (2010) Bionik als Wissenschaft. Erkennen → Absrahieren → Umsetzen. Springer, Berlin

Nachtigall W, Kresling B (1992) Bauformen der Natur. Teil I: Technische Biologie und Bionik von Knoten-Stab-Tragwerken. Naturwissenschaften 79:193–201

Nachtigall W, Kresling B (1993) Bauformen der Natur. Teil II: Technische Biologie und Bionik von Platten- und Faltkonstruktionen. Naturwissenschaften 79:251–259

Nachtigall W, Pohl G (2013) Bau-Bionik: Natur—Analogien—Technik. Springer, Berlin

Nachtigall W, Wisser A (2012) Bionik in Beispielen. Spektrum, Heidelberg

Olberg G (1959) Das Verhalten der solitären Wespen Mitteleuropas. Verlag der Wissenschaften, Berlin

Olberg G (1960) Bauwerke der Tiere. NBB, Wittenberg

Oligmüller D (1995) Die Wiedereinbindung des Baukörpers in den natürlichen Kreislauf—Ein Brückenschlag zwischen Bionik und Architektur. In: Nachtigall W (ed) BIONA-report 9, Akad Wiss Lit. Fischer, Stuttgart, pp 75–90

Oligmüller D (2001) Ganzheitliche Betrachtungsweise bei der Weiterentwicklung bionischer Systeme in der Architektur, im Städtebau und im Verkehrswesen. In: Wisser A, Nachtigall W (eds) BIONA-report 15, Akad Wiss Lit. Fischer, Mainz, pp 254–262

Oligmüller D (2003) How to scratch skies. Wie man Wolken kratzt. Eine zeitgemäße Hochhausstruktur im Spannungsfeld vielfältiger bionischer Aspekte. In: Wisser A (ed) BIONA-report 16, Akad Wiss Lit. Fischer, Mainz (in press)

Olszewski J, Skozen S (1965) The airing of burrows of the mole, *Talpa europaea Linnaeus*, 1758. Acta Teriol 10:181–193

O'Regan B, Grätzel M (1991) A low cost, high efficiency solar cell based on dye-sensitised colloidal TiO2 films. Nature 353:737–740

Otto F (1978) Der Pneu—Bauprinzip des Lebens. Bild der Wissenschaft 10:124–135

Pedretti A, Steingruber P, Pedretti M, Luchsinger RH (2004) The new structural concept Tensairity: FE-modeling and applications. In: Zingoni A (ed) Proceedings of the second international conference on structural engineering, mechanics and computation. Lisse (The Netherlands): A.A. Balkema/Swets ZeitlingerPedretti M, Pedretti A, Steingruber P and Luchsinger R H (2004), "First applications of Tensairity®"

Penzlin H (1991) Lehrbuch der Tierphysiologie, 5th edn. Fischer, Jena

Philippi U, Nachtigall W (1996) Functional morphology of regular echinoid tests (Echinodermata, Echinoida): a finite element study. Zoomorpology 116:35–50

Piano R (1997) Mein Architektur-Logbuch. Hatje, Ostfildern-Ruit

PlasticS Cells http://www.esqsec.unibe.ch/pub_11.htm

Pohl G (ed) (2010) Textiles, polymers and composites for buildings. Woodhead Publishing, in association with the Textile Institute, Philadelphia

Pohl G (2015) Fibre reinforced architecture inspired by nature: COCOON_FS. In: Hamm C (ed) Evolution of lightweight structures—biomechanic adaption and biodiversity of plankton shells: analyses and technical applications. Springer, Berlin
Rainer R (1974) pers. Mitt. B. Lötsch, Wien
Rasdorsky W (1911) Bull. de la Societé des Naturalistes des Moscou. Sect Biol 4:351–405
Reflektivität Schmetterlingsschuppen http://www.tufts.edu/as/tampl/museum/butterfly.html
Remane A, Storch V, Welsch U (1981) Kurzes Lehrbuch der Zoologie. Gustav Fischer, Stuttgart
Rennie J (1833) Die Baukunst der Vögel. Leipzig
Rensch B (1957) Ästhetische Faktoren bei Farb- und Formbevorzugungen von Affen. Zeitschrift f. Tierpsych 14:71–99
Rensch B (1958) Die Wirksamkeit ästhetischer Faktoren bei Wirbeltieren. Zeitschrift f. Tierpsych 15:447–461
Richard PB (1967) Le determinisme de la construction des barrages chez le Castor du Rhone. La Terre et la Vie 4/1967:339–470
Ricolais Le R (1940–1941) Essai sur les systèmes réticulés à trois dimensions. Ann Ponts Chaussées 63–70, 153–165
Ronacher A (1998) Architektur und Zeitgeist. Irrwege des Bauens unserer Zeit—Auswege für das neue Jahrtausend. Mit Projekten von Herwig und Andrea Ronacher. Verlag Johannes Heyn, Klagenfurt
Rüegg H, Völlmin C (2002) Transparente Wärmedämmung (TWD). http://www.energie.ch/themen/bautechnik/twd/index.htm
Rummel G (2001) Bionische Strategie zur Entwicklung eines neuen audiovisuellen Raumkonzepts für die Oper in Oslo (Teil I). In: Wisser A, Nachtigall W (eds) BIONA-report 15 Akad Wiss Lit, Fischer, Mainz, pp 40–51
Sariciftci NS, Smilowitz L, Heeger AJ, Wudl F (1992) Photoinduced electron-transfer from a conducting polymer to buckminsterfullerene. Science 258:1474
Sattler W (1958) Beiträge zur Kenntnis von Lebensweise und Köcherbau der Larve und Puppe von Hydropsyche Pict. (Trichoptera) mit besonderer Berücksichtigung des Netzbaus. Z Morph Ökol Tiere 47:115–192
Sattler W (1962) Die Meisterweber. Umschau 62:205–208
Schmidt RS (1964) Apicotermes Nests. Am Zool 4:221–225
Schmitz H (1994) Thermal characterization of butterfly wings: 1. Absorption in relation to different colour, surface structure and basking type. J Therm Biol 19:403–412
Schmitz H, Tributsch H (1994) Die Eigenschaften von Schmetterlingsflügeln als Solarabsorber. Verh Dtsch Zool Ges 87:112
Schnitzspahn G (2002) Die "Baukultur" und das "Schöne im Bauen". Vortragssammlung zum AIV-Symposium:"Das Stadtquartier der Zukunft", Stuttgart, 25–27 April 2002
Schober W, Grimmberger E (1987) Die Fledermäuse Europas. Franck, Stuttgart
Schwendener S (1874) Das mechanische Prinzip im anatomischen Bau der Monocotylen ... Engelmann, Leipzig
Sedlmayr H (1948) Verlust der Mitte. Die bildende Kunst des 19. und 20. Jahrhunderts als Symptom und Symbol der Zeit. O. Müller (Salzburg) 1951 (1–7th edn). Ullstein Buch Nr. 39, Frankfurt, Berlin (1965)
Selbstbau mit Lehm (1991) Saarbrücker Zeitung 148 vom 30.6.1991, p L3
Smith F (1997) Eastgate—Harare, Zimbabwe. In: Brown DJ (ed) The ARUP-journal 1/1997, 3–8
Snell FM, Shulmann S, Spencer RP, Moos C (1968) Biophysikalische Grundlagen von Struktur und Funktion, vol I. Hirzel, Stuttgart
Spatz H-C, Beismann H, Brüchert F, Emanns A, Speck T (1997) Biomechanics of the giant reed arundo donax. Phil Trans Roy Soc B 352:1–10
Speck O, Speck T, Spatz H-C (1998) Viscoelastizität und Plastizität—oder wie vermeiden Pflanzen destruktive Oszillationen? Eine biomechanisch—funktionsanatomische Analyse des Rhizoms von Arundo donax. In: Nachtigall W, Wisser A (eds) BIONA-report 12, Akad Wiss Lit. Fischer, Mainzt, pp 91–106

Speck T, Masselter T, Prüm B, Speck O, Luchsinger R (2004) Smart materials: light-weight structures with variable stiffness an self-repair mechanisms. In Boblan I, Bannasch R (eds) First international incustrial conference bionik 2004. VDI Reihe Umwelttechnik, Nr. 249

Speck T, Luchsinger R, Busch S, Rüggeberg M, Speck O (2006) Self-healing processes in nature and engineering: self-repairing biomimetic membranes for pneumatic structures. In: Brebbia CA (ed) Design and nature III. WIT Press, Southampton, pp 105–114

Szent-Györgi A (1957) Bioenergetics. Academic, New York

Tada H, Mann SE, Miaoulis IN, Wong PY (1998) Microscale radiative effects in complex microstructures of iridescent butterfly wing scales. MRS Fall 1997 Meeting, Symposium on Materials Science of the Cells, poster presentation. Mat Res Soc Symp Proc 489:173–180

Tautz J (2007) Phänomen Honigbiene. Mit Fotos von H. Heilmann. Spektrum, Heidelberg

Tautz J, Casas J, Sandeman DC (2001) Phase reversal of vibratory signals in honeycomb may assist dancing honeybees to attract their audience. J Exp Biol 204:3737–3746

Thaler E (1976) Nest und Nestbau von Winter- und Sommergoldhähnchen (*Regulus regulus* und *R. ignicapillus*). J Ornithol. 117:121–144

Thallemer http://de.mastersportal.eu/students/browse/programme/31331/industrial-design-scionic.html

Thallemer A (2010) Scionic. Arnoldsche Art Publishers, Stuttgart. http://www.arnoldsche.com/Design/SCIONIC.html

Thallemer Scionic® http://www.facebook.com/pages/Scionic/179549155405507?sk=info,www.scionic.at http://www.scionic.at

Thallemer A, Reese J (2010) Visual permutations. Arnoldsche Art Publishers, Stuttgart. http://www.arnoldsche.com/Design/VISUAL-PERMUTATIONS.html

Tributsch H (1995) Bionik solarer Energiesysteme. In: Nachtigall W, Wisser A (eds) BIONA-report 9, Akad Wiss Lit. Fischer, Mainz, pp 147–170

Tributsch H (2001a) Bionische Vorbilder für eine solare Energietechnik. In: v. Gleich A (ed) Bionik. Ökologische Technik nach dem Vorbild der Natur? 2nd edn. Teubner, Stuttgart, pp 247–261

Tributsch H (2001b) Regenerative Energienutzung: Solarenergie. Unterricht-Arbeit + Technik 10, 62–64

Tributsch H, Goslowsky H, Küppers U, Wetzel H (1990) Light collection and solar sensing through the polar bear pelt. Solarnergy Mat 21:219–236

Vogel S, Ellington C, Kilgorek D (1973) Wind-induced ventilation of the burrow of the prairie-dog, *Cynomys ludovicianus*. J Comp Physiol 85:1–14

Wasserabweisende Haut (1992) Australian science and technology news letter 7, p 5

Weir JS (1973) Air flow, evaporation and mineral accumulation in molds of *Macrotermes subhyalinus* (Rambur). J Anim Ecol 42:509–520

Wester T (1984) Structural order in space: the plate-lattice dualism. Plate Laboratory, Royal Academy of Arts, Copenhagen

Wester T (1997) Some remarkable statical qualities of structures in nature. In: Chilton JC et al (eds) Proc. Inst. Coll. Struct. Morph, Nottingham, 12–21

Weyrauch W (1935) Die Nester der sozialen Wespen. Das Aquarium 2/1935:145–150

Wichard W (1978) Die Köcherfliegen. NBB, Wittenberg

Wickler W, Seibt U (1977) Prinzip Eigennutz

Wöhrle D, Meissner D (1991) Organic solar cells. Rev Article Adv Mater 3:129–138

Wunder W (1931) Brutpflege und Nestbau bei Fischen. Erg der Biol 7:118–192

Wunder W (1932) Nestbau und Brutpflege bei Amphibien. Erg der Biol 8:180–220

Wunder W (1934) Nestbau und Brutpflege bei Reptilien. Erg der Biol 10:1–36

Yanda B, Fisher R (1983) Energie und Nahrung aus dem Solar-Gewächshaus. H. Hu-gendübel, München

Index

G. Pohl, W. Nachtigall, *Biomimetics for Architecture & Design*,
DOI 10.1007/978-3-319-19120-1

MIX
Papier aus verantwortungsvollen Quellen
Paper from responsible sources
FSC® C105338

FSC
www.fsc.org

If you have any concerns about our products,
you can contact us on
ProductSafety@springernature.com

In case Publisher is established outside the EU,
the EU authorized representative is:
Springer Nature Customer Service Center GmbH
Europaplatz 3, 69115 Heidelberg, Germany

Printed by Libri Plureos GmbH
in Hamburg, Germany